An Infantryman's Guide to Combat in Built-Up Areas

An Infantryman's Guide to Combat in Built-Up Areas

Copyright © 1994 by Paladin Press

ISBN 0-87364-800-5
Printed in the United States of America

Published by Paladin Press, a division of
Paladin Enterprises, Inc.,
Gunbarrel Tech Center
7077 Winchester Circle
Boulder, Colorado 80301 USA
+1.303.443.7250

Direct inquiries and/or orders to the above address.

PALADIN, PALADIN PRESS, and the "horse head" design
are trademarks belonging to Paladin Enterprises and
registered in United States Patent and Trademark Office.

All rights reserved. Except for use in a review, no
portion of this book may be reproduced in any form
without the express written permission of the publisher.

Neither the author nor the publisher assumes
any responsibility for the use or misuse of
information contained in this book.

Visit our Web site at www.paladin-press.com

FIELD MANUAL
No. 90-10-1

*FM 90-10-1

HEADQUARTERS
DEPARTMENT OF THE ARMY
Washington, DC, 12 May 1993

AN INFANTRYMAN'S GUIDE TO COMBAT IN BUILT-UP AREAS

CONTENTS

	Page
Preface	vii
Chapter 1. INTRODUCTION	1-1
Section I. **Background**	1-1
1-1. AirLand Battle	1-1
1-2. Definitions	1-1
1-3. Cities	1-2
1-4. The Threat in Built-Up Areas	1-3
Section II. **Characteristics and Categories of Built-Up Areas**	1-4
1-5. Characteristics	1-4
1-6. Categories	1-5
Section III. **Special Considerations**	1-5
1-7. Battles in Built-Up Areas	1-5
1-8. Target Engagement	1-5
1-9. Small-Unit Battles	1-5
1-10. Munitions and Special Equipment	1-6
1-11. Communications	1-6
1-12. Stress	1-6
1-13. Restrictions	1-6
1-14. Fratricide Avoidance	1-7

DISTRIBUTION RESTRICTION: Approved for public release; distribution is unlimited.

*This publication supersedes FM 90-10-1, 30 September 1982.

iii

FM 90-10-1

                                                                                                     page

Chapter 2. URBAN ANALYSIS

    Section I. **Models of Built-Up Areas** 2-1
        2-1. Regional Urban Characteristics 2-1
        2-2. Specific Characteristics of Urban Areas 2-2
        2-3. Characteristics of Urban Areas 2-2
    Section II. **Terrain and Weather Analysis** 2-6
        2-4. Special Terrain Considerations 2-6
        2-5. Special Weather Considerations 2-7
    Section III. **Threat Evaluation and Integration** 2-8
        2-6. Operational Factors 2-8
        2-7. Urban Counterinsurgency, Counterguerrilla,
            and Counterterrorist Operations 2-9
        2-8. Projected Threat Capabilities 2-11

Chapter 3. OFFENSIVE OPERATIONS

    Section I. **Offensive Considerations** 3-1
        3-1. Reasons for Attacking Built-Up Areas 3-1
        3-2. Reasons for Not Attacking a Built-Up Area 3-1
    Section II. **Characteristics of Offensive Operations in a Built-Up Area** . . . 3-2
        3-3. Troop Requirements 3-2
        3-4. Maneuver 3-2
        3-5. Use of Equipment 3-2
    Section III. **Types of Offensive Operations** 3-3
        3-6. Hasty Attack 3-3
        3-7. Deliberate Attack 3-3
    Section IV. **METT-T Factors** 3-7
        3-8. Mission 3-7
        3-9. Enemy 3-8
        3-10. Terrain 3-8
        3-11. Troops 3-9
        3-12. Time 3-16
    Section V. **Command and Control** 3-17
        3-13. Command 3-17
        3-14. Control 3-17
    Section VI. **Battalion Task Force Attack on a Built-Up Area** 3-17
        3-15. Conduct of Deliberate Attack 3-17
        3-16. Seizure of Key Objective 3-18
        3-17. Infiltration 3-19
        3-18. Route Security 3-20

FM 90-10-1

	page
Section VII. **Company Team Attack of a Built-Up Area**	3-21
3-19. Attack of a Block	3-21
3-20. Attack of an Enemy Outpost	3-23
3-21. Seizure of a Traffic Circle	3-24
3-22. Seizure of Key Terrain	3-26
3-23. Reconnaissance	3-27
Section VIII. **Platoon Attack of a Built-Up Area**	3-29
3-24. Attack of a Building	3-29
3-25. Movement Down a Street	3-29
3-26. Counterattacks	3-31

Chapter 4. DEFENSIVE OPERATIONS

Section I. **Defensive Considerations**	4-1
4-1. Reasons for Defending Built-Up Areas	4-1
4-2. Reasons for Not Defending Built-Up Areas	4-2
Section II. **Characteristics of Built-Up Areas**	4-2
4-3. Obstacles	4-3
4-4. Avenues of Approach	4-4
4-5. Key Terrain	4-4
4-6. Observation and Fields of Fire	4-4
4-7. Cover and Concealment	4-4
4-8. Fire Hazards	4-4
4-9. Communications Restrictions	4-4
Section III. **Factors of METT-T**	4-5
4-10. Mission	4-5
4-11. Enemy	4-5
4-12. Terrain	4-5
4-13. Troops Available	4-9
4-14. Time Available	4-13
Section IV. **Command and Control**	4-16
4-15. Command Post Facilities	4-16
4-16. Organization of the Defense	4-16
4-17. Counterattack	4-18
4-18. Defense During Limited Visibility	4-18
Section V. **Defensive Plan at Battalion Level**	4-19
4-19. Defense of a Village	4-19
4-20. Defense in Sector	4-20
4-21. Delay in a Built-Up Area	4-21

	page
Section VI. **Defensive Plan at Company Level**	4-23
4-22. Defense of a Village	4-23
4-23. Defense of a City Block	4-25
4-24. Company Delay	4-26
4-25. Defense of a Traffic Circle	4-27
Section VII. **Defensive Plan at Platoon Level**	4-27
4-26. Defense of a Strongpoint	4-28
4-27. Defense Against Armor	4-29
4-28. Conduct of Armored Ambush	4-32

Chapter 5. FUNDAMENTAL COMBAT SKILLS

Section I. **Movement**	5-1
5-1. Crossing of a Wall	5-1
5-2. Movement Around Corners	5-1
5-3. Movement Past Windows	5-2
5-4. Use of Doorways	5-3
5-5. Movement Parallel to Buildings	5-4
5-6. Crossing of Open Areas	5-6
5-7. Fire Team Employment	5-6
5-8. Movement Between Positions	5-7
5-9. Movement Inside a Building	5-9
Section II. **Entry Techniques**	5-11
5-10. Upper Building Levels	5-11
5-11. Use of Ladders	5-12
5-12. Use of Grappling Hook	5-13
5-13. Scaling of Walls	5-14
5-14. Rappelling	5-16
5-15. Entry at Lower Levels	5-16
5-16. Hand Grenades	5-19
Section III. **Firing Positions**	5-23
5-17. Hasty Firing Position	5-23
5-18. Prepared Firing Position	5-26
5-19. Target Acquisition	5-34
5-20. Flame Operations	5-37
5-21. Employment of Snipers	5-38
Section IV. **Navigation in Built-Up Areas**	5-39
5-22. Military Maps	5-39
5-23. Global Positioning Systems	5-40
5-24. Aerial Photographs	5-40

	page
Section V. **Camouflage**	5-40
5-25. Application	5-40
5-26. Use of Shadows	5-41
5-27. Color and Texture	5-43

Chapter 6. COMBAT SUPPORT

6-1. Mortars	6-1
6-2. Field Artillery	6-3
6-3. Naval Gunfire	6-4
6-4. Tactical Air	6-4
6-5. Air Defense	6-5
6-6. Army Aviation	6-6
6-7. Helicopters	6-6
6-8. Engineers	6-8
6-9. Military Police	6-8
6-10. Communications	6-9

Chapter 7. COMBAT SERVICE SUPPORT AND LEGAL ASPECTS OF COMBAT

Section I. **Combat Service Support**	7-1
7-1. Guidelines	7-1
7-2. Principal Functions	7-1
7-3. Supply and Movement Functions	7-4
7-4. Medical	7-5
7-5. Personnel Services	7-7
Section II. **Legal Aspects of Combat**	7-8
7-6. Civilian Impact in the Battle Area	7-8
7-7. Command Authority	7-8
7-8. Source Utilization	7-9
7-9. Health and Welfare	7-9
7-10. Law and Order	7-9
7-11. Public Affairs Officer and Media Relations	7-9
7-12. Civil Affairs Units and Psychological Operations	7-9
7-13. Provost Marshall	7-10
7-14. Commander's Legal Authority and Responsibilities	7-10

FM 90-10-1

Chapter 8. EMPLOYMENT AND EFFECTS OF WEAPONS

    8-1. Effectiveness of Weapons and Demolitions 8-1
    8-2. M16 Rifle and M249 Squad Automatic Weapon/Machine Gun 8-2
    8-3. Medium and Heavy Machine Guns (7.62-mm and .50-Caliber) 8-4
    8-4. Grenade Launchers, 40-mm (M203 and MK 19) 8-7
    8-5. Light and Medium Recoilless Weapons 8-9
    8-6. Antitank Guided Missiles 8-19
    8-7. Flame Weapons 8-23
    8-8. Hand Grenades 8-26
    8-9. Mortars 8-28
    8-10. 25-mm Automatic Gun 8-31
    8-11. Tank Cannon 8-34
    8-12. Combat Engineer Vehicle Demolition Gun 8-38
    8-13. Artillery and Naval Gunfire 8-38
    8-14. Aerial Weapons 8-40
    8-15. Demolitions 8-41

Appendix A. NUCLEAR, BIOLOGICAL, AND CHEMICAL CONSIDERATIONS A-1

Appendix B. BRADLEY FIGHTING VEHICLE B-1

Appendix C. OBSTACLES, MINES, AND DEMOLITIONS C-1

Appendix D. SUBTERRANEAN OPERATIONS D-1

Appendix E. FIGHTING POSITIONS E-1

Appendix F. ATTACKING AND CLEARING BUILDINGS F-1

Appendix G. MILITARY OPERATIONS IN URBAN TERRAIN (MOUT) UNDER RESTRICTIVE CONDITIONS G-1

Appendix H. URBAN BUILDING ANALYSIS H-1

Appendix I. NIGHT OPERATIONS I-1

Glossary Glossary-1

References References-1

Index Index-1

FOREWORD

This timely, well written manual covers military ground operations in urbanized terrain. Major cities and surrounding areas are increasingly the centers of industry, political power, culture, finance, communications, and commerce. These urban centers are not only destined to be the military battlegrounds of the future but are also critical areas where civil law enforcement must continue to be able to maintain control in the face of potential criminal violence, riots, and insurgency.

Due to changing political and social conditions, the role of our military will probably now take on new dimensions in maintaining long range domestic stability. Local police agencies who are primarily responsible for law and order also are likely to find themselves in situations where it will probably be necessary to employ paramilitary tactics and light weaponry to maintain public order.

The current—and predicted—deterioration of our great cities because of crime, gangs, racial conflict, social and economic conditions combined with civil disturbance, and acts of terrorism will call for permanent standby control forces. These will either be military, or, preferably, civil law enforcement units organized, trained, and appropriately armed along military lines. It's an either/or situation.

This military manual is an update of previous FM 90-10 issued in 1982. Current and super SWAT and similar units yet to be created by civil law enforcement will benefit greatly from its tactical information and other contents. For unknown reasons, new manual FM 19-10-1 (May 1993) has not yet been made available or issued in quantity through customary military channels to those regular Army and National Guard units with a need to know. Paladin Press is doing the U.S. military and civil law enforcement a real service by publication of this nonrestricted text.

—Col. Rex Applegate

PREFACE

This manual provides the infantryman with guidelines and techniques for fighting against a uniformed enemy in built-up areas who may or may not be separated from the civilian population. Some techniques for dealing with insurgents, guerrillas, and terrorists are included; however, the manual which best addresses these issues is FM 7-98. This manual does not address any techniques for missions that require the restoration of order to urban areas. Information and techniques to accomplish this mission are addressed in FM 19-15. The probability is great that United States forces will become engaged by enemy forces who are intermingled with the civilian population. Therefore, units using the techniques outlined in this manual under these conditions must obey the rules of engagement issued by their headquarters and the laws of land warfare. Infantry commanders and staffs should concentrate on the skills contained in Chapters 3 through 5 as they train their units.

The urban growth in all areas of the world has changed the face of the battlefield. Military operations in urbanized terrain (MOUT) constitute the battlefield in the Eurasian continent. It includes all man-made features (cities, towns, villages) as well as natural terrain. Combat in built-up areas focuses on fighting for and in those cities, towns, and villages.

The proponent of this publication is the US Army Infantry School. Send comments and recommendations on DA Form 2028 directly to Commandant, US Army Infantry School, ATTN: ATSH-ATD, Fort Benning, Georgia 31905-5410.

Unless this publication states otherwise, masculine nouns and pronouns do not refer exclusively to men.

CHAPTER 1

INTRODUCTION

The increased population and accelerated growth of cities have made the problems of combat in built-up areas an urgent requirement for the US Army. This type of combat cannot be avoided. The make up and distribution of smaller built-up areas as part of an urban complex make the isolation of enemy fires occupying one or more of these smaller enclaves increasingly difficult. MOUT is expected to be the future battlefield in Europe and Asia with brigade- and higher-level commanders focusing on these operations. This manual provides the infantry battalion commander and his subordinates a current doctrinal source for tactics, techniques, and procedures for fighting in built-up areas.

Section I. BACKGROUND

Friendly and enemy doctrine reflect the fact that more attention must be given to urban combat. Expanding urban development affects military operations as the terrain is altered. Although the current doctrine still applies, the increasing focus on operations short of war, urban terrorism, and civil disorder emphasizes that combat in built-up areas is unavoidable.

1-1. AIRLAND BATTLE

AirLand Battle doctrine describes the Army's approach to generating and applying combat power at the operational and tactical levels. It is based on securing or retaining the initiative and exercising it aggressively to accomplish the mission. The four basic AirLand Battle tenets of initiative, agility, depth, and synchronization are constant. During combat in built-up areas, the principles of AirLand Battle doctrine still apply—only the terrain over which combat operations will be conducted has changed.

1-2. DEFINITIONS

MOUT is defined as all military actions that are planned and conducted on terrain where man-made construction affects the tactical options available to the commander. These operations are conducted to defeat an enemy that may be mixed in with civilians. Therefore, the rules of engagement (ROE) and use of combat power are more restrictive than in other conditions of combat. Due to political change, advances in technology, and the Army's role in maintaining world order, MOUT now takes on new dimensions that previously did not exist. These new conditions affect how units will fight or accomplish their assigned missions. The following definitions provide clarity and focus for commanders conducting tactical planning for MOUT. The terms "surgical MOUT operations" and "precision MOUT operations" are descriptive in nature only. These are conditions of MOUT, not doctrinal terms.

    a. **Built-Up Area.** A built-up area is a concentration of structures, facilities, and people that forms the economic and cultural focus for the surrounding area. The four categories of built-up areas are large cities, towns and small cities, villages, and strip areas.

    b. **Surgical MOUT.** These operations are usually conducted by joint special operation forces. They include missions such as raids, recovery operations, rescues, and other special operations (for example, hostage rescue).

c. **Precision MOUT.** Conventional forces conduct these operations to defeat an enemy that is mixed with noncombatants. They conduct these operations carefully to limit noncombatant casualties and collateral damage. Precision MOUT requires strict accountability of individual and unit actions through strict ROE. It also requires specific tactics, techniques, and procedures for precise use of combat power (as in Operation Just Cause). (See Appendix G for more detailed information.)

1-3. CITIES
Cities are the centers of finance, politics, transportation, communication, industry, and culture. Therefore, they have often been scenes of important battles (Table 1-1).

CITY	YEAR	CITY	YEAR
RIGA	1917	BUDAPEST	1956
MADRID	1936	* BEIRUT	1958
WARSAW	1939	* SANTO DOMINGO	1965
ROTTERDAM	1940	* SAIGON	1968
MOSCOW	1942	* KONTUM	1968
STALINGRAD	1942	* HUE	1968
LENINGRAD	1942	BELFAST	1972
WARSAW	1943	MONTEVIDEO	1972
* PALERMO	1944	QUANGTRI CITY	1972
* BREST	1944	AN LOC	1972
WARSAW	1944	XUAN LOC	1975
* AACHEN	1944	SAIGON	1975
ORTONA	1944	BEIRUT	1975-1978
* CHERBOURG	1944	MANAGUA	1978
BRESLAU	1945	ZAHLE	1981
* WEISSENFELS	1945	TYRE	1982
BERLIN	1945	* BEIRUT	1983
* MANILA	1945	* PANAMA CITY	1989-1990
* SAN MANUEL	1945	* COLON	1989-1990
* SEOUL	1950	* KUWAIT CITY	1991

*Direct US Troop Involvement

Table 1-1. Cities contested during twentieth century conflicts.

a. Operations in built-up areas are conducted to capitalize on the strategic and tactical advantages of the city, and to deny those advantages to the enemy. Often, the side that controls a city has a psychological advantage, which can be enough to significantly affect the outcome of larger conflicts.

b. Even in insurgencies, combat occurs in cities. In developing nations, control of only a few cities is often the key to control of national resources. The city riots of the 1960's and the guerrilla and terrorist operations in Santo Domingo, Caracas, Belfast, Managua, and Beirut indicate the many situations that can result in combat operations in built-up areas.

c. Built-up areas also affect military operations because of the way they alter the terrain. In the last 40 years, cities have expanded, losing their well-defined boundaries as they extended into the countryside. New road systems have opened areas to make them passable. Highways, canals, and railroads have been built to connect population centers. Industries have grown along those connectors, creating "strip areas." Rural areas, although retaining much of their farm-like character, are connected to the towns by a network of secondary roads.

d. These trends have occurred in most parts of the world, but they are the most dramatic in Western Europe. European cities tend to grow together to form one vast built-up area. Entire regions assume an unbroken built-up character, as is the case in the Ruhr and Rhein Main complex. Such growth patterns block and dominate the historic armor avenues of approach, or decrease the amount of open maneuver area available to an attacker. It is estimated that a typical brigade sector in a European environment will include 25 small towns, most of which would lie in the more open avenues of approach (Figure 1-1).

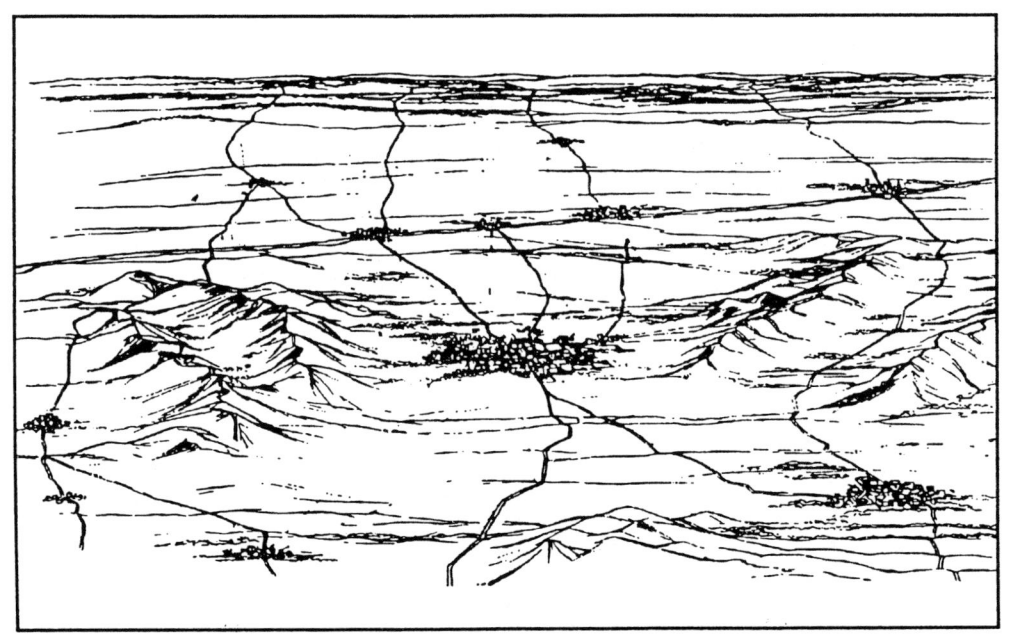

Figure 1-1. Urban areas blocking maneuver areas.

e. Extensive urbanization provides conditions that a defending force can exploit. Used with mobile forces on the adjacent terrain, antitank forces defending from built-up areas can dominate avenues of approach, greatly improving the overall strength of the defense.

f. Forces operating in such areas may have elements in open terrain, villages, towns, or small and large cities. Each of these areas calls for different tactics, task organization, fire support, and CSS.

1-4. THE THREAT IN BUILT-UP AREAS
The Commonwealth of Independent States and other nations that use Soviet doctrine have traditionally devoted much of their training to urban combat exercises. Indications are that they believe such combat is unavoidable in

future conflicts. But, the threat of combat in built-up areas cannot be limited to former Soviet doctrine. Throughout many Third World countries, the possibility of combat in built-up areas exists through acts of insurgents, guerrillas, and terrorists. (Information on operations in this environment is found in the reference list.)

Section II. CHARACTERISTICS AND CATEGORIES OF BUILT-UP AREAS

One of the first requirements for conducting operations in built-up areas is to understand the common characteristics and categories of such areas.

1-5. CHARACTERISTICS

Built-up areas consist mainly of man-made features such as buildings. Buildings provide cover and concealment, limit fields of observation and fire, and block movement of troops, especially mechanized troops. Thick-walled buildings provide ready-made, fortified positions. Thin-walled buildings that have fields of observation and fire may also be important. Another important aspect is that built-up areas complicate, confuse and degrade command and control.

   a. Streets are usually avenues of approach. However, forces moving along streets are often canalized by the buildings and have little space for off-road maneuver. Thus, obstacles on streets in towns are usually more effective than those on roads in open terrain since they are more difficult to bypass.

   b. Subterranean systems found in some built-up areas are easily overlooked but can be important to the outcome of operations. They include subways, sewers, cellars, and utility systems (Figure 1-2).

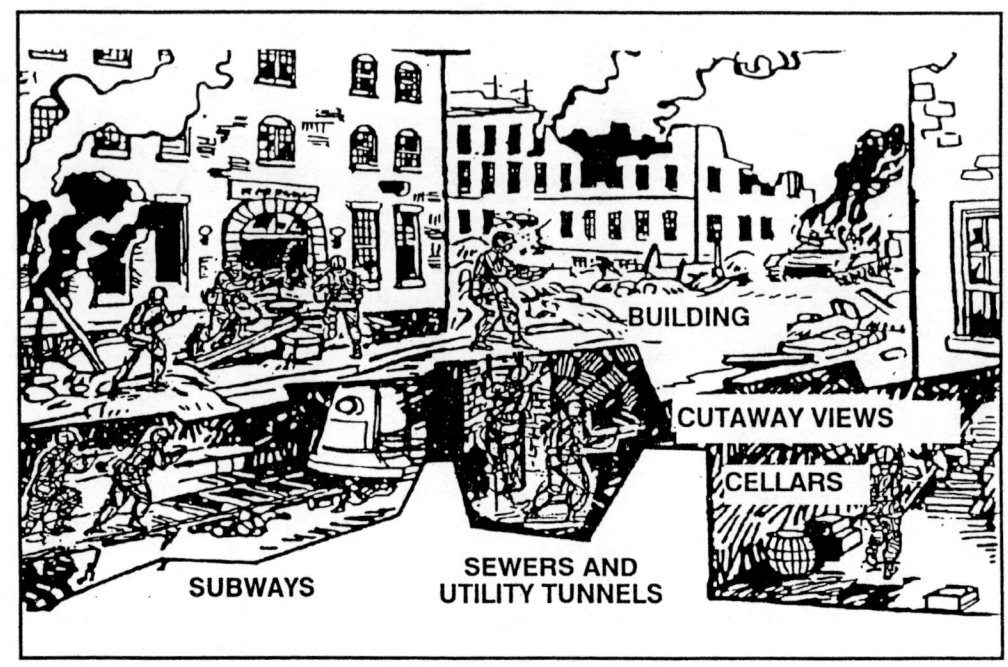

Figure 1-2. Underground systems.

1-6. CATEGORIES
Built-up areas are classified into four categories:
- Villages (population of 3,000 or less).
- Strip areas (urban areas built along roads connecting towns or cities).
- Towns or small cities (population up to 100,000 and not part of a major urban complex).
- Large cities with associated urban sprawl (population in the millions, covering hundreds of square kilometers).

Each area affects operations differently. Villages and strip areas are commonly encountered by companies and battalions. Towns and small cities involve operations of entire brigades or divisions. Large cities and major urban complexes involve units up to corps size and above.

Section III. SPECIAL CONSIDERATIONS

Several considerations are addressed herein concerning combat in built-up areas.

1-7. BATTLES IN BUILT-UP AREAS
Battles in built-up areas usually occur when—
- A city is between two natural obstacles and there is no bypass.
- The seizure of a city contributes to the attainment of an overall objective.
- The city is in the path of a general advance and cannot be surrounded or bypassed.
- Political or humanitarian concerns require the seizure orretention of a city.

1-8. TARGET ENGAGEMENT
In the city, the ranges of observation and fields of fire are reduced by structures as well as by the dust and smoke of battle. Targets are usually briefly exposed at ranges of 100 meters or less. As a result, combat in built-up areas consists mostly of close, violent combat. Infantry troops will use mostly light and medium antitank weapons, automatic rifles, machine guns, and hand grenades. Opportunities for using antitank guided missiles are rare because of the short ranges involved and the many obstructions that interfere with missile flight.

1-9. SMALL-UNIT BATTLES
Units fighting in built-up areas often become isolated, making combat a series of small-unit battles. Soldiers and small-unit leaders must have the initiative, skill, and courage to accomplish their missions while isolated from their parent units. A skilled, well-trained defender has tactical advantages over the attacker in this type of combat. He occupies strong static positions, whereas the attacker must be exposed in order to advance. Greatly reduced line-of-sight ranges, built-in obstacles, and compartmented terrain require the commitment of more troops for a given frontage. The troop density for both an attack and defense in built-up areas can be as much as three to five

times greater than for an attack or defense in open terrain. Individual soldiers must be trained and psychologically ready for this type of operation.

1-10. MUNITIONS AND SPECIAL EQUIPMENT

Forces engaged in fighting in built-up areas use large quantities of munitions because of the need for reconnaissance by fire, which is due to short ranges and limited visibility. LAWs or AT-4s, rifle and machine gun ammunition, 40-mm grenades, hand grenades, and explosives are high-usage items in this type of fighting. Units committed to combat in built-up areas also must have special equipment such as grappling hooks, rope, snaplinks, collapsible pole ladders, rope ladders, construction material, axes, and sandbags. When possible, those items should be either stockpiled or brought forward on-call, so they are easily available to the troops.

1-11. COMMUNICATIONS

Urban operations require centralized planning and decentralized execution. Therefore, communications plays an important part. Commanders must trust their subordinates' initiative and skill, which can only occur through training. The state of a unit's training is a vital, decisive factor in the execution of operations in built-up areas.

    a. Wire is the primary means of communication for controlling the defense of a city and for enforcing security. However, wire can be compromised if interdicted by the enemy.

    b. Radio communication in built-up areas is normally degraded by structures and a high concentration of electrical power lines. Many buildings are constructed so that radio waves will not pass through them. The new family of radios may correct this problem, but all units within the built-up area may not have these radios. Therefore, radio is an alternate means of communication.

    c. Visual signals may also be used but are often not effective because of the screening effects of buildings, walls, and so forth. Signals must be planned, widely disseminated, and understood by all assigned and attached units. Increased noise makes the effective use of sound signals difficult.

    d. Messengers can be used as another means of communication.

1-12. STRESS

A related problem of combat in built-up areas is stress. Continuous close combat, intense pressure, high casualties, fleeting targets, and fire from a concealed enemy produce psychological strain and physical fatigue for the soldier. Such stress requires consideration for the soldiers' and small-unit leaders' morale and the unit's esprit de corps. Stress can be reduced by rotating units that have been committed to heavy combat for long periods.

1-13. RESTRICTIONS

The law of war prohibits unnecessary injury to noncombatants and needless damage to property. This may restrict the commander's use of certain weapons and tactics. Although a disadvantage at the time, this restriction may be necessary to preserve a nation's cultural institutions and to gain the support of its people. Units must be highly disciplined so that the laws of land warfare and ROE are obeyed. Leaders must strictly enforce orders against looting and expeditiously dispose of violations against the UCMJ.

1-14. FRATRICIDE AVOIDANCE

The overriding consideration in any tactical operation is the accomplishment of the mission. Commanders must consider fratricide in their planning process because of the decentralized nature of execution in the MOUT environment. However, they must weigh the risk of fratricide against losses to enemy fire when considering a given course of action. Fratricide is avoided by doctrine; by tactics, techniques, and procedures; and by training.

    a. **Doctrine.** Doctrine provides the basic framework for accomplishment of the mission. Commanders must have a thorough understanding of US, joint, and host nation doctrine.

    b. **Tactics, Techniques, and Procedures.** TTP provide a "how-to" that everyone understands. TTP are disseminated in doctrinal manuals and SOPs.

        (1) *Tactics.* Tactics is the employment of units in combat or the ordered arrangement and maneuver of units in relation to each other and or the enemy in order to use their full potential.

        (2) *Techniques.* Techniques are the general and detailed methods used by troops or commanders to perform assigned missions and functions. Specifically, techniques are the methods of using weapons and personnel. Techniques describe a method, but not the only method.

        (3) *Procedures.* Procedures are standard, detailed courses of action that describe how to accomplish a task.

        (4) *Planning.* A simple, flexible maneuver plan that is disseminated to the lowest level of command will aid in the prevention of fratricide. Plans should make the maximum possible use of SOPs and battle drills at the user level. They should incorporate adequate control measures and fire support planning and coordination to ensure the safety of friendly troops and allow changes after execution begins.

        (5) *Execution.* The execution of the plan must be monitored, especially with regard to the location of friendly troops and their relationship to friendly fires. Subordinate units must understand the importance of accurately reporting their position.

    c. **Training.** The most important factor in the prevention of fratricide is individual and collective training in the many tasks that support MOUT.

        (1) *Situational awareness.* Well-trained soldiers accomplish routine tasks instinctively or automatically. This allows them to focus on what is happening on the battlefield. They can maintain an awareness of the relative location of enemy and friendly forces.

        (2) *Rehearsal.* Rehearsal is simply training for the mission at hand. Commanders at every level must allow time for this critical task.

        (3) *Train to standard.* Soldiers that are trained to the Army standard are predictable. This predictability will be evident to any NCO or officer who may be required to lead them at a moments notice or who is observing their maneuvers to determine if they are friend or foe.

CHAPTER 2

URBAN ANALYSIS

Intelligence preparation of the battlefield (IPB) is key to all operations conducted in built-up areas—intelligence is an important part of every combat decision. To succeed as fighters in built-up areas, commanders and leaders must know the nature of built-up areas. They must analyze its effect on both enemy and friendly forces. The focus of the material presented in this chapter will be on those issues of urban analysis that commanders and their staffs must be aware of before conducting the IPB process. (For a detailed explanation of IPB in the urban battle, see FM 34-130.)

Section I. MODELS OF BUILT-UP AREAS

Each model of an urban area has distinctive characteristics. Most urban areas resemble the generalized model shown in Figure 2-1.

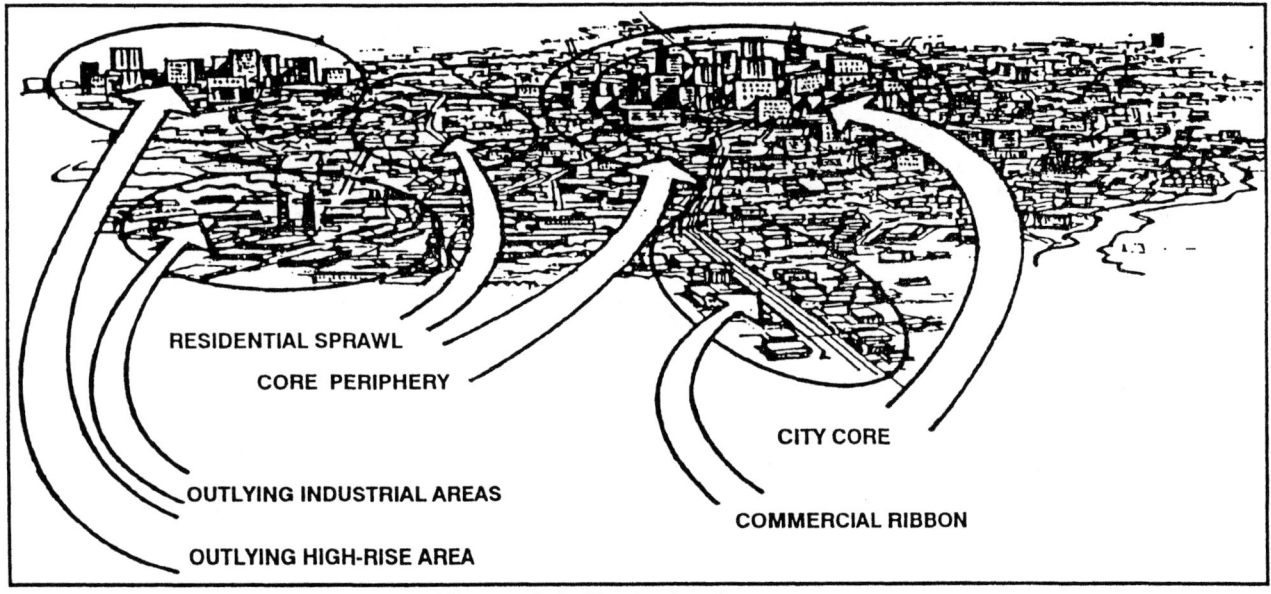

Figure 2-1. Typical urban area.

2-1. REGIONAL URBAN CHARACTERISTICS

Cities of the world are characterized by density of construction and population, street patterns, compartmentalization, affluent and poor sections, modernization, and presence of utility systems. The differences in built-up areas are in size, level of development, and style.

　　a. Due to colonization, most major cities throughout the world have European characteristics. They have combination street patterns, distinct economic and ethnic sections, and areas known as shanty towns. All of which present obstacles to vehicles. Also, concrete and steel high-rise structures hinder wall breaching and limit radio communications.

　　b. Variations in cities are caused mainly by differences in economic development and cultural needs. Developed and developing countries differ more in degree and style rather than in structure and function. Major urban trends are: high-rise apartments, reinforced concrete construction, truck-related industrial storage, shopping centers, detached buildings, suburbs at outer edges, and apartment complexes.

c. The spatial expanse of cities throughout the world in the last three decades presents problems for MOUT. The increased use of reinforced concrete framed construction is only one example of the trend to use lighter construction, which affects how forces will attack or defend such an area. Another example is the growing apartment complexes, shopping centers, and truck-related industrial storage that lie on the outskirts of towns and cities. This change in style causes offensive action to be more difficult and enhances the defense of such an area.

2-2. SPECIFIC CHARACTERISTICS OF URBAN AREAS

A summary of regional urban characteristics is as follows:

a. **Middle East and North Africa.** All nations in the region can be reached by sea and urbanization rates are high. This region has long, hot, dry summers and mild winters, making life outside cities difficult. In spite of its vast deserts, greater urban congestion has resulted. Ancient cities have expanded into their current metropolises, and many new cities have been created because of the petroleum industry (mainly in the Persian Gulf). European influence and petroleum revenues have resulted in urban centers with modern sections of multistory buildings.

b. **Latin America.** Most urban centers can be reached by sea with many capitals serving as ports. This is a region that has mainly a tropical climate. It has a strong Spanish influence characterized by broad avenues that radiate outward from a central plaza with a large church and town hall. Upper and middle class sections combine with the urban centers, while the lower class sections are located farther out. The poor sections are located in slums at the outer edges of the city.

c. **Far East.** Except for Mongolia, all nations in this region can be reached by sea. Urbanization is dense, especially in coastal cities where modern commercial centers are surrounded by vast industrial developments and residential districts.

d. **South Asia.** This region has great European influence with wide busy streets that are overcrowded. Urban centers may be composed mainly of poorer native sections with few or no public services and alleys no more than a yard wide.

e. **Southeast Asia.** This region also has strong European influences with all capitals and major cities serving as seaports. Urban centers contain both the older, high-density native quarters with temples or religious shrines, and modern sections with boulevards, parks, and warehouses.

f. **Sub-Sahara Africa.** In contrast to other regions, this region cannot be accessed by sea and has impassable terrain. Except for a few kingdoms, towns did not exist before the arrival of the Europeans. As a result, urban areas are relatively modern and without "an old quarter," although many do have shanty towns.

2-3. CHARACTERISTICS OF URBAN AREAS

A typical urban area consists of the city core, commercial ribbon, core periphery, residential sprawl, outlying industrial areas, and outlying high-rise areas.

a. In most cities, the core has undergone more recent development than the core periphery. As a result, the two regions are often quite different. Typical city cores of today are made up of high-rise buildings, which vary greatly in height. Modern planning for built-up areas allows for more open spaces between buildings than in old city cores or in core peripheries. Outlying high-rise areas are dominated by this open construction style more than city cores (Figures 2-2 and 2-3).

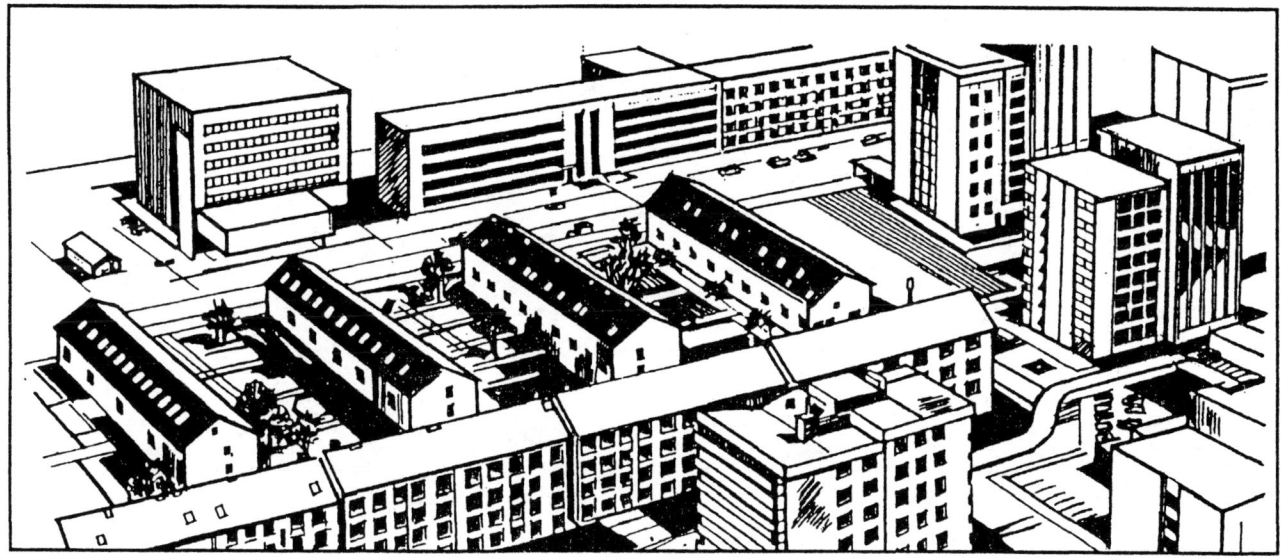

Figure 2-2. City core.

Figure 2-3. Outlying high-rise area.

b. Commercial ribbons are rows of stores, shops, and restaurants that are built along both sides of major streets through built-up areas. Usually, such streets are 25 meters wide or more. The buildings are uniformly two to three stories tall—about one story taller than the dwellings on the streets behind them (Figure 2-4).

Figure 2-4. Commercial ribbons.

c. The core periphery consists of streets 12 to 20 meters wide with continuous fronts of brick or concrete buildings. The building heights are fairly uniform—2 or 3 stories in small towns, 5 to 10 stories in large cities (Figure 2-5).

Figure 2-5. Core periphery.

d. Residential sprawl and outlying industrial areas consist of low buildings that are 1 to 3 stories tall. Buildings are detached and arranged in irregular patterns along the streets with many open areas (Figures 2-6 and 2-7).

Figure 2-6. Residential sprawl.

Figure 2-7. Outlying industrial areas.

Section II. TERRAIN AND WEATHER ANALYSIS

Terrain analysis for urban combat differs significantly from that of open country, whereas a weather analysis does not. Although special considerations peculiar to the urban environment must be considered, a weather analysis for urban combat is mostly the same as for other operations. (See Appendix H for more detailed information.)

2-4. SPECIAL TERRAIN CONSIDERATIONS

Several special considerations have implications in a terrain analysis and must be considered when developing the tactical plan for combat. Special terrain products must be developed to include specialized overlays, maps, and plans augmented by vertical or hand-held imagery. The depiction of GO, SLOW-GO, NO-GO, obstacles, avenues of approach, key terrain, observation and fields of fire, and cover and concealment must focus on the terrain analysis.

   a. Military maps, normally the basic tactical terrain analysis tool, do not provide sufficient detail for a terrain analysis in built-up areas. Due to growth, towns and cities are constantly adding new structures and demolishing existing ones. Therefore, any map of a built-up area, including city maps or plans published by the city, state, or national government, will be inaccurate and obsolete.

   b. The nature of combat can radically alter the terrain in a built-up area in a short period. Incidental or intentional demolition of structures can change the topography of an area and destroy reference points, create obstacles to mobility, and provide additional defensive positions for defenders.

   c. Maps and diagrams of sewer systems, subway systems, underground water systems, elevated railways, mass transit routes, fuel and gas supply and storage facilities, electric power stations and emergency systems, and mass communications facilities (radio, telephone) are of key importance during urban operations. Sewer and subway systems provide covered infiltration and small-unit approach routes. Elevated railways and mass transit routes provide mobility between city sectors, and point to locations where obstacles might be expected. Utility facilities are key targets for insurgents, guerrillas, and terrorists, and their destruction can hinder the capabilities of a defending force.

   d. Certain public buildings must be identified during the terrain-analysis phase of an IPB. Hospitals, clinics, and surgical facilities are critical because the laws of war prohibit their attack when not being used for military purposes other than medical support. As command and control breaks down during urban operations, hospitals become an important source of medical support to combat forces. The locations of civil defense, air raid shelters, and food supplies are critical in dealing with civilian affairs. The same is true during insurgency, guerrilla, or terrorist actions.

   e. Stadiums, parks, sports fields, and school playgrounds are of high interest during both conventional and unconventional operations in built-up areas. They provide civilian holding areas, interrogation centers, insurgent segregation areas, and prisoner of war holding facilities. These open areas also provide helicopter landing sites. They provide logistic support areas and offer air resupply possibilities because they are often centrally located within a city or city district.

f. Construction sites and commercial operations, such as lumberyards, brickyards, steelyards, and railroad maintenance yards, serve as primary sources of obstacle and barrier construction materials when rubble is not present or is insufficient. They can also provide engineers with materials to strengthen existing rubble obstacles or with materials for antitank hedgehogs or crib-type roadblocks.

g. Roads, rivers, streams, and bridges provide high-speed avenues of movement. They also provide supporting engineer units locations to analyze demolition targets and to estimate requirements for explosives.

h. Public baths, swimming facilities, and cisterns are useful in providing bathing facilities. They also provide an alternate water source when public utilities break down.

i. A close liaison and working relationship should be developed with local government officials and military forces. In addition to information on items of special interest, they may provide information on the population, size, and density of the built-up area; fire fighting capabilities; the location of hazardous materials; police and security capabilities; civil evacuation plans; and key public buildings. They may also provide English translators, if needed.

2-5. SPECIAL WEATHER CONSIDERATIONS

Some weather effects peculiar to an urban environment are discussed herein.

a. Rain or melting snow often floods basements and subway systems. This is especially true when automatic pumping facilities that normally handle rising water levels are deprived of power. Rain also makes storm and other sewer systems hazardous or impassable. Chemical agents are washed into underground systems by precipitation. As a result, these systems contain agent concentrations much higher than surface areas and become contaminated "hot spots." These effects become more pronounced as agents are absorbed by brick or unsealed concrete sewer walls.

b. Many major cities are located along canals or rivers, which often creates a potential for fog in the low-lying areas. Industrial and transportation areas are the most affected by fog due to their proximity to waterways.

c. Air inversion layers are common over cities, especially cities located in low-lying "bowls" or in river valleys. Inversion layers trap dust, chemical agents, and other pollutants, reducing visibility, and often creating a greenhouse effect, which causes a rise in ground and air temperature.

d. The heating of buildings during the winter and the reflection and absorption of summer heat make built-up areas warmer than surrounding open areas during both summer and winter. This difference can be as great as 10 to 20 degrees, and can add to the already high logistics requirements of urban combat.

e. Wind chill is not as pronounced in built-up areas. However, the configuration of streets, especially in closed-orderly block and high-rise areas, can cause wind canalization. This increases the effects of the wind on streets that parallel the wind direction, while cross-streets remain relatively well protected.

f. Light data have special significance during urban operations. Night and periods of reduced visibility favor surprise, infiltration, detailed reconnaissance, attacks across open areas, seizure of defended strongpoints, and reduction of defended obstacles. However, the difficulties of night navigation in restrictive terrain, without reference points and near the enemy, forces reliance on simple maneuver plans with easily recognizable objectives.

Section III. THREAT EVALUATION AND INTEGRATION

Threat evaluation for urban combat uses a three-step process: developing a threat data base, determining enemy capabilities, and developing a doctrinal template file as threat evaluation for open terrain. Due to the unique aspects of urban combat, certain operational factors and future threat capabilities must be recognized. These factors must be considered before preparing the required templates during threat integration of the IPB process.

2-6. OPERATIONAL FACTORS

The basic tenets of AirLand Battle doctrine are the rapid deployment and employment of US forces across the operational spectrum to achieve national and strategic objectives. This doctrinal concept, and recent changes in the international security environment, presupposes the increasing chance of conflict with regional threats. These conflicts will be with the conventional forces of one or more Third World nations, to include the possibility of a regional war or, at the lower end of the operational spectrum, combat operations against insurgent forces. Because of the political and socioeconomic structures within the Third World, urban combat will be a greater probability in the future.

    a. Most regular armies emphasize managing combined arms operations in built-up areas. Among the conventional force structures, the poorer the nation, the less likely it is to field, maneuver, and sustain forces beyond logistic centers. Also, the extreme environment in some regions restricts operations beyond urban centers.

    b. Urban structural characteristics are shaped by social, cultural, and economic factors. These elements are the prime reasons that MOUT doctrine and tactics differ between nations. Coupled with the restrictive nature of urban combat, the differences in tactics may be superficial. More than any other factor, the advent of high technology, precision weapons has enabled nations to modify and update their MOUT doctrine and tactics. Research has revealed many factors to consider in the planning and execution of MOUT. Some key factors are—

    (1) Urban combat is only combat in different terrain. Urban combat consumes **time.** A well-planned defense, even if cut off or lacking in air, armor, or artillery weapons, can consume a great deal of an attacker's time.

    (2) The ability to control military operations in highly decentralized circumstances remains the priority for both attacker and defender. Personnel training and motivation continue to be as important as equipment or force balance factors.

    (3) The required size of the attacking force depends on the quality of intelligence, degree of surprise, and degree of superior firepower the attacker can achieve rather than the degree of sophistication with which the defender has prepared the city.

    (4) The degree of a defender's resistance depends on whether or not he is separated from the local population, is wholly or partly cut off from external support, or has effective communication systems.

(5) The belief that armor has no role in city fighting is wrong. Tanks and APCs have proven vital to the attacker inside the city as long as they are protected by dismounted infantry.

(6) If the attacker is subject to any constraints, the defender has a good chance of winning or prolonging the battle, thus raising the cost for the attacker.

(7) The defender has three tactical options: defense in depth, key sector defense, and mobile defense. Defense in depth suggests an outer and inner defense combination; key sector defense means strongpoint defense of vital positions, mainly those controlling major avenues of approach; and mobile defense is based on counterattacks. These are not mutually exclusive options.

(8) Exfiltration and movement within a city by small groups are easy at night.

(9) The prevention of the reentry of cleared buildings by the enemy will be a significant challenge to both the attacker and defender.

(10) Mortars may be used more heavily than other artillery in MOUT due to their immediate response and high-angle fire capabilities.

(11) The employment of snipers in urban combat can prove to be extremely effective for both the attacker and defender. Snipers are usually found two to three stories below the top floor in high buildings.

(12) Ammunition consumption is five to ten times greater in urban environments than in field environments. (See Chapter 7 for more information.)

2-7. URBAN COUNTERINSURGENCY, COUNTERGUERRILLA, AND COUNTERTERRORIST OPERATIONS

During urban counterinsurgency, counterguerrilla, and counterterrorist operations, threat evaluation is similar to that for low-intensity conflict. When conducting these operations, the five low-intensity imperatives (political dominance, unity of effort, adaptability, legitimacy, and perserverance) must be followed. (See FM 7-98 for more information.)

a. Population status overlays are prepared for the city, showing potential neighborhoods or districts where a hostile population could be encountered. Overlays are also prepared showing insurgent or terrorist safe houses, headquarters, known operating areas, contact points, and weapons supply sources. These overlays must include buildings that are known, or could become, explosives, ammunition, or weapons storage sites.

b. Underground routes are of primary concern when considering insurgent and terrorist avenues of approach and lines of communications. Sewers, subways, tunnels, cisterns, and basements provide mobility, concealment, cover, and storage sites for insurgents and terrorists. Elevated railways, pedestrian overpasses, rooftops, fire escapes, balconies, and access ladders provide mobility and concealment, and can serve as relatively good fighting or sniper positions.

c. Although doctrinal templates are not developed for urban insurgency and terrorist operations, pattern analysis reveals how the insurgent or terrorist group operates, and what its primary targets are. Once the group's method of operation is determined, insurgent situation maps can be developed. These maps pinpoint likely sabotage targets, kidnap or assassination targets, ambush points, and bombing targets. When developing these maps, electric power generation and transmission facilities, gas production and holding facilities, water and sewer pumping and treatment plants, telephone exchanges and facilities, and radio and television stations should be considered as primary insurgent and terrorist targets.

d. If the enemy has, for whatever reason, become intermingled with the population, a greater degree of control is required for military operations. While detection is more difficult, the enemy forces operating without uniforms share some common characteristics with guerrillas, insurgents, and terrorists.

(1) As with any operation of this type, intelligence, rather than force, plays the dominant role. Known members of the armed forces, their auxiliaries, and the underground must be identified and arrested and or removed from the populace. Use of minimum force is critical. As a last resort, cordon and search into suspected or known hostile areas may be used. This is the least preferred method since it will cause moderate to severe casualties for both the friendly forces and the local civilian population.

(2) The local population's support to the enemy may be either forced or given willingly. In either case, an effort must be made to separate the enemy from the local population base. A population can be forced into giving support by a combination of terrorism (either by coercion or intimidation) and harassment. The friendly force commander must be observant and sensitive to the local population's concerns before the population may be willing to help the friendly forces.

(3) Logistical support will be in smaller packages. The enemy must rely on the local population to support the distribution of logistics so that identification and destruction of the logistics base is more difficult. To curb resupply operations totally, the friendly forces would have to stop all movement within the built-up area. For obvious reasons, this is not an option. Therefore, a priority for intelligence should be to identify and destroy the enemy's logistics base.

(4) Soldiers must remember the political and psychological impact of their actions if they use force. The local population may be neutral or have luke-warm support for the friendly forces, but excessive use of force will cause the local civilians to support the enemy. Of special concern is the media (newspapers, television, magazines, and so forth). Due to the large numbers of journalists and amateur and or professional photographers in built-up areas, any negative images of friendly forces will probably be published. This negative publicity could have a serious adverse effect on both civilian opinion and United States political interests. Conversely, positive publicity can greatly enhance friendly operations and morale. This also can sway the local population away from the enemy. Therefore, all media members should be accompanied.

(5) While not officially part of doctrine, infantry forces have historically been used as a part of the effort to separate the enemy from the local civilian populace. Some units may be detailed to provide civil services such as law enforcement patrols, trash pick-up, and or the restoration and maintenance of power, telephone, and water services.

2-8. PROJECTED THREAT CAPABILITIES

The wealth of some Third World nations will be used to modernize their armed forces through the acquisition of new technologies. Future conflicts may be against Third World forces armed better than or equal to US weapon systems. Projected future threat force capabilities are—

- New munitions such as fuel air explosives (FAE), enhanced blast, intense light, and other improved ballistic technologies.
- Systems with interchangeable warheads, some designed for MOUT combat.
- Precision-guided munitions.
- Robotics.
- Day or night target acquisition systems.
- Elevated gun systems.
- Improved engineering abilities to breach or emplace obstacles.
- Soft-launch hand-held AT and flame weapons.
- Nonlethal incapacitating chemical or biological agents by conventional forces.
- Lethal chemical or biological agents by insurgent forces.
- Improved self-protection (body armor).
- Improved communications.

CHAPTER 3

OFFENSIVE OPERATIONS

Good cover and concealment in a built-up area gives the defender an advantage. Attackers must fight from the outside into a well-defended position. While a decision to attack a major built-up area usually rests at a level higher than battalion, commanders at all levels must be prepared to fight in such areas.

Section I. OFFENSIVE CONSIDERATIONS

A commander must decide if attacking a built-up area is needed to accomplish his mission. He should consider those issues discussed in this section.

3-1. REASONS FOR ATTACKING BUILT-UP AREAS

A commander should consider the following reasons for attacking a built-up area.

a. Cities control key routes of commerce and provide a tactical advantage to the commander who controls them. Control of features, such as bridges, railways, and road networks, can have a significant outcome on future operations. The requirement for a logistics base, especially for a port or airfield, may play a pivotal role during a campaign.

b. The political importance of some built-up areas may justify the use of time and resources to liberate it. Capturing the city could deal the threat a decisive psychological blow and or lift the moral of the people within the city.

c. Though the terrain around a built-up area may facilitate its bypass, the enemy within that urban area may be able to interdict lines of communications. Therefore, the situation may require the enemy force to be contained. Also, the urban area itself may sit on dominating terrain that would hinder bypassing for CS and CSS elements.

d. The results of the commander's and staff's estimate may preclude bypassing as an option. The mission itself may dictate an attack of a built-up area.

3-2. REASONS FOR NOT ATTACKING A BUILT-UP AREA

The unit's mission may allow it to bypass an urban area. The commander should consider the following reasons for not attacking a built-up area.

a. The commander may decide to bypass if he determines that no substantial threat exists in the built-up area that could interdict his unit's ability to accomplish its mission. Also, the commander's intent may dictate that speed is essential to the mission. Since combat in an urban area is time consuming, the commander may choose to bypass the urban area to save time.

b. During the estimate process, the commander and staff may realize that a sufficient force is not available to seize and clear the built-up area. A situation may exist where more than enough forces are available to accomplish the mission but logistically the attack cannot be supported. If the tactical and political situation allow it, the commander should avoid attacks on the built-up area.

c. The built-up area is declared an "open city" to prevent civilian casualties or to preserve cultural or historical sites. An open city, by the law of land warfare, is a city that cannot be defended or attacked. The defender must immediately evacuate the open city and cannot distribute weapons to the city's inhabitants. The attacker assumes administrative control of the city and must treat its citizens as noncombatants in an occupied country.

Section II. CHARACTERISTICS OF OFFENSIVE OPERATIONS IN A BUILT-UP AREA

Offensive operations in urban areas are based on offensive doctrine modified to conform to the area. Urban combat also imposes a number of demands that are different from ordinary field conditions such as problems with troop requirements, maneuver, and use of equipment. As with all offensive operations, the commander must retain his ability to fix and maneuver against enemy positions.

3-3. TROOP REQUIREMENTS
Due to the nature of combat in built-up areas, more troops are normally needed than in other combat situations. This is mainly due to the requirement to clear buildings in a given zone or objective, refuge control, and the possible increase in the number of friendly casualties.

  a. Because of the need to clear buildings and provide security for forces in the attack, the number of troops required to accomplish an offensive mission will be much greater. Some forces must be left behind in a building once it has been cleared to prevent enemy forces from repositioning on or counterattacking friendly forces.

  b. Commanders must also consider the soldiers' fatigue. Room clearing techniques are highly physical and will quickly tire a force. Commanders must plan for the relief of their forces before they reach the point of exhaustion.

  c. Additional forces may be needed to control the civilians in the built-up area. These forces must protect civilians, provide first aid, and prevent them from interfering with the tactical plan.

  d. Fighting in a built-up area normally results in a greater number of friendly casualties than does conventional fighting. The ability to see the enemy is fleeting and confined to very short ranges compared to ordinary field combat. Fratricide can become a serious problem and must be addressed in detail by the commander. Evacuating casualties from the MOUT environment also presents a problem.

3-4. MANEUVER
Combat operations in a built-up area have a slower pace and tempo, and an increase in methodical, synchronized missions. Unlike open terrain, commanders cannot maneuver quickly due to the close, dense environment. Clearing buildings and looking for antiarmor ambushes degrade speed, thus increasing the duration of enemy contact. Due to the dense environment and the restricted ability to use all available weapon systems, synchronization of combat power will be one of the commander's main challenges.

3-5. USE OF EQUIPMENT
Commanders attacking a built-up area must recognize some important limitations in the use of available assets.

a. Normally, the use of indirect fires is much more restricted in built-up areas than in open terrain. Consideration must be given to the effects of the indirect fire on the urban area and the civilian population. When indirect fires are authorized, they must be fired in greater mass to achieve the desired effect. The rubbling caused by massive preparatory indirect fires will adversely affect a unit's maneuvers later on in the attack.

b. Communications equipment may not function properly because of the massive construction of buildings and the environment. More graphic control measures and understanding of the commander's intent at all levels become even more important to mission accomplishment.

c. The commander and his staff must consider the effect city lights, fires, and background illumination have on night vision devices. These elements "white out" NVGs and make thermal imagery identification difficult.

Section III. TYPES OF OFFENSIVE OPERATIONS

Offensive operations in a built-up area are planned and implemented based on the factors of METT-T and established doctrine. At battalion level, the offense takes the form of either a hasty or deliberate attack. Both the hasty and deliberate attacks are characterized by as much planning, reconnaissance, and coordination as time and the situation permit.

3-6. HASTY ATTACK

Battalions and companies conduct hasty attacks as a result of a movement to contact, a meeting engagement, or a chance contact during a movement; after a successful defense or part of a defense; or in a situation where the unit has the opportunity to attack vulnerable enemy forces. When contact is made with the enemy, the commander immediately deploys; suppresses the enemy; attacks through a gap, flank, or weak point; and reports to his higher commander. The preparation for a hasty attack is similar to that of a deliberate attack, but time and resources are limited to what is available. The hasty attack in a built-up area differs from a hasty attack in open terrain because the close nature of the terrain makes command, control, and communications difficult. Also, massing fires to suppress the enemy may be difficult.

a. In built-up areas, incomplete intelligence and concealment may require the maneuver unit to move through, rather than around, the friendly unit fixing the enemy in place. Control and coordination become important to reduce congestion at the edges of the built-up area.

b. On-order missions, be-prepared missions, or fragmentary orders may be given to a force conducting a hasty attack so it can react to a contingency once its objective is secured.

3-7. DELIBERATE ATTACK

A deliberate attack is a fully synchronized operation that employs all available assets against the enemy's defense. It is necessary when enemy positions are well prepared, when the built-up area is large or severely congested, or when the element of surprise is lost. Deliberate attacks are characterized by precise planning based on detailed information, thorough reconnaissance, preparation, and rehearsals.

Given the nature of urban terrain, the deliberate attack of a built-up area is similar to the techniques employed in assaulting a strong point. Attacking the enemy's main strength is avoided and combat power is focused on the weakest point of his defense. A deliberate attack of a built-up area is usually conducted in the following phases:

    a. **Reconnoiter the Objective.**

    b. **Move to the Objective.**

    c. **Isolate the Objective.** Isolating the objective involves seizing terrain that dominates the area so that the enemy cannot supply or reinforce its defenders. This step may be taken at the same time as securing a foothold. If isolating the objective is the first step, the following steps should be enacted quickly so that the defender has no time to react (Figure 3-1).

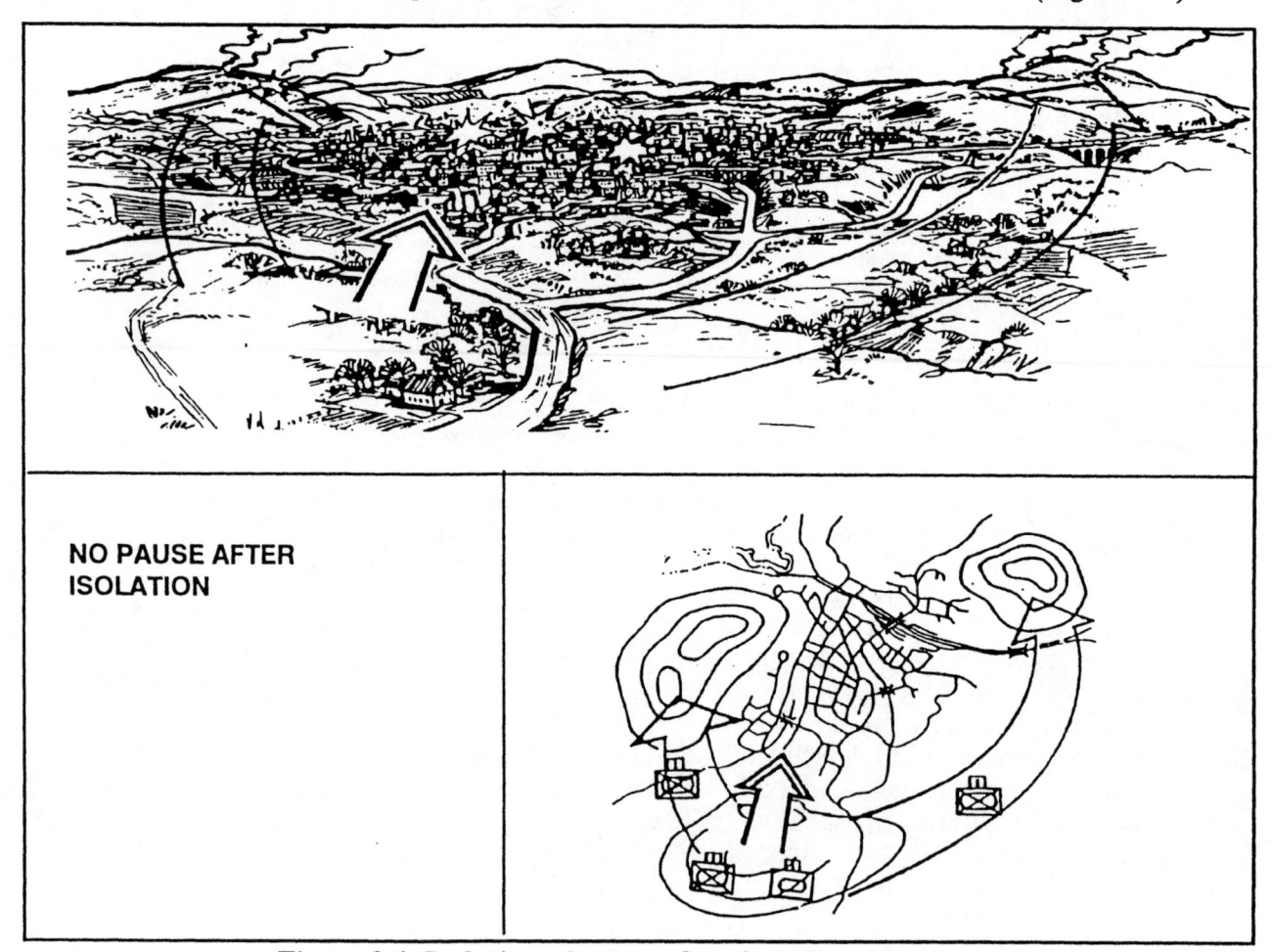

NO PAUSE AFTER ISOLATION

Figure 3-1. Isolation of an area by a battalion task force.

    d. **Secure a Foothold.** Securing a foothold involves seizing an intermediate objective that provides cover from enemy fire and a place for attacking troops to enter the built-up area. A foothold is normally one to two city blocks and is an intermediate objective of a company. As the company attacks to secure the foothold, it should be supported by suppressive fire and smoke (Figure 3-2).

FM 90-10-1

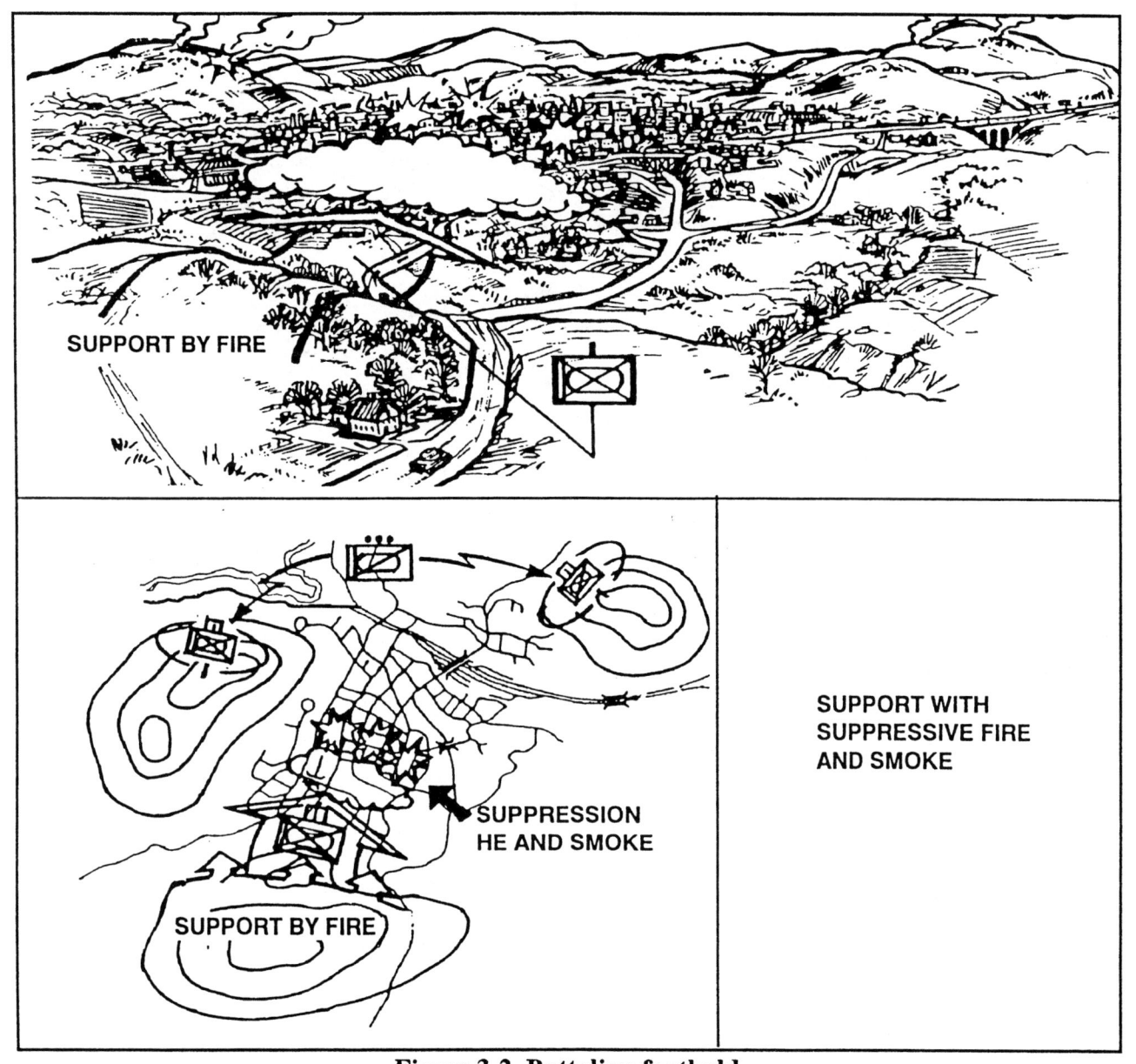

Figure 3-2. Battalion foothold.

e. **Clear a Built-up Area.** Before determining to what extent the built-up area must be cleared, the factors of METT-T must be considered. The commander may decide to clear only those parts necessary for the success of his mission if—
- An objective must be seized quickly.
- Enemy resistance is light or fragmented.
- The buildings in the area are of light construction with large open areas between them. In that case, he would clear only those buildings along the approach to his objective, or only those buildings necessary for security (Figure 3-3, page 3-6).

3-5

Figure 3-3. Clearing buildings along the route of an attack.

A unit may have a mission to systematically clear an area of all enemy. Through detailed analysis, the commander may anticipate that he will be opposed by a strong, organized resistance or will be in areas having strongly constructed buildings close together. Therefore, one or two companies may attack on a narrow front against the enemy's weakest sector. They move slowly through the area, clearing systematically from room to room and building to building. The other company supports the clearing units and is prepared to assume their mission (Figure 3-4).

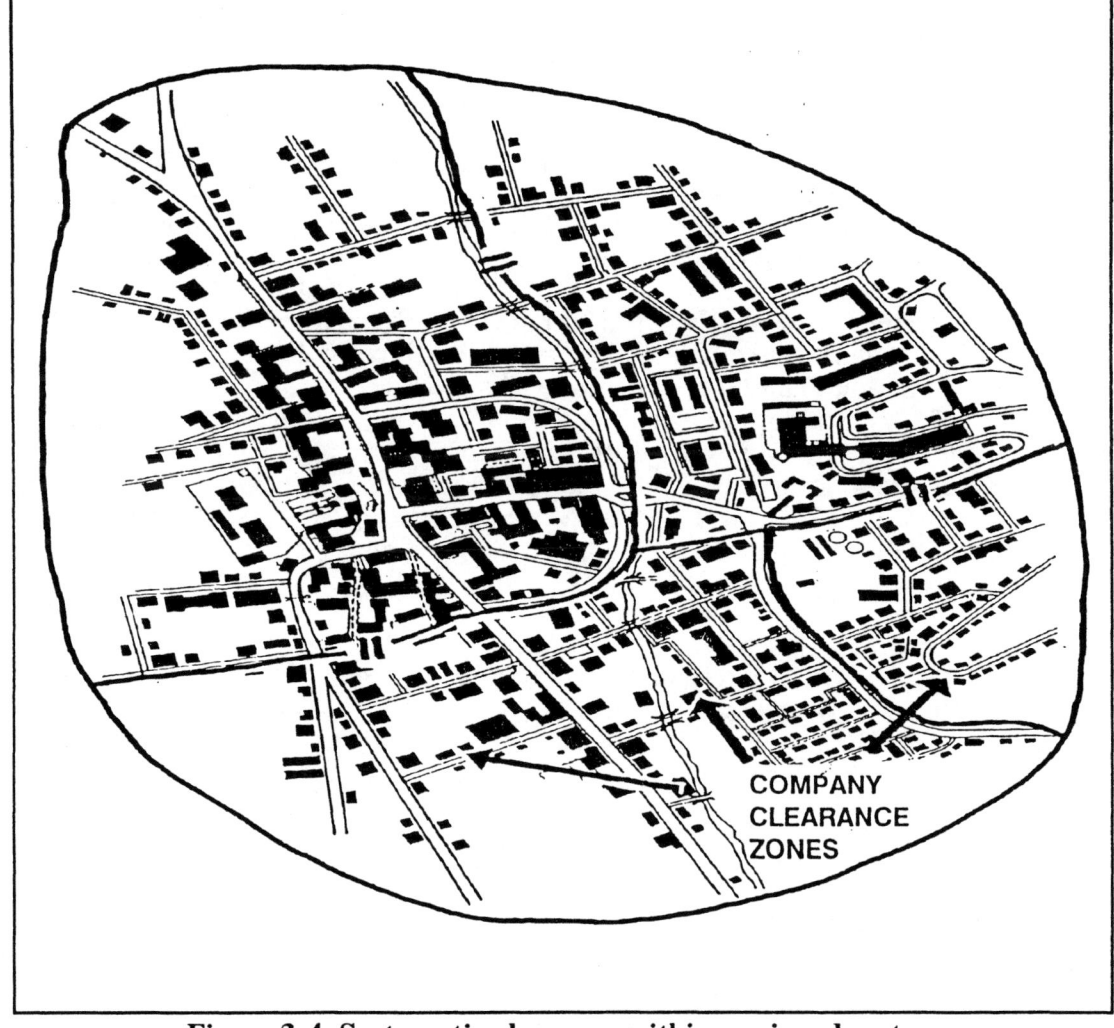

Figure 3-4. Systematic clearance within assigned sectors.

Section IV. METT-T FACTORS

The planning, preparation, and conduct of offensive operations in an urban area are the same as any other offensive operation. An attack plan against a well-defended built-up area must be based on METT-T factors. Commanders must focus on the synchronization of maneuver forces and the fire support plan to accomplish the assigned mission. Combat support and combat service support will play a critical role in the offense. (See Chapters 6 and 7 for further details on CS and CSS.)

3-8. MISSION
When conducting the estimate, commanders and staffs must consider the overall intent of the operation in regard to the requirement for clearance of the urban area. The commander must determine if clearance means every building, block by block, or the seizure of a key objective, which may only require clearing along the axis of advance.

3-9. ENEMY
The enemy is analyzed in detail using the IPB process (FM 34-130). The unique factor the commander must decide on to complete the IPB process is the type threat he is attacking. He must determine if the threat forces are conventional or unconventional. This determines how the battalion or company will task-organize and how combat power will be synchronized to accomplish the mission.

a. **Conventional Forces.** Most third world countries have adopted techniques of urban combat from either the United States or the Commonwealth of Independent States. Therefore, the future threat will consider the motorized or mechanized rifle battalion the most effective unit for urban combat because of its inherent mobility, armor protection, and ability to quickly adapt buildings and other structures for defense.

(1) Threat defenses are organized into two echelons to provide greater depth and reserves. Company strongpoints are prepared for perimeter defense and form the basis for the battalion defensive position. The reserve is located in a separate strongpoint. Ambush locations are established in the gaps of the strongpoints, and dummy strongpoints are constructed to deceive the attacker. Positions for securing and defending the entrances to and exits from underground structures and routes are established. Security positions are prepared forward of first echelon defensive positions.

(2) Within a built-up area, a motorized/mechanized rifle company may defend several buildings with mutually supporting fires or a single large building. Each platoon defends one or two buildings, or one or two floors of a single building.

b. **Unconventional Forces.** Enemy analysis is similar to that for LIC during urban counterinsurgency, counterguerrilla, and counterterrorist operations. (See FMs 34-130 and 7-98 for details of IPB in counterinsurgency operations.)

3-10. TERRAIN
Offensive operations must be tailored to the urban environment based on a detailed analysis of each urban terrain setting, its types of built-up areas, and existing structural form. (See FM 34-130 for details of urban terrain analysis.) Commanders and subordinate leaders must incorporate the following special planning considerations for an urban environment when conducting an offensive operation:

- Military maps that do not provide enough detail for urban terrain analysis or reflect the underground sewer system, subways, underground water system, mass transit routes, and utility generation.
- Natural terrain surrounding the built-up area.
- Key and decisive terrain (stadiums, parks, sports fields, school play grounds, public buildings, and industrial facilities).
- Confined spaces that limit observation, fields of fire, and maneuver, which also prevents the concentration of fires at critical points.
- Covered and concealed routes to the urban area.
- Covered and concealed routes within the built-up area.

- Limited ability to employ maximum combat power due to the need to minimize damage and rubbling effects.
- A greater demand for ammunition and rations, thus imposing unusual strains on logistics elements.
- Problems with conducting effective reconnaissance during conventional operations. (Reconnaissance by force becomes the most effective reconnaissance means. This method involves probing a defense with successively larger units until the enemy positions are disclosed and can be successfully attacked. During unconventional operations, the opposite is true. Reconnaisance and security are easily accomplished by both sides and may be unstoppable.)

3-11. TROOPS

In an attack on a large built-up area, a battalion would probably participate as part of an attacking brigade. In that case, the battalion may have to isolate the objective or seize a foothold. If the objective is a smaller built-up area, a battalion or company may accomplish the entire mission independently, assigning subordinate tasks to its companies or platoons. In either case, the maneuver platoons accomplish their entry and clearance tasks as explained in Appendix F.

   a. When attacking to seize a foothold, the battalion normally assigns a forward company the first block of buildings as its first objective. When an objective extends to a street, only the near side of the street is included. The company's final objective may be buildings at the far edge of the built-up area or key terrain on the far side. Key buildings or groups of buildings also may be assigned as intermediate objectives. Buildings along the route of attack should be identified by numbers to simplify assigning objectives and reporting (Figure 3-5).

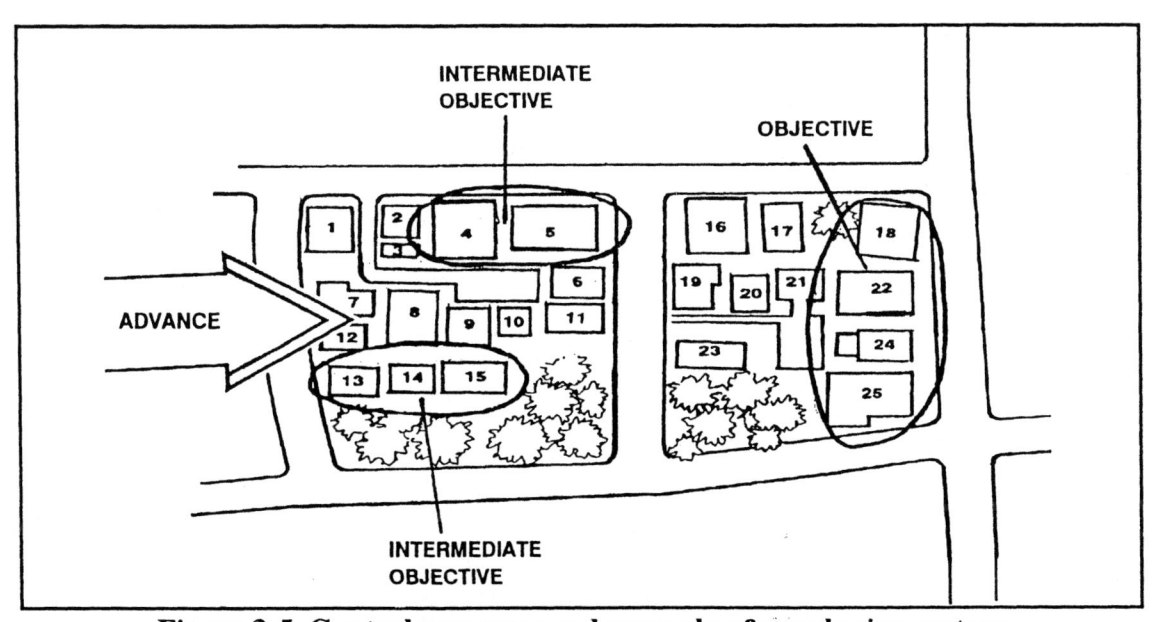

Figure 3-5. Control measures and example of numbering system

FM 90-10-1

b. When the unit is involved in clearing, bypassing buildings increases the risk of attack from the rear or flank. Thus, the clearing unit must enter, search, and clear each building in its zone of action. A single building may be an objective for a rifle squad, or if the building is large, for a rifle platoon or even a company. When the commander's concept is based on speed or when conducting a hasty attack, a battalion may be directed not to clear its entire zone.

c. Phase lines can be used to report progress or to control the advance of attacking units. Principal streets, rivers, and railroad lines are suitable phase lines, which should be on the near side of the street or open area. In systematic clearing, a unit may have the mission to clear its zone of action up to a phase line. In that case, the unit commander chooses his own objectives when assigning missions to his subunits.

d. Battalion and company boundaries are usually set within blocks so that a street is included in a company zone. Boundaries must be placed to ensure that both sides of a street are included in the zone of one unit (Figure 3-6).

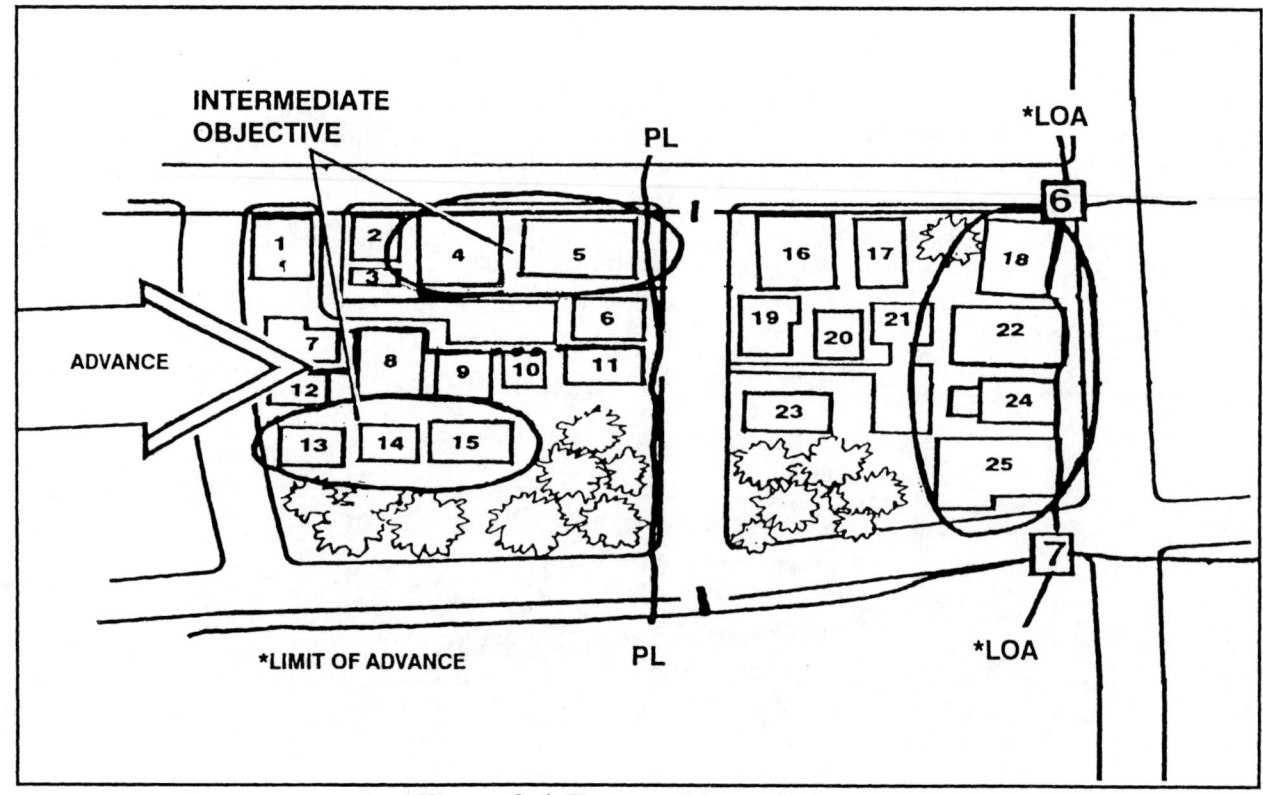

Figure 3-6. Zone and boundaries.

e. Checkpoints and contact points are planned at street corners, buildings, railway crossings, bridges, or any other easily identifiable feature. Checkpoints aid in reporting locations and controlling movement. Contact points are used to designate specific points where units make physical contact.

3-10

f. An attack position may be occupied by forward units for last-minute preparation and coordination. The attack position is often behind or inside the last large building before crossing the LD. The LD should be the near side of either a street or rail line.

g. A unit's assigned frontage for the attack of a built-up area depends on the size of buildings and the resistance anticipated. A company normally attacks on a one- to two-block front, and a battalion on a two- to four-block front, based on city blocks averaging 175 meters in width.

h. The first phase of the attack should be conducted when visibility is poor. Troops should exploit poor visibility to cross open areas, to gain access to rooftops, to infiltrate enemy areas, and to gain a foothold. If the attack must be made when visibility is good, units should use smoke to conceal movement.

i. The formation used in an attack depends on the width and depth of the zone to be cleared, the character of the area, enemy resistance, and the formation adopted by the next higher command.

j. The reserve should be mobile and prepared for commitment. Because of the available cover in built-up areas, the reserve can stay close to forward units. Battalion reserves normally follow one to two blocks to the rear of the lead company. If a company reserve is available, it follows within the same block so that it can immediately influence the attack. A unit with a reserve mission may be called upon to perform one or more of the following tasks:

- Attacking from another direction.
- Exploiting an enemy weakness or friendly success.
- Clearing bypassed enemy positions.
- Securing the rear or a flank.
- Maintaining contact with adjacent units.
- Supporting or counterattacking by fire.

k. The reconnaissance platoon is normally employed to reconnoiter the battalion's flanks and rear. Its capability for reconnaissance and security is somewhat reduced in built-up areas. The reconnaissnce platoon can also help isolate a village or small town. They must be prepared to dismount and enter buildings for reconnaissance or for setting up OPs. Infantry platoons and squads conduct reconnaissance patrols and man OPs to supplement the reconnaissance platoon effort.

l. Leading companies may have engineers attached for providing immediate support. Engineers equipped with the M728 combat engineer vehicle (CEV) can quickly clear rubble and other obstructions using the blade or the 165-mm demolition gun. Other tasks given the engineers include:

- Preparing and using explosives to breach walls and obstacles.
- Finding and exploding mines in place or helping to remove them.
- Destroying fortifications to a maximum range of 925 meters with the CEV (165-mm demolition gun).
- Clearing barricades and rubble to ease movement.
- Cratering roads and other countermobility measures.

m. Security in a built-up area presents special problems. All troops must be alert to an enemy that may appear from the flanks, from above, or from underground passages (Figure 3-7).

Figure 3-7. Enemy firing from flank.

n. The fire support plan may require extensive air and artillery bombardment to precede the ground attack on a built-up area. This supporting fire suppresses the defender's fire, restricts his movement, and possibly destroys his position. However, use of indirect fire in built-up areas with heavily clad construction creates rubble. This can be used effectively for cover but may also restrict the movements of attacking troops. For that reason, an artillery preparation should be short and violent. Assaulting troops must closely follow the artillery fire to exploit its effect on the defenders. While the enemy is suppressed by the supporting fire, maneuver units move near the FCL. As the attacking force assaults the objective, fires are lifted or shifted to block enemy withdrawal or to prevent the enemy from reinforcing their position.

o. Prior coordination is made to determine the techniques and procedures to use for communication, target identification, and shifting of fires. Additional consideration must be given to the civilian population, houses of worship, medical centers, schools, public services, and historical monuments. The fire support plan can include the integration of tanks, infantry weapons, artillery, CEVs, and dismounted fires. Fire support can be categorized into indirect and direct fires.

(1) Indirect fire is employed in its normal role of support to the maneuver units.

(a) Indirect artillery fire is planned to isolate objectives, to prevent reinforcement and resupply, to neutralize known and suspected command and observation posts, and to suppress enemy defenders. Due to the restricted nature of urban terrain, most indirect artillery fires will be high-angle.

(b) Mortars are the most responsive indirect fire that can hit targets of opportunity at the close ranges typical of combat in built-up areas. Forward observers move with the forward units to adjust fire on targets as requested by the supported troops.

(2) The direct-fire system is the most effective fire support in built-up areas. Once a target can be located in a building, one or two direct-fire rounds can accomplish what entire salvos of indirect-fire artillery cannot. Direct fire support is key to success in fighting in built-up areas. The best direct fire support is provided by BFVs, but it can also be provided by tanks, howitzers, and or CEVs. (See Chapter 8 for specific weapons effects.) Tanks, howitzers, and CEVs may create rubble and building and street damage that could restrict movement for the attacking force.

(a) Tanks may support by fire when lead units are seizing a foothold. During the attack of a built-up area, tanks overwatch the infantry's initial assault until an entry into the area has been secured. Tanks must be supported by infantry organic weapons to suppress enemy strongpoints and by ATGMs while they move into overwatch positions. The commander must employ tanks to take advantage of the long range of their main armament. This can usually be achieved with tanks employed outside the built-up area, where they remain for the duration of the attack to cover high-speed armor avenues of approach. This is especially true during the isolation phase.

(b) In house-to-house and street fighting, tanks and or BFVs move down the streets protected by the infantry, which clears the area of enemy ATGM weapons. Tanks and BFVs in turn support the infantry by firing their main guns and machine guns from a safe stand-off range to destroy enemy positions. Tanks are the most effective weapon for heavy fire against structures and may be used to clear rubble with dozer blades (Figure 3-8, page 3-14). The BFV can provide sustained, accurate suppressive fires with its 25-mm gun.

(c) Large-caliber artillery rounds that are shot by direct fire are effective for destroying targets in buildings. If available, self-propelled 155-mm howitzers can use direct fire to destroy or neutralize bunkers, heavy fortifications, or enemy positions in reinforced concrete buildings (Figure 3-9, page 3-14). The self-propelled 155-mm can also be used to clear or create avenues of approach. The 105-mm and 203-mm artillery can also be used in this role. However, due to the exposed positions of the gun crew, these are not the preferred artillery pieces used in MOUT operations. In any case, whenever artillery is used in the direct fire role, it must be close to the infantry who will provide security against enemy ground attack. Prior coordination must be accomplished so the bulk of the field artillery unit's shells are switched to HE.

Figure 3-8. Tank in direct fire supported by infantry.

Figure 3-9. Artillery in direct-fire role.

(d) Tanks, self-propelled artillery, and BFVs are vulnerable in built-up areas where streets and alleys provide ready-made fire lanes for defenders. Motorized traffic is greatly restricted, canalized, and vulnerable to ambush and close-range fire. Tanks are at a further disadvantage because their main guns cannot be depressed sufficiently to fire into basements or be elevated to fire into upper floors of buildings at close range (Figure 3-10).

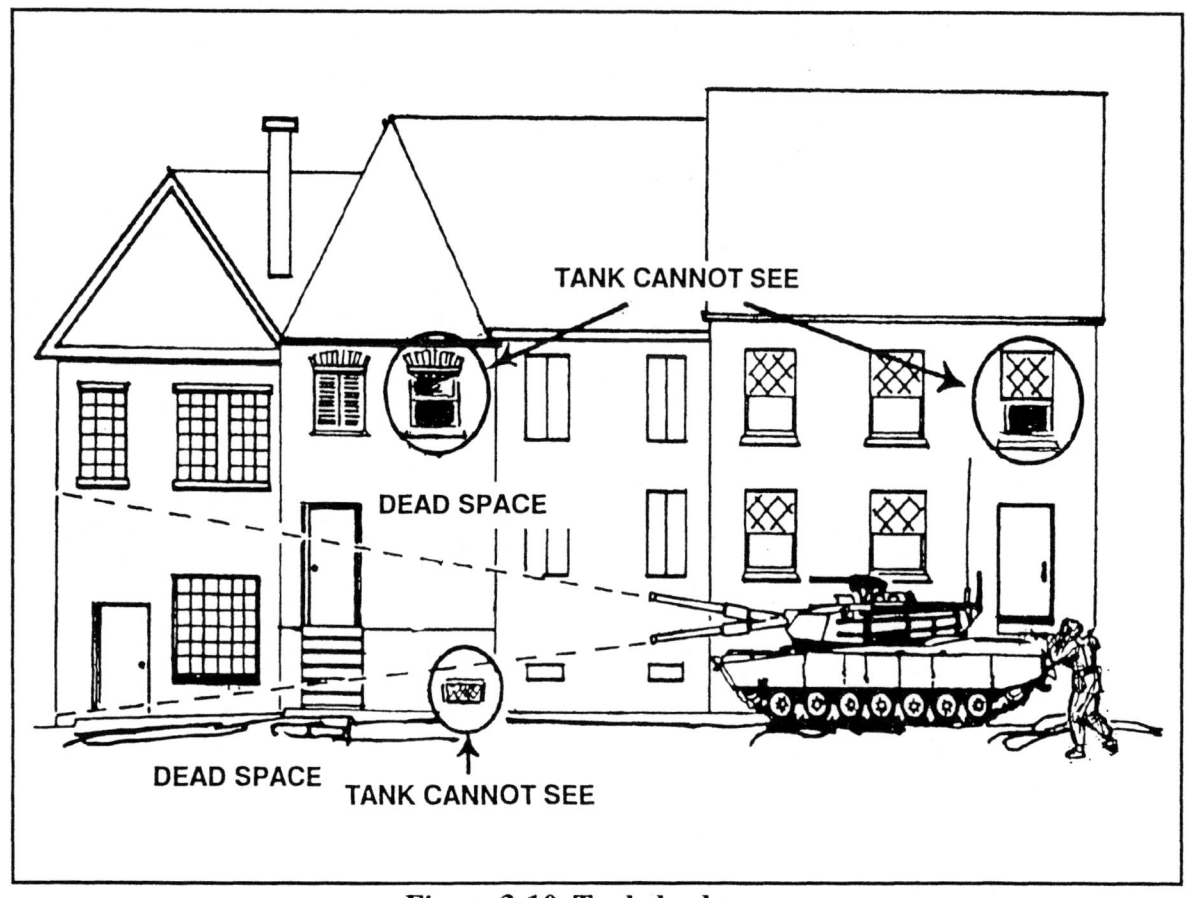

Figure 3-10. Tank dead space.

(e) In movement down narrow streets, or down wider streets with narrow paths through the debris, infantry should move ahead of the tanks, clearing the buildings on each side. The movement of personnel across open areas must be planned with a specific destination in mind. Suppression of enemy positions and smoke to cover infantry movement should also be included in the plan. When needed, tanks move up to places secured by the infantry to hit suitable targets. When that area is cleared, the infantry again moves forward to clear the next area. Due to the restricted movement and limited observation of buttoned-up tanks, the infantry must clear the route in advance of the tanks. The tanks and infantry should use the traveling overwatch movement technique. Infantrymen can communicate with the tank crews by using arm-and-hand signals and radio.

(f) For movement down wider streets, infantry platoons normally have a section of attached tanks with one tank on each side of the street—tanks should not be employed singly. Other tanks of the attached tank platoon should move behind the infantry and fire at targets in the upper stories of buildings. In wide boulevards, commanders may employ a tank platoon secured by one or more infantry platoons. The infantry can secure the forward movement of the lead tanks, while the rearward tanks overwatch the movement of the lead units.

(g) If an infantry unit must travel streets that are too narrow for this type of tank support, it uses tanks in single file for support. The tanks move and fire to cover each other's approach while the infantry provides ATGM fire from buildings.

(h) Where feasible, tanks may drive inside buildings or behind walls for protection from enemy antitank missile fire. Buildings should first be cleared by the infantry. Ground floors should be checked to ensure they will support the tank or that there is no basement into which the tank could fall and become trapped. When moving, all bridges and overpasses should be checked for mines and booby traps, and for load capacity. Specific infantry elements should be assigned to protect specific tanks.

(i) Direct-fire systems organic to infantry battalions—mainly ATGMs, recoilless rifles (in some units), and LAWs—are initially employed to support the seizure of a foothold. Then, if necessary, they are brought forward to fight enemy armor within the town. Positioning of antitank weapons in buildings must allow for enough space for backblasts. Antitank weapons are not as effective as tank rounds for neutralizing targets behind walls. They neutralize a target only if that target is located directly behind the point of impact. ATGMs are at a greater disadvantage because of their 65-meter arming distance and the possibility of their guiding wires becoming caught on the ground clutter. These factors limit employment in close engagements like those in built-up areas.

p. Snipers are a valuable asset during MOUT operations. In situations where the ROE permit the use of destructive force, snipers can be used as part of the security element to provide accurate, long-range fires. Depending on the commander's concept, snipers can be dedicated to the counter-sniper role or assigned priority targets. If a restrictive ROE is in effect, the sniper may be the best asset the battalion or company commander has to prevent collateral damage. Snipers can also overwatch breaching operations and call for indirect artillery fires. Regardless of the mission, snipers must be equipped with effective observation devices and placed in a key area to be effective. (For more information on the offensive employment of snipers, see Change 1 to FM 71-2, FM 7-20, and TC 23-14.)

3-12. TIME
Offensive operations in built-up areas have a slower pace and tempo of operation. The following issues must be considered when analyzing time available for an attack in urban terrain.

a. Due to the dense environment of urban terrain, more time is required for clearing buildings, blocks, or axes of advance.

b. Troops tire quicker because of stress and additional physical exertion related to clearing.

c. More time must be allowed for thorough reconnaissance and rehearsals. This saves time in the execution of the commander's plan.

Section V. COMMAND AND CONTROL

Units fight separated and isolated from one another in built-up areas. Planning is centralized but execution is decentralized.

3-13. COMMAND
Soldiers and units require mission-type orders that are restrictive in nature. They use detailed control measures to ease decentralized execution. Increased difficulties in command, control, and communications from higher headquarters demand increased responsibility and initiative from junior leaders.

3-14. CONTROL
In built-up areas, radio communications are less effective than field telephones and messengers. Units often fight without continuous communications from higher headquarters, since dependable communications above company level are uncertain. Pyrotechnic signals are hard to see because of buildings and smoke. Voice commands are degraded by the high noise level of battles within and around buildings.

Section VI. BATTALION TASK FORCE ATTACK ON A BUILT-UP AREA

The following are techniques that might be employed by a battalion. These may be independent operations but are normally part of a brigade operation.

3-15. CONDUCT OF DELIBERATE ATTACK
Because companies or company teams may become isolated during the attack, the task force commander should attach some support elements to ensure the success of his plan. Mechanized vehicles (tanks, self-propelled artillery, BFVs, or ITVs) attached to light units must have their own logistics packages. Tanks, BFVs, and ITVs can be used to clear or isolate hardened targets protected by buildings or rubble. Engineers can neutralize obstacles hindering the attack. All of these actions could be modified for use by any type of infantry. The TF commander plans to conduct a deliberate attack by performing the following actions.

    a. **Reconnoiter the Objective.** The commander conducts a thorough reconnaissance of the objective with subordinate leaders to complete the attack plan.

    b. **Move to the Objective.** The TF moves to the objective using covered and concealed routes to approach gaps or lightly held areas or the enemy's flanks and rear. Reconnaissance elements and security elements detect enemy forces, positions, and obstacles to prevent them from interfering with the attack plan. Obstacles encountered are either breached or bypassed. Enemy elements encountered en route are defeated by subordinate elements.

    c. **Isolate the Objective.** The TF commander positions direct and indirect fire elements where they can best support the attack. OPSEC is employed to deceive the enemy as to the time, location, and strength of the attack. The battalion support element provides support to the assault element. The TF commander uses direct and indirect fire support to suppress and kill the enemy, screen the assault element, protect breaching actions, and isolate the enemy by blocking reinforcements and counterattacks.

d. **Secure a Foothold.** The TF assault element kills, captures, destroys, or forces the withdrawal of all enemy on objectives as required by the commander's intent.

e. **Clear the Built-up Area.** The assault force or other designated forces clear the built-up area using the appropriate technique based on commander's intent.

3-16. SEIZURE OF KEY OBJECTIVE
Many built-up areas are built around key features such as road junctions or bridges. The key feature could be a bridge over a river. A normal deliberate attack would not succeed because it might allow the enemy time to destroy the bridge. Instead, the commander must plan a rapid advance through the built-up area, leaving the task of clearing to following units (Figure 3-11).

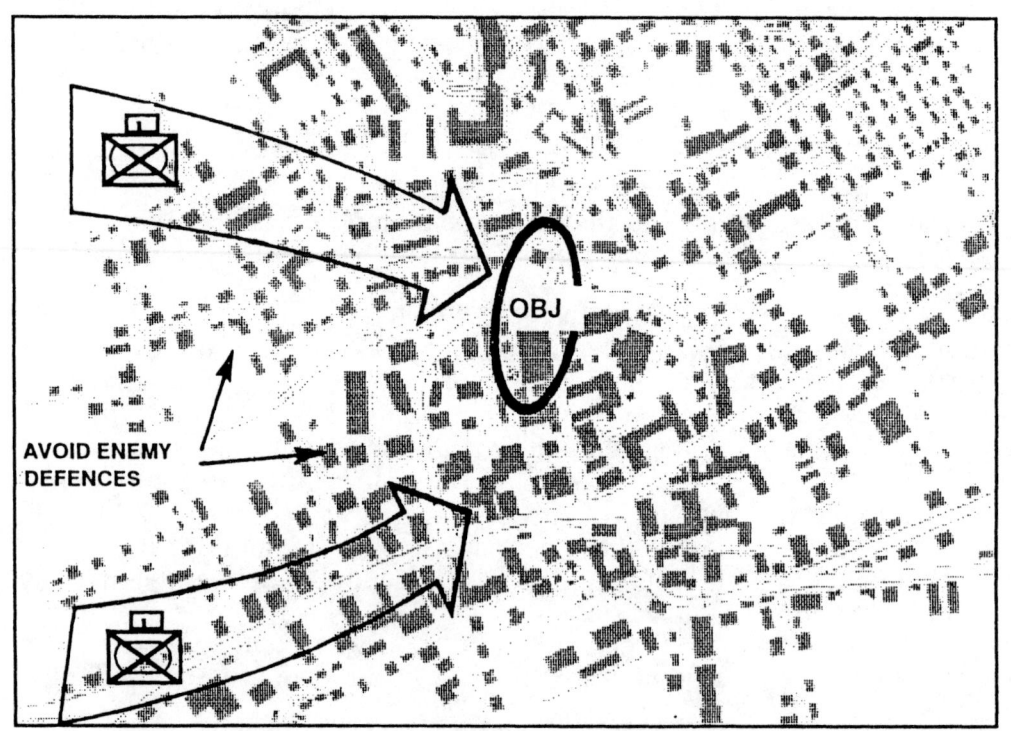

Figure 3-11. Seizure of a key objective.

a. This type of operation has the highest chance of success when the enemy has not had time to set up a well-established defense. Because of the importance of the objective, the prime considerations are to get through the area fast before the enemy can react and to seize the objective while it is still intact.

b. The TF should avoid contact with the enemy. If enemy resistance is encountered, it should be bypassed. Time-consuming combat must be avoided so that the TF can arrive at the bridge as quickly as possible.

c. The TF commander organizes his TF for movement on two axes to allow for more flexibility in reacting to enemy contact. The lead unit on each axis reconnoiters as it moves. Lead units must find enemy positions, fix them by fire, and quickly bypass them.

   d. The units move mounted toward the built-up area. On reaching the edge of the built-up area, troops stay mounted until they meet enemy resistance so as not to slow the advance. Platoons are dropped off to assume blocking positions and to secure the TF advance.

   e. Once the objective is seized, the TF establishes a perimeter defense. The companies clear buildings and expand the size of the perimeter until it is large enough to secure the bridge against enemy action. Attached engineers examine the bridge and clear it of any explosives (Figure 3-11).

3-17. INFILTRATION
The following is an example that describes the actions of a light infantry battalion conducting an infiltration with engineers attached. With some modification, it could also apply to a dismounted mechanized infantry battalion.

   a. The outskirts of a town may not be strongly defended. Its defenders may have only a series of antiarmor positions, security elements on the principal approach, or positions blocking the approaches to key features in the town. The strongpoints and reserves are deeper in the city.

   b. A battalion may be able to seize a part of the town by infiltrating platoons and companies between those enemy positions on the outskirts. Moving by stealth on secondary streets by using the cover and concealment of back alleys and buildings, the battalion may be able to seize key street junctions or terrain features, to isolate enemy positions, and to help following units pass into the built-up area. Such an infiltration should be performed when visibility is poor and no civilians are in the area.

   c. The light infantry battalion is best organized into infiltration companies with engineers attached to each company in platoon strength and a reserve consistent with METT-T. Each company should have an infiltration lane based on the commander's estimate of the situation. Depending on the construction of the built-up areas and streets, the infiltration lane may be 500 to 1,500 meters wide.

   d. The infiltrating companies advance on foot, with stealth, using available cover and concealment. Mortar and artillery fire can be used to divert the enemy's attention and cover the sound of infiltrating troops.

   e. BFVs or TOWs are positioned to cover likely avenues of approach for enemy armored vehicles. The battalion commander may position his antiarmor platoon (light) or company (airborne, air assault) to cover the likely avenues of approach if no BFVs or tanks area available. The reconnaissance platoon and antiarmor company screen the battalion's more vulnerable flanks. Also, the antiarmor company can support by fire if the situation provides an adequate position.

   f. As the companies move into the built-up area, they secure their own flanks. Security elements may be dropped off along the route to warn of a flank attack. Engineers assist in breaching or bypassing minefields or obstacles encountered. Enemy positions are avoided but reported.

   g. The infiltrating companies proceed until they reach their objective. At that time, they consolidate and reorganize and arrange for mutual support. They patrol to their front and flanks, and establish contact with each other. The company commander may establish a limit of advance to reduce chances of enemy contact or to ensure safety from friendly forces.

h. If the infiltration places the enemy in an untenable position and he must withdraw, the rest of the battalion is brought forward for the next phase of the operation. If the enemy does not withdraw, the battalion must clear the built-up area before the next phase of the operation (Figure 3-12).

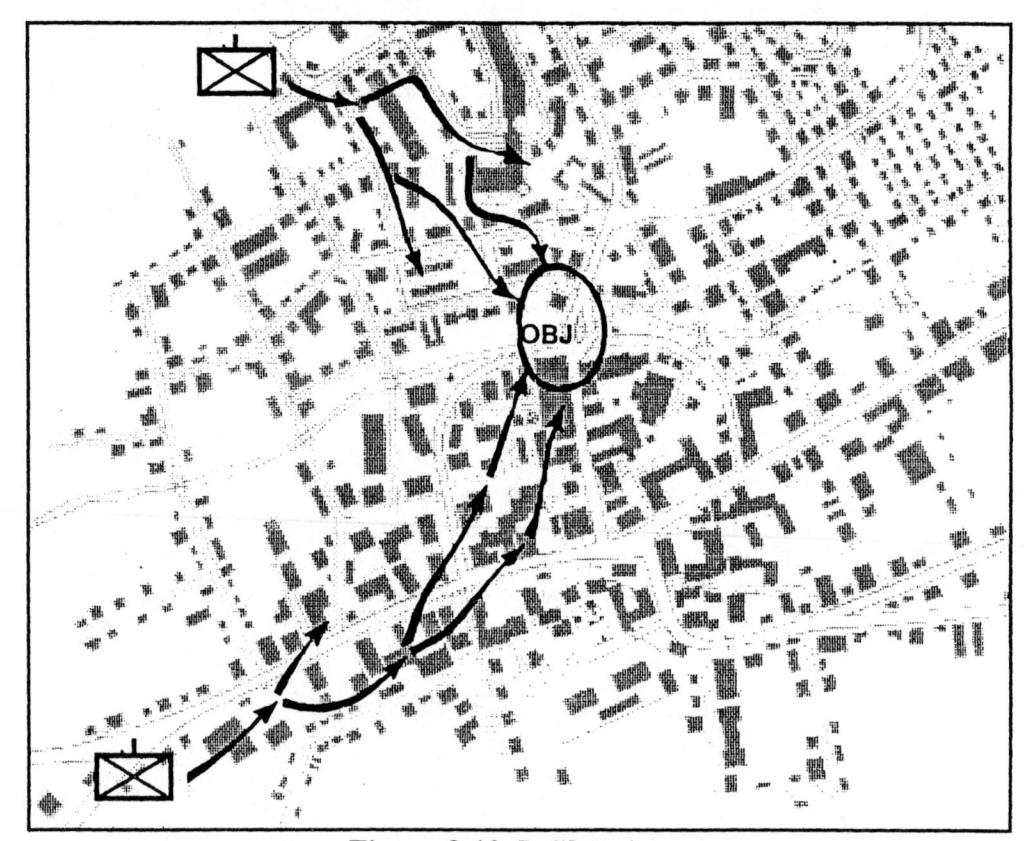

Figure 3-12. Infiltration.

3-18. ROUTE SECURITY
An infantry battalion may have to clear buildings to secure a route through a city. How quickly the battalion can clear the buildings depends on the enemy resistance and the size and number of the buildings. In the outlying area, the forward units proceed by bounds from road junction to road junction. Other platoons provide flank security by moving down parallel streets and by probing to the flanks.

a. Depending on the required speed and enemy situation, the infantry may either move mounted or dismounted. The platoons move down the widest streets, avoiding narrow streets. Each section overwatches the squad to its front, keeping watch on the opposite side of the street. Sections provide their wingman with mutual support. Combat vehicles providing overwatch should be secured by dismounted troops. The rest of the infantry should stay mounted to maximize speed and shock effect until required to dismount due to the enemy situation or upon reaching the objective.

b. When contact with the enemy is made, tanks support as usual. Supporting fire fixes and isolates enemy positions, which dismounted troops maneuver to attack.

c. Phase lines can be used to control the rate of the company's advance and other action. For example, at each phase line, the forward companies might reestablish contact, reorganize, and continue clearing (Figure 3-13).

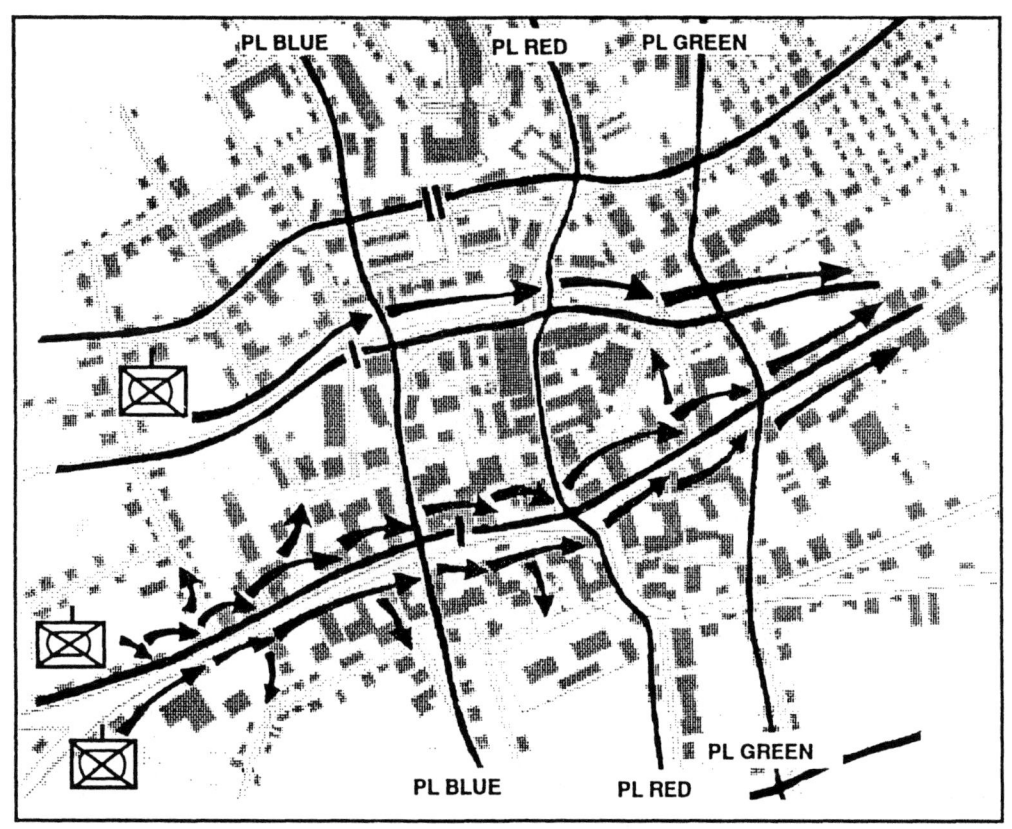

Figure 3-13. Clearing along a route.

Section VII. COMPANY TEAM ATTACK OF A BUILT-UP AREA

The following are techniques that might be employed by a company. These may be independent operations but are normally part of a battalion operation and apply to any type of infantry.

3-19. ATTACK OF A BLOCK

To attack a block in a built-up area, a company should be reinforced with tanks and engineers.

a. This operation is characterized by platoon attacks supported by both direct and indirect fires. Success depends on isolating the enemy positions (which often become platoon objectives), suppressing enemy weapons, seizing a foothold in the block, and clearing the block's buildings room by room.

(1) Task organization of the company team varies because of the nature of the built-up area. For example, a nonmechanized infantry company fighting in the outskirts of a city might organize as follows:
- Two rifle platoons reinforced with engineers—to assault.
- One rifle platoon—reserve.
- One tank platoon—in support of the assaulting rifle platoons.

(2) In a core or core periphery area, that same company might be organized as follows:
- Two rifle platoons, each with engineers and tanks under the platoon leader's operational control (OPCON)—to assault. (The engineers and tanks are placed under the platoon leader's OPCON due to the independent, isolated combat that can be expected in those areas.)
- One platoon—in reserve.
- All available direct and indirect fire weapons should be used to isolate objective buildings. Direct fire down streets and indirect fire in open areas between buildings helps in the objective isolation.

b. Tanks, machine guns, and other direct fire support weapons fire on the objective from covered positions. These weapons should not be fired for prolonged periods from one position. The gunners should use a series of positions and displace from one to another to gain better fields of fire and to avoid being targeted by the enemy. Direct fire support tasks are assigned as follows:
- Machine guns fire along streets and into windows, doors, and so forth.
- BFVs, tanks, TOWs, and Dragons fire at enemy tanks and other armored vehicles.
- Tanks fire at targets protected by walls, make entrances in buildings, and provide backup against enemy tanks, as required.
- Riflemen engage targets of opportunity.

c. Before an assault, the company commander should employ smoke to conceal the assaulting platoons. He secures their flanks with direct fire weapons and by employment of the reserve, if necessary.

(1) Concealed by smoke and supported by direct fire weapons, an assaulting platoon attacks the first isolated building. The platoon must close on the building quickly while the enemy is still stunned by supporting fire. The company commander must closely coordinate the assault with its supporting fire so that the fire is shifted at the last possible moment.

(2) The squads and platoons clear each building as described in Appendix E. After seizing the block, the company consolidates and reorganizes to repel a counterattack or to continue the attack.

(3) A mechanized infantry company team would be organized on similar lines. The assault platoons should be dismounted. The BFV and tanks can provide direct fire support (Figure 3-14).

(4) The company commander may or may not use the technique of numbering the buildings in the area of the attack. In the assault of a strongpoint, the strongpoint itself may have the corners lettered to identify enemy forces.

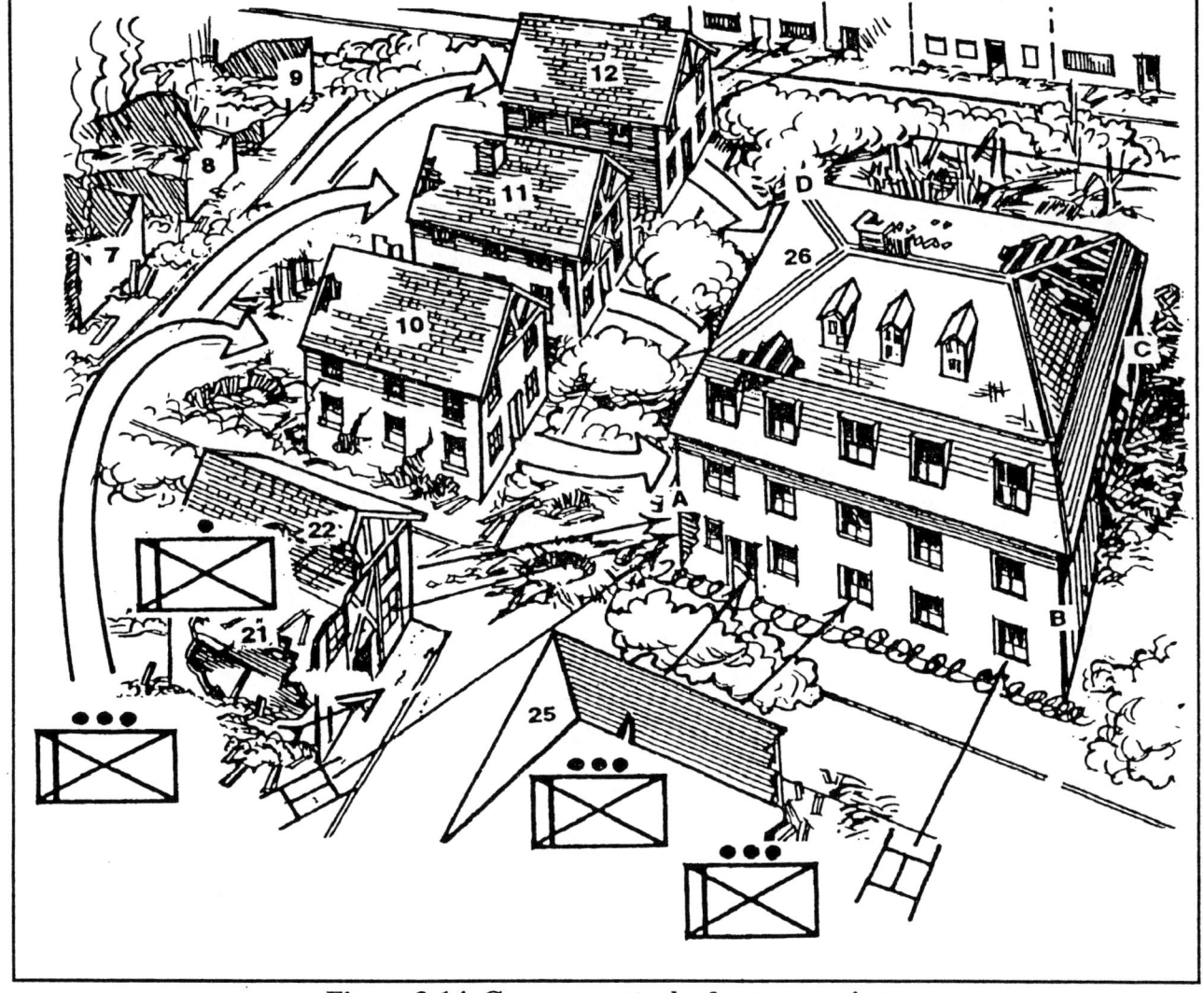

Figure 3-14. Company attack of a strongpoint.

3-20. ATTACK OF AN ENEMY OUTPOST

The following discussion provides a technique for conduct of a hasty attack on an enemy outpost. The company team commander makes a quick assessment of the factors of METT-T and reacts appropriately to support the commander's intent.

   a. The company team commander uses a form of fire and movement. His tanks, BFVs, MK 19s or M2HBs mounted on HMMWVs, and TOWs assume support-by-fire positions from which they can fire on the outpost, keep the enemy from escaping, and destroy any reinforcements.

   b. The rifle platoons then move into the area. They do not attack head on, but from a covered route so as to hit the outpost at a vulnerable point. As the platoons approach the outpost, smoke is employed to screen their movement and supporting fires are shifted. Once the platoons close on the outpost, they clear the buildings quickly and consolidate. The company is then ready to continue operations (Figure 3-15, page 3-24).

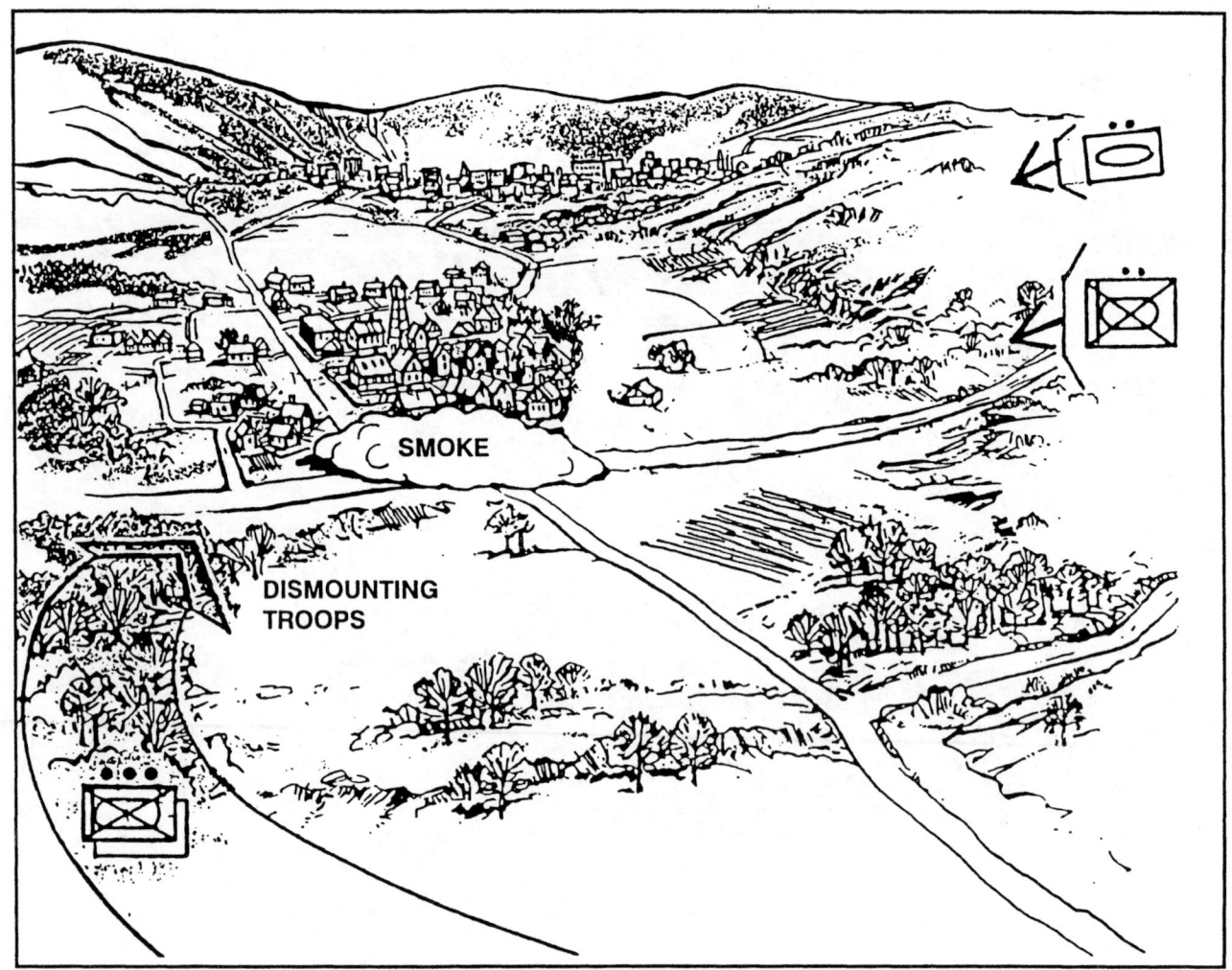

Figure 3-15. Hasty attack of an outpost.

3-21. SEIZURE OF A TRAFFIC CIRCLE
A company may have to seize a traffic circle either to secure it for friendly use or to deny it to the enemy (Figure 3-16). This operation consists of seizing and clearing the buildings that control the traffic circle bringing direct-fire weapons into position to cover it. Routes to overwatch positions for direct fire weapons may have to first be cleared of mines. Enemy avenues of approach into the flanks of the position may have to be mined to prevent enemy use.

    a. After gathering all available intelligence on the terrain, enemy, and population, the commander plans for the following steps:

- Isolate the objectives.
- Seize and clear the buildings along the traffic circle under cover of tanks, ATGMs, and machine guns.
- Consolidate and prepare for counterattack.

    b. Friendly troops should not venture into the traffic circle until it is under friendly control. A traffic circle is a natural kill zone.

c. The company should be organized as follows:
- A security element (charged with isolating the traffic circle).
- An assault element reinforced with engineers.
- A support element (providing direct fire support for the assault element) made up of the company's BFVs, TOWs, MK 19s or M2HBs mounted on HMMWVs, and attached tanks occupying an attack-by-fire position.
- A reserve.

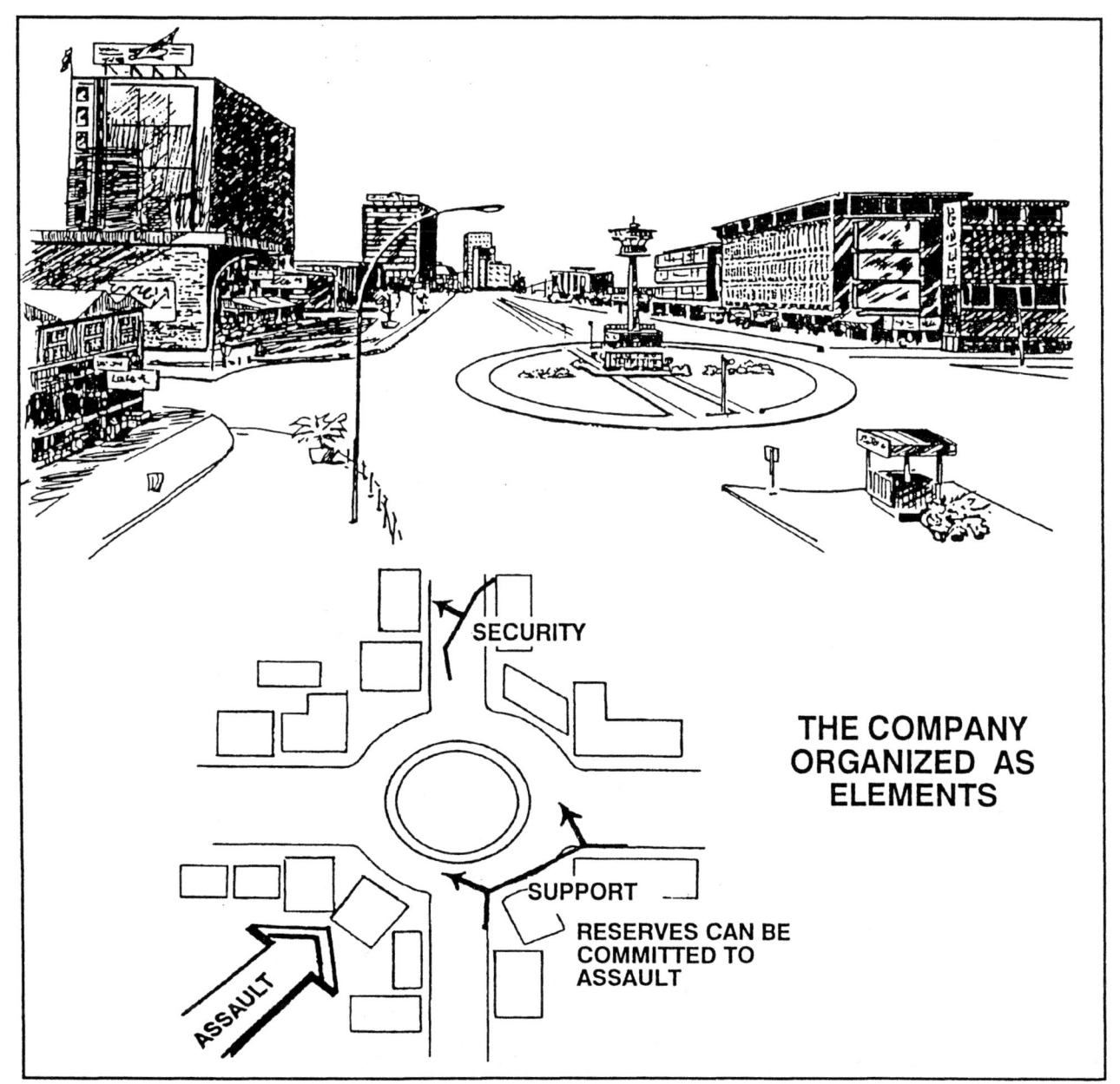

Figure 3-16. Seizure of a traffic circle.

d. At various stages in this operation, those roles may change. For example, the assault element may clear buildings until the support element can no longer support it. Then the reserve can be committed to the assault. It may also happen that one of the assault elements is in a better position to isolate the traffic circle. At that time, the isolating element would become part of the assault element.

3-22. SEIZURE OF KEY TERRAIN
Key terrain dominates an avenue of approach or is a location which, if held by either friendly forces or the enemy, will directly affect the operation. A bridge or overpass that spans a canal, a building complex, or, in some cases, the population itself are examples of key terrain in a city. Therefore, seizing such a crossing point intact and securing it for friendly use is a likely mission for an infantry company.

  a. For this mission, an infantry company should—
  - Clear the buildings on the near bank that permit a clear view of the bridge and provide good fields of fire for supporting weapons.
  - Quickly suppress enemy weapons on the far bank with direct fire.
  - Use screening smoke to limit enemy observation and reduce interference with friendly direct fires.
  - Seize a bridgehead (buildings that dominate the bridge) on the far bank by an assault across the bridge.
  - Secure a perimeter around the bridge so that the engineers can clear any obstacle and remove demolitions on the bridge.

  b. The first step in seizing a bridge is to clear the buildings on the near bank. The commander must find out which buildings dominate the approaches to the bridge. Buildings that permit him to employ LAWs, Dragons, machine guns, and riflemen are cleared while supporting fire prevents the enemy from reinforcing his troops on the far bank and keeps enemy demolition parties away from the bridge.

  c. In suppressing the enemy's positions on the far bank, priority is given to those positions from which the enemy can fire directly down the bridge. Tanks, BFVs, TOWs, and machine guns mounted on HMMWVs in the light infantry antiarmor platoon or the airborne or air assault antiarmor company are effective in this role. TOWs, Dragons, and, in some cases, LAWs can be used against enemy tanks covering the bridge. The company FSO should plan artillery and mortar fires to suppress infantry and antitank weapons.

  d. The objectives of the assaulting platoons are buildings that dominate the approaches to the bridge. One or two platoons assault across the bridge using all available cover while concealed by smoke. They are supported by the rest of the company and attached tanks. Once on the other side, they call for the shift of supporting fire and start clearing buildings. When the first buildings are cleared, supporting fire is shifted again and the assault continues until all the buildings in the objective area are cleared.

  e. At this point, the engineers clear the bridge and its approaches of all mines, demolitions, and obstacles. The company commander may expand his perimeter to prepare for counterattack. Once the bridge is cleared, the tanks and other support vehicles are brought across to the far bank (Figure 3-17).

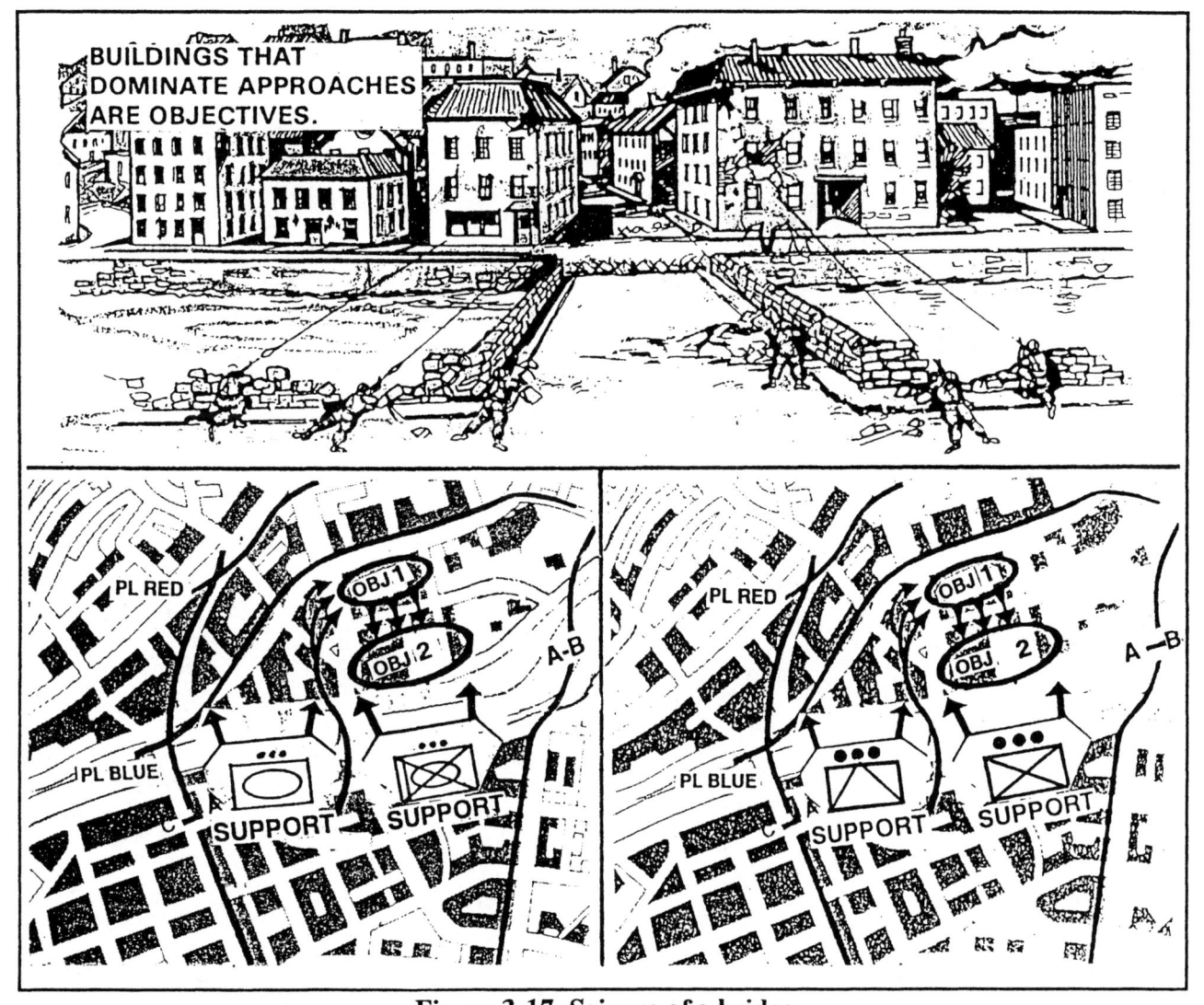

Figure 3-17. Seizure of a bridge.

3-23. RECONNAISSANCE

In a fast-moving situation, a company may have a movement to contact through a built-up area along a highway. Similarly, a company may have to reconnoiter such a route to prepare for a battalion task force attack. This type of mission can be accomplished by an infantry company of any type with an attached tank platoon, if available.

    a. This operation is characterized by alternating periods of rapid movement to quickly cover distances and much slower movement for security. The speed of movement selected depends on the terrain and enemy situation.

    b. In open areas where rapid movement is possible, a tank section should lead. In closer terrain, the infantry should lead while overwatched by the tanks. Another infantry platoon and the other tank section should move on a parallel street. Artillery fire should be planned along the route. Engineers accompany the lead platoon on the main route to help clear obstacles and mines.

c. The team should seize the key points on the highway (crossroads, bridges, and overpasses, and so forth) by a combination of actions:
- Between key points, the team moves with the infantry mounted when contact is not likely.
- At key points or when enemy contact is likely, the team moves dismounted to clear enemy positions or to secure the key point. Tanks and other combat vehicles support the dismounted troops.

d. In peripheral or strip areas, this advance should be on one axis with the lead unit well forward and security elements checking side streets as they are reached. In the city core, this operation is conducted as a coordinated movement on two or three axes for more flank security.

e. Enemy positions can be either destroyed by the team itself or, if the need for speed is great, bypassed, reported, and left to following units.

f. The subunits of the team must coordinate their action. The company commander reports all information collected to the battalion task force (Figure 3-18).

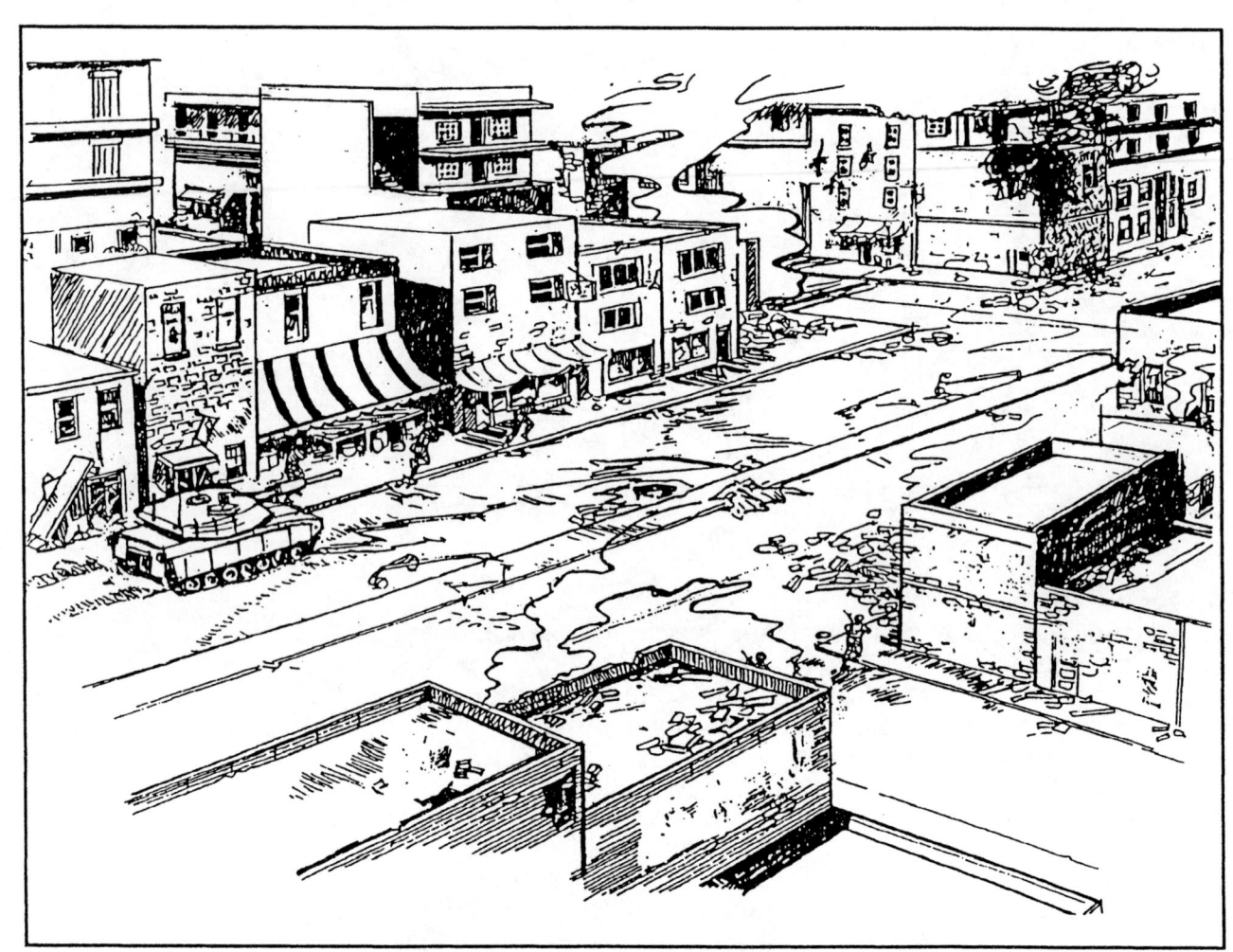

Figure 3-18. Route reconnaissance for a movement to contact along a highway through a city (commercial ribbon).

Section VIII. PLATOON ATTACK OF A BUILT-UP AREA

Platoons seldom perform independent operations in combat in built-up areas, but because of the type of combat to be expected, they can become isolated and seem to be alone. This section discusses techniques that might be employed by a platoon under such conditions. These operations are conducted as part of a company operation.

3-24. ATTACK OF A BUILDING

The most common platoon offensive mission in a built-up area is the attack of a building. The platoon must kill the defenders and secure the building.

a. The attack involves isolating the building to prevent the escape or reinforcement of its defenders (normally coordinated at company level); suppressing the defenders with BFV 25-mm main gun, tank, machine gun, and mortar fire; entering the building at the least-defended point or through a hole breached by tank fire; and clearing the building. To clear it, troops normally go quickly to the top floor and clear from the top down. There must be close coordination between the assault and support elements of the platoon using radios, telephones, arm-and-hand signals, or pyrotechnics.

(1) If a platoon is attacking a building independently, it should be organized with an assault element, support element, and security element to cover its flanks and rear. In addition to its own support element, the platoon can be supported by BFVs or tanks and other elements of the company.

(2) If one platoon is attacking, supported by the rest of the company, security may be provided by the other rifle platoons. The assault has three steps:

STEP 1: Isolate the building.
STEP 2: Enter the building (secure a foothold).
STEP 3: Clear the building methodically room by room and floor by floor.

(3) The clearing is performed by the rifle squads, which pass successively through each other (leapfrogging) as rooms and floors are secured. Platoons that clear buildings should be reinforced with engineers to help with demolition clearing (Figure 3-19, page 3-30).

3-25. MOVEMENT DOWN A STREET

When moving in built-up areas, a platoon follows the same principles of movement as in other areas. However, some movement techniques must be modified to adjust to a built-up area. This discussion focuses on the movement down the street of the lead platoon of a rifle company, either mechanized or nonmechanized.

a. The platoon members must be prepared to return fire immediately. They must also be alert for any signs of the enemy and report this information promptly.

b. The speed of movement depends on the type of operation, terrain, and degree of enemy resistance. In outlying or lightly defended areas, a mechanized infantry platoon proceeds along the street mounted, but sends dismounted men forward to reconnoiter key points (crossroads, bridges). In the center of a built-up area or in situations when there is heavy fighting, the platoon moves on foot with two squads leading—one on each side of the road, using all cover. They move through the buildings, if feasible, to avoid exposure on the streets. The squads give each other mutual support.

Figure 3-19. Attack of a building.

    c. Enemy action against the platoon might consist of an ambush on the street, enfilade fire down the streets, sniper fire from rooftops, or artillery or mortar fire.

    d. For protection from those dangers, the platoon should move through buildings and along walls, use tanks for fire support and station men on the roofs or upper stairs for overwatch, and search for defenders in all three dimensions.

    e. The platoon should move in two elements: a maneuver element (one squad on narrow streets, two squads on wide streets), which moves forward, scouts danger areas, and closes with the enemy; and an overwatch element (the rest of the platoon and its supporting weapons), which moves behind the maneuver element, secures the flanks and rear, and provides fire support. These two elements, or parts of them, can exchange roles (Figure 3-20).

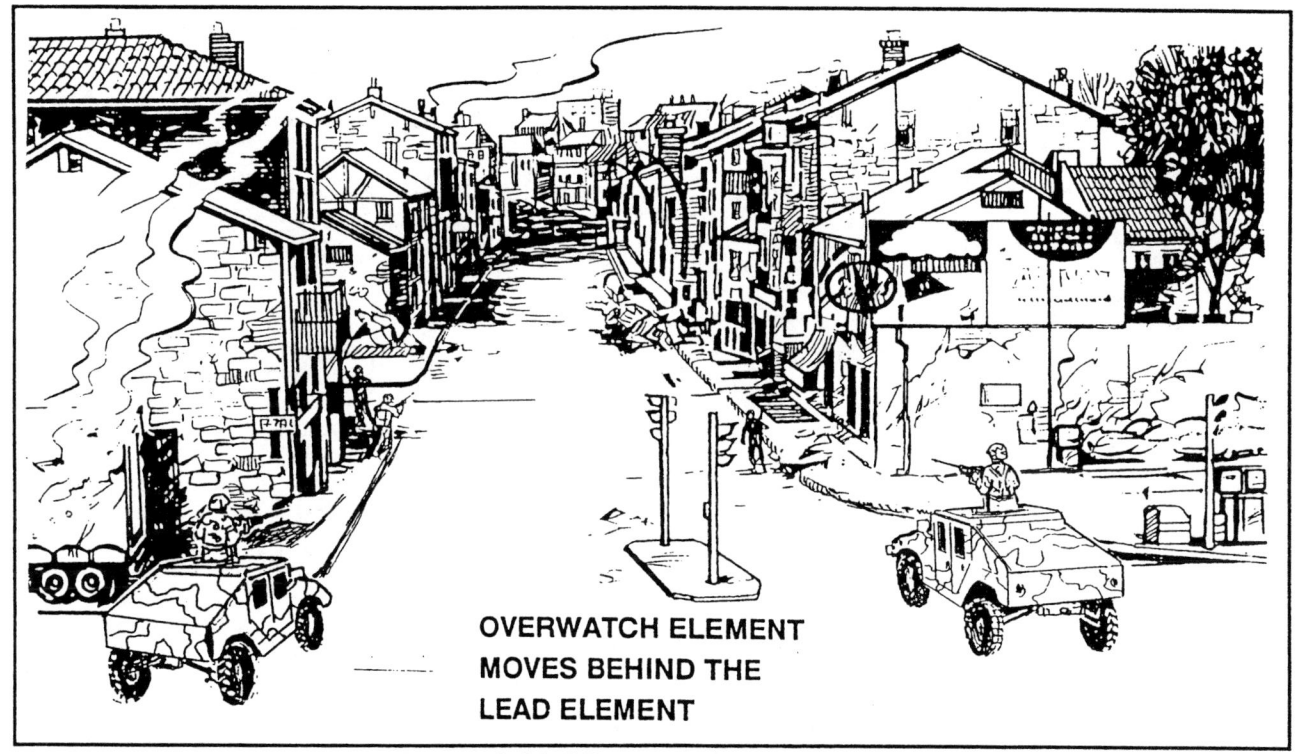

Figure 3-20. Movement down a street.

3-26. COUNTERATTACKS

A platoon may be given the mission of counterattacking for one of two reasons: to recapture a defensive position or a key point, destroying or ejecting an enemy foothold; or to stop an enemy attack by striking his flank, forcing him to stop and adopt a hasty defense.

    a. Platoon counterattacks are planned at company level to meet each probable enemy penetration. They must be well coordinated and executed violently. Preferably, counterattacks should be directed at an enemy flank and supported with direct and indirect fire.

    b. In outlying areas, where the terrain is relatively open, a mechanized infantry platoon accompanied by tanks can approach the counterattack objective mounted for speed. The tanks destroy the enemy's tanks and heavy weapons while the infantry dismounts to clear the objective. In central or more congested areas, the tanks progress deliberately, from point to point, providing close support to the dismounted troops. Counterattacks require the following:

- An analysis of the probable avenues of enemy approach.
- Reconnaissance and rehearsal along each counterattack route and of each proposed overwatch position.
- Construction of obstacles and fighting positions to canalize or block the enemy.
- Gaps or lanes through these obstacles if the counterattacks are to be quick enough to affect the action.

- Rapid and aggressive execution—leaders must set the example.
- Flexibility to react to unforeseen circumstances.
- An analysis of the probable counter-counterattack routes by the enemy.
- A fire support plan for the counterattack and possible counter-counterattack.

CHAPTER 4
DEFENSIVE OPERATIONS

Of the two patterns of defense, area and mobile, the area defense is the pattern most used since most of the reasons for defending a city are focused on retaining terrain. The mobile defense pattern is more focused on the enemy and the commander may decide to use it based on his estimate of the situation. In a built-up area, the defender must take advantage of the abundant cover and concealment. He must also consider restrictions to the attacker's ability to maneuver and observe. By using the terrain and fighting from well-prepared and mutually supporting positions, a defending force can inflict heavy losses on, delay, block, or fix a much larger attacking force.

Section I. DEFENSIVE CONSIDERATIONS

A commander must decide whether defending a built-up area is needed to successfully complete his mission. Before making his decision, the commander should consider the issues discussed herein.

4-1. REASONS FOR DEFENDING BUILT-UP AREAS

A commander should consider the following reasons for defending built-up areas.

   a. Certain built-up areas contain strategic industrial, transportation, or economic complexes that must be defended. Capitals and cultural centers can be defended for strictly psychological or national morale purposes even if they do not offer a tactical advantage to the defender. Because of the sprawl of such areas, significant combat power is required for their defense. Thus, the decision to defend these complexes is made by political authorities or the theater commander.

   b. The defender's need to shift and concentrate combat power, and to move large amounts of supplies over a wide battle area require that he retain vital transportation centers. Since most transportation centers serve large areas, the commander must defend all of the built-up area to control such centers.

   c. The worldwide increase in sprawling built-up areas has made it impossible for forces conducting combat operations to avoid cities and towns. Most avenues of approach are straddled by small towns every few kilometers and must be controlled by defending forces. These areas can be used as battle positions or strongpoints. Blocked streets covered by mortar and or artillery fire can canalize attacking armor into mined areas or zones covered by antiarmor fire. If an attacker tries to bypass a built-up area, he may encounter an array of tank-killing weapons. To clear such an area, the attacker must sacrifice speed and momentum, and expend many resources. A city or town can easily become a major obstacle.

   d. Forces can be concentrated in critical areas. Due to the tactical advantages to the defender, a well-trained force defending a built-up area can inflict major losses on a numerically superior attacker. The defender can conserve the bulk of his combat power so that it is available for use in open terrain. The defenders remaining in built-up areas perform an economy-of-force role.

   e. Forces can be well concealed in built-up areas. Aerial photography, imagery, and sensory devices cannot detect forces deployed in cities. CPs, reserves, CSS complexes, and combat forces emplaced well in built-up areas make them hard to detect.

4-2. REASONS FOR NOT DEFENDING BUILT-UP AREAS
The commander should consider the following reasons for not defending built-up areas.

    a. The location of the built-up area does not support the overall defensive plan. If the built-up area is too far forward or back in a unit's defensive sector, is isolated, or is not astride an enemy's expected avenue of approach, the commander may choose not to defend it.

    b. Nearby terrain allows the enemy to bypass on covered or concealed routes. Some built-up areas, mainly smaller ones, are bypassed by main road and highway systems. A built-up area that can be easily bypassed normally will be.

    c. Structures within the built-up area do not adequately protect the defenders. Extensive areas of lightly built or flammable structures offer little protection to the defender. Built-up areas near flammable or hazardous industrial areas, such as refineries or chemical plants, may not be defended.

    d. Dominating terrain is close to the built-up area. If the built-up area can be dominated by an enemy force occupying close terrain, the commander may choose to defend from there rather than the built-up area. This applies mainly to small built-up areas such as a village.

    e. Better fields of fire exist outside the built-up area. The commander may choose to base all or part of his defense on the long-range fields of fire that exist outside a built-up area. This applies mainly to armor-heavy forces defending sectors with multiple, small, built-up areas surrounded by farm areas.

    f. The built-up area has cultural, religious, or historical significance. The area may have been declared an "open city," in which case, by international law, it is demilitarized and must be neither defended nor attacked. The attacking force must assume civil administrative control and treat the civilians as noncombatants in an occupied country. The defender must immediately evacuate and cannot arm the civilian population. A city can be declared open only before it is attacked. The presence of large numbers of noncombatants, hospitals, or wounded personnel may also affect the commander's decision not to defend a built-up area.

Section II. CHARACTERISTICS OF BUILT-UP AREAS

The defense of a built-up area should be organized around key terrain features, buildings, and areas that preserve the integrity of the defense and that provide the defender ease of movement. The defender must organize and plan his defense by considering obstacles, avenues of approach, key terrain, observation and fields of fire, cover and concealment, fire hazards, and communications restrictions.

4-3. OBSTACLES
A city itself is an obstacle since it canalizes and impedes an attack. Likely avenues of approach should be blocked by obstacles and covered by fire. Barriers and obstacles should be emplaced in three belts.

    a. The first obstacle belt is at the nearest buildings across from and parallel to the main defensive position (MDP). This belt consists of wire and improvised barriers (to include inside buildings, in subterranean avenues of approach, and outside in open areas), danger areas, and dead space. These barriers and obstacles should be heavily booby trapped and covered by long-range fires as appropriate. This belt impedes enemy movement, breaks up and disorganizes attack formations, and inflicts casualties.

   b. The second obstacle belt is placed between the first belt and the MDP buildings, but out of hand grenade range from defensive positions. It impedes movement, channelizes the enemy into the best fields of fire, breaks up attack formations, and inflicts casualties. This belt is not meant to stop enemy soldiers permanently. It should be constructed efficiently to give the most benefit—not to be an impenetrable wall. It consists mainly of wire obstacles, improvised barriers, road craters, and mine fields. It should be booby trapped heavily (including trip-wire-activated Claymores). Triple-strand concertina is placed along the machine-gun FPL (as designated earlier with engineer tape) to slow the enemy on the FPL and allow the machine gun to be used effectively.

   c. The third obstacle belt is the defensive positions denial belt. It consists of wire obstacles placed around, through, and in the defensive buildings and close-in mine fields as well as in subterranean accesses. It impedes and complicates the enemy's ability to gain a foothold in the defensive area. It should be booby trapped, and Claymores should be used extensively, both trip wire activated and command detonated. The booby traps and Claymores should be placed where they will not cause friendly casualties.

   d. All avenues of approach (surface and subsurface) must be denied. Units must not overlook the use of field-expedient obstacles such as cars, light poles, and so on (Figure 4-1), or the emplacement of antipersonnel and antitank mines.

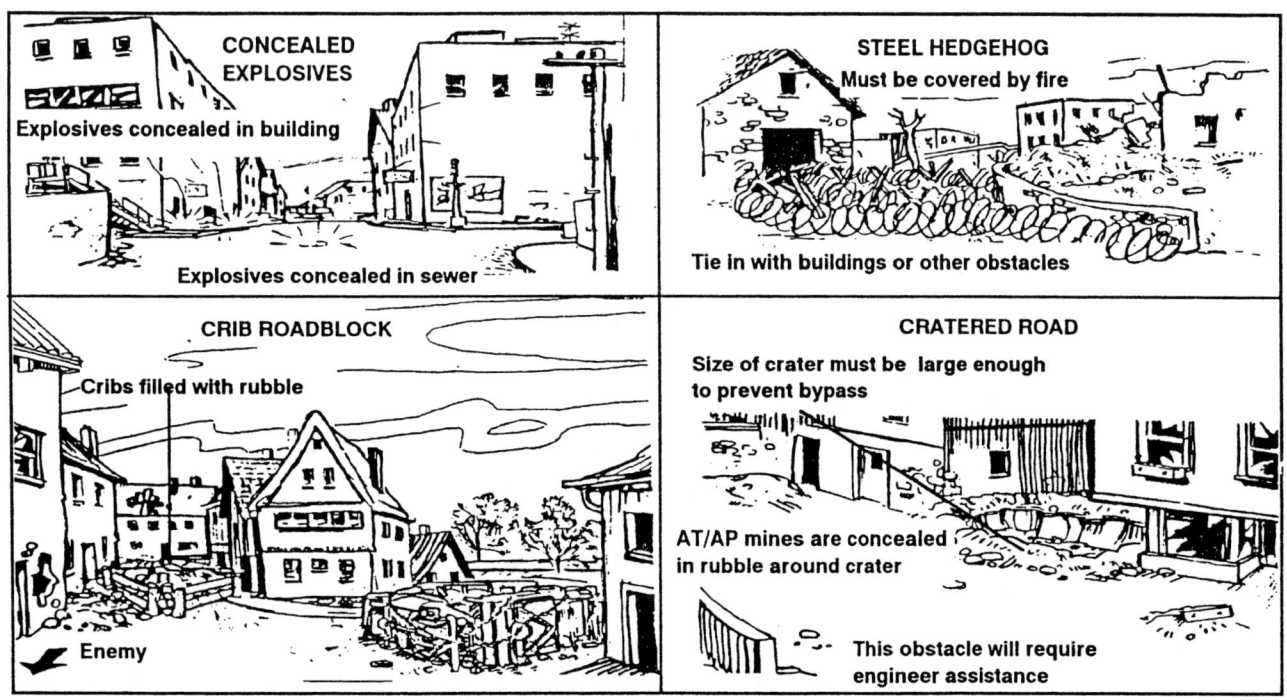

Figure 4-1. Example of field-expedient obstacles.

4-4. AVENUES OF APPROACH

The defender must not only consider the conventional avenues of approach into and out of the city but also the avenues within built-up areas that are above and below ground level. The defender normally has the advantage. He knows the city and can move rapidly from position to position through buildings and underground passages.

4-5. KEY TERRAIN

Key terrain is any place where seizure, retention, or control affords a marked advantage to either enemy or friendly forces. Primary examples of key terrain are bridges over canals or rivers, building complexes, public utilities or services, or parks. Built-up areas are unusual in that the population of the area itself may be considered key terrain. The identification of key terrain allows the defender to select his defensive positions and assists in determining the enemy's objectives.

4-6. OBSERVATION AND FIELDS OF FIRE

The defender must position weapons to obtain maximum effect and mutual supporting fire. This allows for long-range engagements out to the maximum effective ranges. Artillery FOs should be well above street level to adjust fires on the enemy at maximum range. Fires and FPFs should be preregistered on the most likely approaches to allow for their rapid shifting to threatened areas.

4-7. COVER AND CONCEALMENT

The defender should prepare positions using the protective cover of walls, floors, and ceilings. Soldiers should always improve positions using materials at hand. When the defender must move, he can reduce his exposure by—

- Using prepared breaches through buildings.
- Moving through reconnoitered and marked underground systems.
- Using trenches and sewage systems.
- Using the concealment offered by smoke anddarkness to cross open areas.

To accomplish his mission, the attacker must advance by crossing streets and open areas between buildings where he is exposed to fires from concealed weapons positions.

4-8. FIRE HAZARDS

The defender's detailed knowledge of the terrain permits him to avoid areas that are likely to be fire hazards. All cities are vulnerable to fire, especially those with many wooden buildings. The defender can deliberately set fires—

- To disrupt and disorganize the attackers.
- To canalize the attackers into more favorable engagement areas.
- To obscure the attacker's observation.

4-9. COMMUNICATIONS RESTRICTIONS

Wire is the primary means of communication for controlling the defense of a city and for enforcing security. However, wire can be compromised if interdicted by the enemy. Radio communication in built-up areas is normally degraded by structures and a high concentration of electrical power lines.

The new family of radios may correct this problem, but all units within the built-up area may not have these radios. Therefore, radio is an alternate means of communication. Messengers can be used well as another means of communication. Visual signals may also be used but are often not effective because of the screening effects of buildings, walls, and so forth. Signals must be planned, widely disseminated, and understood by all assigned and attached units. Increased noise makes the effective use of sound signals difficult.

Section III. FACTORS OF METT-T

Procedures and principles for planning and organizing the defense of a built-up area are the same as for other defensive operations. In developing a defensive plan, the defender considers METT-T factors with emphasis on fire support, preparation time, work priorities, and control measures. Planning for the defense of a city must be detailed and centralized.

4-10. MISSION
The commander must receive, analyze, and understand the mission before he begins planning. He may receive the mission as a FRAGO or formal OPORD, and must analyze all specified and implied tasks.

4-11. ENEMY
The commander must also analyze the type of enemy he may encounter. If the attacker is mostly dismounted infantry, the greatest danger is allowing him to gain a foothold. If the attacker is mostly armor or mounted motorized infantry, the greatest danger is that he will mass direct fire and destroy the defender's positions.

Intelligence gathering for defensive operations is not limited to only studying the enemy. Commanders must emphasize obtaining and using all intelligence. The items of intelligence peculiar to combat in built-up areas are discussed in Chapter 2. They include:

- Street, water, and sewer plans.
- Key installations and facilities.
- Key civilians.
- Civilian police and paramilitary forces.
- Sources of food.
- Communications facilities and plans.
- Power stations.

4-12. TERRAIN
Terrain in built-up areas is three-dimensional: ground level (streets and parks), above ground (buildings), and below ground (subways and sewers). Analysis of all man-made and natural terrain features is critical when planning to defend on built-up terrain. The commander's defensive plan is affected by the type of built-up area he will be operating in. (See Chapter 1.)

   a. **Villages.**

   (1) Villages are often on chokepoints in valleys, dominating the only high-speed avenue of approach through the terrain. If the buildings in such a village are well constructed and provide good protection against both

direct and indirect fires, a formidable defense can be mounted by placing a company in the town, while controlling close and dominating terrain with other battalion elements.

(2) If the terrain allows easy bypass and there are no other villages on defendable terrain within a mutually supporting distance, units may be unwise to defend it. This would allow friendly forces to be easily bypassed and cut off (Figure 4-2).

Figure 4-2. Villages.

(3) Villages on the approaches to large towns or cities may be used by commanders to add depth to the defense or to secure the flanks. These villages are often characterized by clusters of stone, brick, or concrete houses and buildings. Company-sized battle positions can be established in these small villages to block approaches into the main defensive positions.

b. **Strip Areas.**

(1) Strip areas consist of houses, stores, and factories and are built along roads or down valleys between towns and villages. They afford the defender the same advantages as villages.

(2) If visibility is good and enough effective fields of fire are available, a unit acting as a security force need occupy only a few strong positions spread out within the strip. This will deceive the enemy, when engaged at long ranges, into thinking the strip is an extensive defensive line. Strip areas often afford covered avenues of withdrawal to the flanks once the attacking force is deployed and before the security force becomes decisively engaged (Figure 4-3).

Figure 4-3. Strip areas.

c. **Towns and Cities.**

(1) A small force can gain combat power advantage when defending a small city or town that is a chokepoint if it places tanks, BFVs, TOWs, and Dragons on positions dominating critical approaches. To deny the enemy the ability to bypass the town or city, the defending force must control key terrain and coordinate with adjacent forces. Reserve forces should be placed where they can quickly reinforce critical areas. Obstacles and minefields assist in slowing and canalizing the attacker.

(2) Finding positions in towns and cities that provide both good fields of fire and cover is often difficult. The forward edges of a town usually offer the best fields of fire but can be easily targeted by enemy overwatch and supporting fire. These areas often contain residential buildings constructed of light material. Factories, civic buildings, and other heavy structures, which provide adequate cover and are more suitable for a defense, are more likely to be found deeper in the town and have limited fields of fire on likely avenues of approach.

(3) Since the forward edge of a town is the obvious position for the defender, it should be avoided. However, the defender can set up his position there if the terrain limits the enemy's ability for engagement or it contains strongly constructed buildings that give defending units adequate protection.

(4) A force may initially be assigned battle positions on the forward edge of the town. Its mission is to provide early warning of the enemy's advance, to engage the enemy at long range, and to deceive the enemy as to the true location of the defense. This force should withdraw in time to avoid decisive engagement. If there is limited observation from the forward edge, a force should be positioned on more favorable terrain forward or to the flanks of the town to gain better observation and to engage the enemy at long range.

(5) To prevent airmobile or airborne landings within the city or town, the commander must cover probable LZs and DZs, such as parks, stadiums, or large rooftops and heliports with obstacles or fire (Figure 4-4).

Figure 4-4. Towns and cities.

d. Large Built-Up Areas.

(1) In large built-up areas the commander must consider that the terrain is restrictive due to large buildings that are normally close together. This requires a higher density of troops and smaller defensive sectors than in natural open terrain. Units occupy defensive frontages about one-third the size of those in open areas. An infantry company, which might occupy 1,500 to 2,000 meters in open terrain, is usually restricted to a frontage of 300 to 800 meters in built-up areas. The density of buildings and rubble and street patterns will dictate the frontage of the unit (Table 4-1).

UNIT	FRONTAGES	DEPTHS
Battalion or Battalion TF	4 to 8 blocks	3 to 6 blocks
Company or Company Team	2 to 4 blocks	2 to 3 blocks
Platoon	1 to 2 blocks	1 block
NOTE: An average city block has a frontage of about 175 meters. These minimum figures apply in areas of dense, block-type construction; multistory buildings; and underground passages.		

Table 4-1. Approximate frontages and depths in large built-up areas.

(2) In a large built-up area, a battalion is given a sector to defend and normally establishes a series of defensive positions. Unlike villages or towns, natural terrain close to the built-up area is not usually available for the commander to integrate into his plan. Although mutual support between positions should be maintained, built-up terrain often allows for infiltration routes that the enemy may use to pass between positions. Therefore, the defender must identify the following:

- Positions that enable him to place surprise fires on the infiltrating enemy.
- Covered and concealed routes for friendly elements to move between positions (subways and sewers).
- Structures that dominate large areas.
- Areas such as parks, boulevards, rivers, highways, and railroads where antiarmor weapons have fields of fire.
- Firing positions for mortars.
- Command locations that offer cover, concealment, and ease of command and control.
- Protected storage areas for supplies.

(3) Buildings that add most to the general plan of defense are chosen for occupation. Mutual support between these positions is vital to prevent the attacker from maneuvering and outflanking the defensive position, making it untenable. Buildings chosen for occupation as defensive positions should—

- Offer good protection.
- Have strong floors to keep the structure from collapsing under the weight of debris.
- Have thick walls.
- Be constructed of nonflammable materials (avoid wood).
- Be strategically located (corner buildings and prominent structures).
- Be adjacent to streets, alleys, vacant lots, and park sites. (These buildings usually provide better fields of fire and are more easily tied in with other buildings.)

4-13. TROOPS AVAILABLE

Employment of troops in built-up areas depends on many factors governed by METT-T and on the mission.

a. **Employment of Squads.** Squads are usually employed abreast so that they all can fire toward the expected direction of attack. In a built-up area, squads may be separated by rooms within buildings or be deployed in different buildings. Squad positions must be mutually supporting and allow for overlapping sectors of fire, even if buildings or walls separate the positions (Figure 4-5, page 4-10).

b. **Employment of Platoons.** Once the commander has decided where to defend, he should select platoon battle positions or sectors that block or restrict the enemy's ability to maneuver and control key areas. The frontage for a platoon is about one to two city blocks long. Along with his primary and alternate positions, the platoon leader normally selects one supplementary position to reorient his defense to meet enemy threats from another direction.

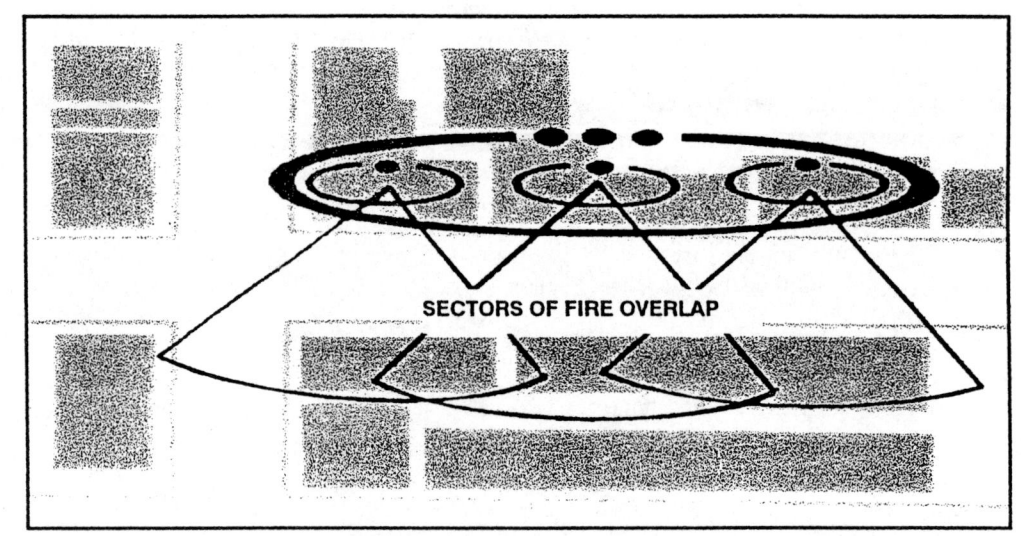

Figure 4-5. Sectors of fire.

c. **Employment of Companies.** Battalion commanders employ their companies in battle positions or sectors. The frontage of a company or company team defending in a built-up area is normally two to four city blocks long (300 to 800 meters). Depending on the type of built-up area, a company may be employed on the forward edge of the flanks of the area. This forces the enemy to deploy early without decisive engagement since it deceives the enemy as to the true location of the main defense. Other companies may then be employed in a series of strongpoints in the center of the city or town. In all cases, mutual support between positions is vital. Companies should also have designated alternate and supplementary positions.

d. **Employment of the Reserve.** The commander's defensive plan must always consider the employment of a reserve. The reserve force should be prepared to counterattack to regain key positions, to block enemy penetrations, to protect the flanks, or to assist by fire in the disengagement and withdrawal of endangered positions. For combat in a built-up area, a reserve force—

- Normally consists of infantry.
- Must be as mobile as possible.
- May be a platoon or squad at company level or one platoon at battalion level.
- May be supported by tanks.

e. **Employment of Tanks and BFVs.** The commander should employ tanks and BFVs to take advantage of their long-range fires and mobility. Built-up areas restrict the mobility of tanks and BFVs and make them vulnerable to the antiarmor weapons of the enemy infantry.

(1) When tanks and BFVs are employed in the defense of a city, infantry should be positioned to provide security against close antitank fires and to

detect targets for the armored vehicles. Tanks and BFVs should be assigned primary, alternate, and supplementary positions, as well as primary and alternate sectors. BFVs and antitank weapons should supplement tank fires.

(2) Tanks and BFVs should be located on likely avenues of approach to take advantage of their long-range fires. They may be—

- Positioned on the edge of the city in mutually supporting positions.
- Positioned on key terrain on the flanks of towns and villages.
- Used to cover barricades and obstacles by fire.
- Part of the reserve.

(3) Tanks and BFVs are normally employed as a platoon. However, sections and individual tanks and BFVs may be employed with infantry platoons or squads. This provides tanks and BFVs with the close security of the infantry. Tanks and BFVs provide the commander with a mobile force to respond quickly to enemy threats on different avenues of approach.

f. **Employment of Fire Support.** Fire planning must be comprehensive due to the proximity of buildings to targets, minimum range restrictions, and repositioning requirements. Mortar and artillery fires are planned on top of and immediately around defensive positions for close support.

(1) Artillery fire support may be used in the direct or indirect fire role. Artillery fire should be used—

- To suppress and blind enemy overwatch elements.
- To disrupt or destroy an assault.
- To provide counterbattery fire.
- To support counterattacks.
- To provide direct fire when necessary.

(2) Mortars at battalion and company level are employed to maximize the effect of their high-angle fires. They should be used to engage—

- Enemy overwatch positions.
- Enemy infantry before they seize a foothold.
- Targets on rooftops.
- Enemy reinforcements within range.

(3) Final protective fires are planned to stop dismounted assaults in front of the defensive positions. Fires within the city are planned along likely routes of advance to destroy the enemy as he attempts to deepen a penetration.

(4) At battalion level, the commander should establish priorities of fire based on enemy avenues of approach and threat systems that present the greatest danger to the defense. For example, during the attacker's initial advance, tanks, BMPs, and overwatching elements are the greatest threat to the defense. ATGMs should concentrate on destroying tanks first, then BMPs. Artillery and mortar fires should suppress and destroy enemy ATGMs and overwatch positions and or elements. If enemy formations secure a foothold, priority is shifted to the destruction of enemy forces within the penetration.

(5) As the enemy attack progresses in the city, fires are increased to separate infantry from supporting tanks and fighting vehicles. During this phase, friendly artillery concentrates on attacking infantry, counterfire missions, and the destruction of reinforcements that are approaching the city.

(6) When initiated, counterattacks are given priority of supporting fires. When artillery is firing the missions as mentioned above, it must remain mobile and be prepared to displace to preplanned positions to avoid enemy counterbattery fire.

(7) The battalion mortar platoon may be initially positioned forward in support of the reconnaissance platoon. After withdrawal of the reconnaissance platoon, the mortar platoon is positioned where it can support the whole battalion.

(8) At company and platoon level, fire plans include fires of organic, attached, and supporting weapons. The company commander also plans his own mortar and artillery fires on top of and immediately around his battle positions for close support.

(9) Based on the location of platoon positions in relation to the most likely avenues of advance, the company commander should assign FPFs to platoon leaders. Each rifle platoon leader then assigns his machine guns sectors of fire and FPLs. These positions should be selected to provide interlocking grazing fire and mutual support between adjacent units. FPLs are fired on planned signals from the platoon forward observers. Proposed FPLs must be "walked out" to determine the extent of grazing fire available and to locate dead space, which can be covered by—

- Sniper fire.
- Grenade launchers.
- Mines and booby traps.
- Indirect fires.

(10) Air defense assets available to the commander, such as Stinger and Vulcan, are normally employed to ensure all-round air defense. The lack of good firing positions for long-range air defense missile systems in the built-up area may limit the number of deployed weapons. In the defense, weapons systems may have to be winched or airlifted into positions. Rooftops and parking garages are good firing positions because they normally offer a better line-of-sight. Stingers and Vulcans can be assigned the missions of protecting specific positions or in general support of the battalion.

g. **Employment of Engineers.** Engineers are employed under battalion control or attached to companies and platoons. Normally, one engineer platoon or company supports a battalion or battalion task force. Commanders must consider engineer tasks that enhance survivability, mobility, and countermobility. Tasks that engineers can accomplish in the defense of a built-up area include:

- Constructing obstacles and rubbling.
- Clearing fields of fire.
- Laying mines.
- Preparing routes to the rear.
- Preparing fighting positions.

h. **Employment of the Antiarmor Company.** The antiarmor company normally supports the battalion security force, providing long-range antiarmor fires forward of the main defense. Separate antiarmor sections may be attached to companies to cover likely armor approaches. Once the security force withdraws, the antiarmor company is normally employed in GS of the battalion. If the threat is not armored, or if the terrain prevents the use of the TOW weapons system, antiarmor platoons and companies in light airborne and air assault units can mount MK 19 grenade launchers and .50 caliber machine guns to support the defending units.

i. **Employment of the Reconnaissance Platoon.** Depending on the situation and terrain, the battalion reconnaissance platoon may provide a security force forward of the built-up area to give the commander early warning of enemy activity. Upon withdrawal of the security force, the reconnaissance platoon may be given the mission to ensure flank or rear security, to occupy a defensive sector (or battle position), or to stay in reserve.

j. **Employment of Ground Surveillance Radar.** If attached, GSR is best employed on the outskirts of built-up areas because of the line-of-sight problems within the area. During limited visibility, if suitable avenues exist, GSR can be placed to monitor sectors. Because of the normal ranges found in built-up areas and the likely narrowness of the sector ranges, GSR can be vulnerable to detection and direct fire. Cross vectoring is important in this environment.

4-14. TIME AVAILABLE

The commander must organize and establish priorities of work, depending upon the time available. Many tasks can be accomplished at the same time, but priorities for preparation should be according to the commander's order. In defensive operations, an example priority of work sequence follows.

a. **Establish Security.** The unit should quickly establish all-round security by placing forces on likely approaches. Troop positions should have at least one soldier to provide security during all preparations. The reconnaissance and counterreconnaissance plan should be emphasized.

b. **Assign Sectors of Responsibility.** Boundaries define sectors of responsibility. They include areas where units may fire and maneuver without interference or coordination with other units. Responsibility for primary avenues of approach should never be split. In areas of semidetached construction, where observation and movement are less restricted, boundaries should be established along alleys or streets to include both sides of a street in a single sector. Where buildings present a solid front along streets, boundaries may have to extend to one side of the street (Figure 4-6, page 4-14).

c. **Clear Fields of Fire.** In built-up areas, commanders may need to rubble certain buildings and structures to provide greater protection and fields of fire to the defender. If the ceiling of a lower-story room can support the weight of the rubble, collapsing the top floor of a building before the battle starts may afford better protection against indirect fires. Rubbling an entire building can increase the fields of fire and create an obstacle to enemy movement. Defenders must be careful, however. Rubbling buildings too soon (or too many) may give exact locations and destroy the cover from direct fire. Rubbled buildings may also interfere with planned routes of withdrawal or counterattack.

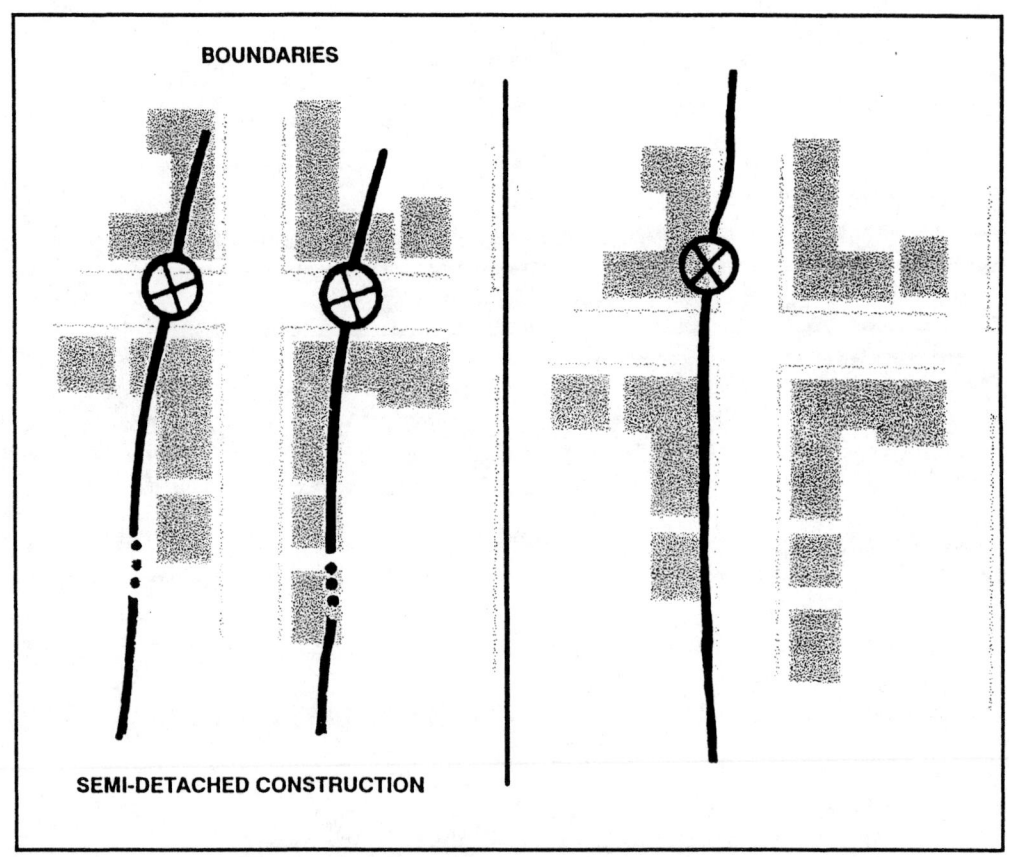

Figure 4-6. Boundaries in built-up areas.

d. Select and Prepare Initial Fighting Positions. The commander should select positions in depth. The unit should prepare positions as soon as troops arrive and continue preparing as long as positions are occupied. Enemy infiltration or movement sometimes occurs between and behind friendly positions. Therefore, each position must be organized for all-round defense. The defender should also—

(1) Make minimum changes to the outside appearance of buildings where positions are located.

(2) Screen or block windows and other openings to keep the enemy from seeing in and tossing in hand grenades. This must be done so that the enemy cannot tell which openings the defenders are behind.

(3) Remove combustible material to limit the danger of fire. Fires are dangerous to defenders and create smoke that could conceal attacking troops. For these reasons, defenders should remove all flammable materials and stockpile firefighting equipment (water, sand, and so forth). The danger of fire also influences the type of ammunition used in the defense. Tracers or incendiary rounds should not be used extensively if threat of fire exists.

(4) Turn off electricity and gas. Both propane and natural gas are explosive. Natural gas is also poisonous and is not filtered by a protective mask. Propane gas, although not poisonous, is heavier than air. If it leaks into an enclosed area, it displaces the oxygen and causes suffocation. Gas mains and electricity should be shut off at the facility that serves the city.

(5) Locate positions so as not to establish a pattern. The unit should avoid obvious firing locations like church steeples.

(6) Camouflage positions.

(7) Reinforce positions with materials available such as beds, furniture, and so forth.

(8) Block stairwells and doors with wire or other material to prevent enemy movement. Create holes between floors and rooms to allow covered movement within a building.

(9) Prepare range cards, fire plans, and sector sketches.

(10) Emplace machine guns in basements. When basements are not used, they should be sealed to prevent enemy entry.

(11) Establish an extra supply of Class V and medical supplies.

e. **Establish Communications.** Commanders should consider the effects of built-up areas on communications when they allocate time to establish communications. Line-of-sight limitations affect both visual and radio communications. Wire laid at street level is easily damaged by rubble and vehicle traffic. Also, the noise of built-up area combat is much louder than in other areas, making sound signals difficult to hear. Therefore, the time needed to establish an effective communications system may be greater than in more conventional terrain. Commanders should consider the following techniques when planning for communications:

(1) If possible, lay wire through buildings for maximum protection.

(2) Use existing telephone systems. Telephones are not always secure even though many telephone cables are underground.

(3) Emplace radios and retransmission sites on the second or third floor of a building.

(4) Use messengers at all levels since they are the most secure means of communications.

f. **Emplace Obstacles and Mines.** To save time and resources in preparing the defense, commanders must emphasize using all available materials (automobiles, railcars, rubble) to create obstacles. Civilian construction equipment and materials must be located and inventoried. This equipment can be used with engineer assets or in place of damaged equipment. Coordination must be made with proper civilian officials before use. Engineers must be able to provide advice and resources as to the employment of obstacles and mines. The principles for employing mines and obstacles do not change in the defense of a built-up area; however, techniques do change. For example, burying and concealing mines in streets are hard due to concrete and asphalt. Obstacles must be tied in to buildings and rubble areas to increase effectiveness and to canalize the enemy. FASCAM may be effective on the outskirts of a city or in parks; however, in a city core, areas may be too restrictive (see Appendix G).

g. **Improve Fighting Positions.** When time permits, all positions, to include supplementary and alternate positions, should be reinforced with sandbags and provide overhead cover. Timely and accurate support from attached engineers helps in this effort (see Appendix E).

h. **Establish and Mark Routes Between Positions.** Reconnaissance by all defending elements should help select routes for use by defenders moving between positions. Movement is crucial in fighting in built-up areas. Early selection and marking of routes adds to the defender's advantages.

Section IV. COMMAND AND CONTROL

In all defensive situations, the commander should position himself well forward so that he can control the action. In a built-up environment, this is even more critical due to obstacles, poor visibility, difficulty in communication, and intense fighting.

Graphic control measures common to other tactical environments are also used in combat in built-up areas. Streets are ideal for phase lines. These and other control measures ensure coordination throughout the chain of command.

4-15. COMMAND POST FACILITIES

Command post facilities should be located underground. Their vulnerability requires all-round security. Since each facility may have to secure itself, it should be near the reserve unit for added security. When collocated with another unit, command post facilities may not need to provide their own security. Also, a simplified organization for command posts is required for ease of movement. Since rubble often hinders movement of tracked and wheeled vehicles, battalion and company headquarters must be prepared to backpack communications and other needed equipment for operations.

4-16. ORGANIZATION OF THE DEFENSE

The battlefield is divided into three operational areas—deep, close, and rear. At the battalion level, operations are conducted in the close operational area. The defense is organized into three areas—the security force area, main battle area, and rear area. A battalion defending in built-up areas may have missions in any one of these areas, depending on the mission of the brigade or division (Figure 4-7).

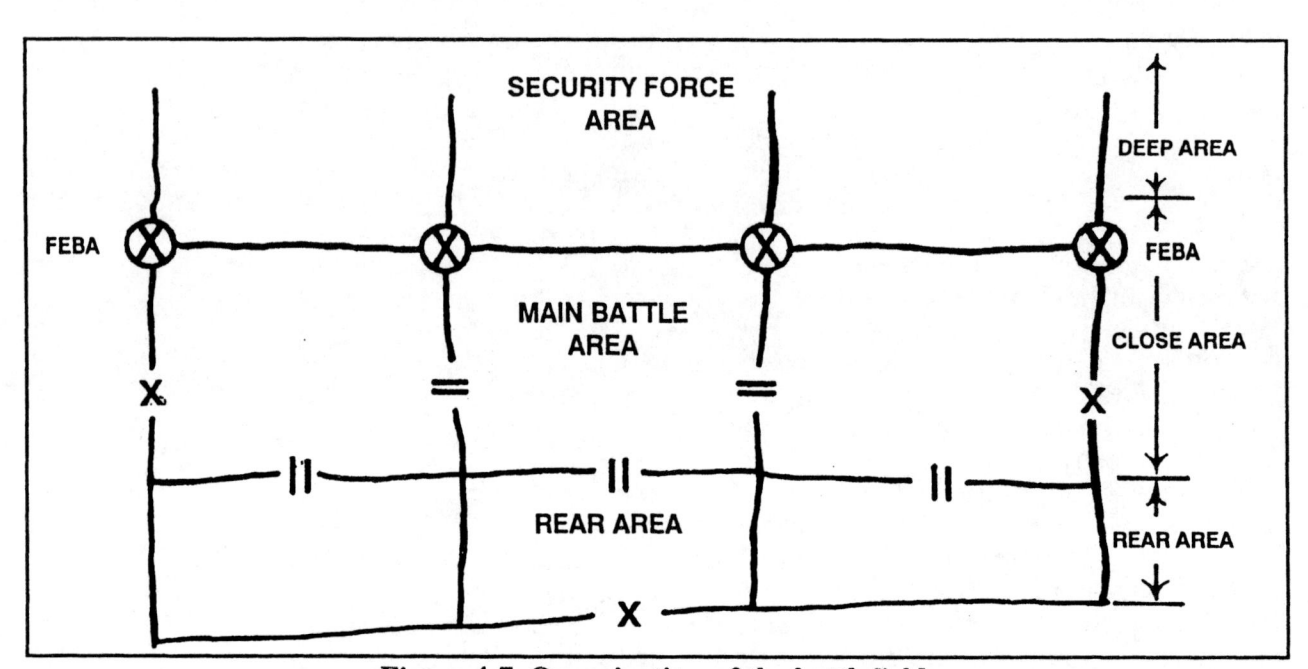

Figure 4-7. Organization of the battlefield.

a. **Security Operations.** The defensive battle begins with a combined arms force conducting security operations well forward of the main body. Security operations include screening operations, guard operations, and when, augmented with additional CS and CSS assets, covering force operations. The commander decides which operation is appropriate based on the factors of METT-T. The designated force uses all available combat power to destroy the enemy and slow his momentum. Artillery, tactical air, and attack helicopters are employed to fight the initial battle.

(1) Security operations add to the defense by—
- Alerting the defense to the strength, location, and general direction of the enemy's main and supporting attacks.
- Delaying enemy first-echelon detachments.
- Initiating early engagement of enemy forces.
- Deceiving the enemy as to the true location of the main defense force.

(2) The withdrawal of the security forces must not result in an easing of pressure on the enemy. The built-up area environment may complicate battle handover from the security force to the main battle area force. However, this transition must be accomplished smoothly to prevent the enemy from gaining momentum.

b. **Main Battle Area.** The decisive battle is fought in the MBA. Depending on the Threat, the battalion commander may deploy companies on the forward edges of the city or in battle positions in depth. In either case, the defense is made stronger by including forces that are defending on close terrain on the flanks into the defensive scheme. The battalion commander normally employs a security force to the front to provide early warning and to deny the enemy intelligence on the battalion's defensive dispositions (counterreconnaissance) (Figure 4-8).

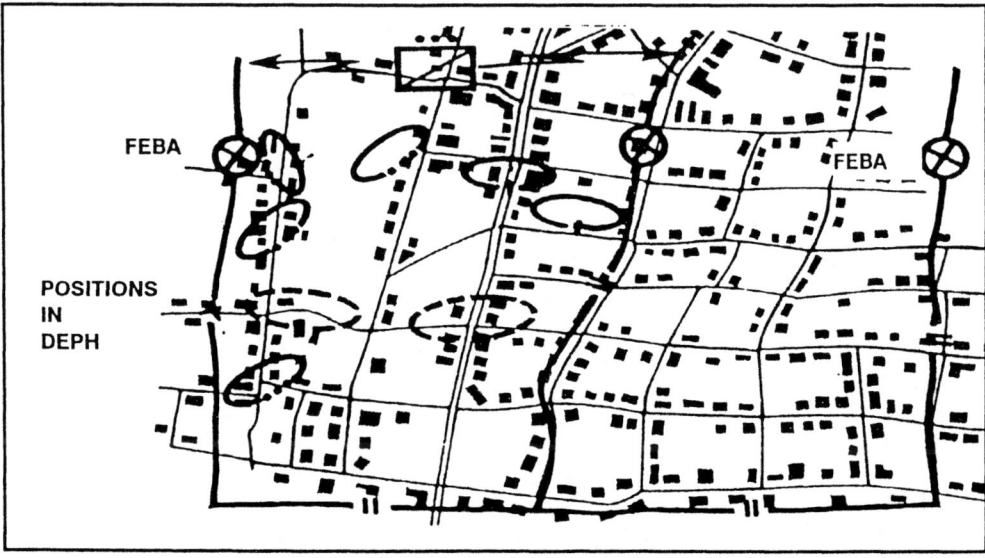

Figure 4-8. Main battle area.

(1) The size and location of battle positions within the battalion's area of operations depends mainly on the type of enemy encountered and the ability to move between positions to block threatened areas. It may be desirable to place small antiarmor elements, secured by infantry, on the forward edges while the main defense is deployed in depth.

(2) A force assigned battle positions on the forward edge of a city or town should—

- Provide early warning of the enemy's advance.
- Engage the enemy at long range.
- Deceive the enemy as to the true location of the defense.

(3) When enemy forces enter and maneuver to seize initial objectives, the defender should employ all available fires to destroy and suppress the direct-fire weapons that support the ground attack. Tanks and BMPs should be engaged as soon as they come within the effective range of antiarmor weapons.

(4) As the enemy attack develops, the actions of small-unit leaders assume increased importance. Squad and platoon leaders are often responsible for fighting independent battles. Thus, it is important that all leaders understand their commander's concept of the defense.

c. **Rear Area.** The rear area is behind the MBA. It is the area from which supply and maintenance support is sent forward. At battalion level, the rear area facilities are in the MBA. They are not organized as combat elements but are critical to the overall defense. Protection of these elements is vital.

4-17. COUNTERATTACK

Small infantry-heavy reserves supported by BFVs and or tanks (if present) should be prepared to counterattack to regain key positions, to block enemy penetrations, to provide flank protection, and to assist by fire the disengagement and withdrawal of endangered positions. When the reserves are committed to counterattack to reinforce a unit, they may be attached to the unit in whose sector the counterattack is taking place. Otherwise, the counterattack becomes the main effort. This makes coordination easier, especially if the counterattack goes through the unit's positions.

4-18. DEFENSE DURING LIMITED VISIBILITY

The TF and TM commander can expect the attacker to use limited visibility conditions to conduct necessary operations to sustain or gain daylight momentum. (See Appendix I for more information.)

a. Commanders should employ the following measures to defend against night attacks.

(1) Defensive positions and crew-served weapons should be shifted just before dark to deceive the enemy as to their exact location. (A squad or fire team can often be shifted to an adjacent building and cover the same avenue of approach.)

(2) Unoccupied areas between units, which can be covered by observed fire during daylight, may have to be occupied or patrolled at night.

(3) Radar, remote sensors, and night observation devices should be emplaced on streets and open areas.

(4) Nuisance mines, noise-making devices, tanglefoot tactical wire, and OPs should be positioned on secondary avenues of approach for early warning.

(5) Observation posts, planned indirect fires, patrols, and anti-intrusion devices should be used to prevent infiltration.

(6) Artificial illumination should be planned, to include the use of street lamps, stadium lights, and so forth.

(7) Indirect fire weapons, grenade launchers, and hand grenades should be used when defenses are probed to avoid disclosure of defensive positions.

b. When the enemy begins his night assault, FPFs should be initiated by a planned signal. Crew-served weapons, tank-mounted weapons, and individual riflemen fire within their assigned sectors. Grenades and command detonated mines should be used to supplement other fires as the enemy approaches the positions.

c. Defenders should move to daylight positions before the BMNT. During attacks in fog, rain, or snowstorms, many of the techniques described for night defense apply. Commanders must rely on OPs and patrolling in these situations.

Section V. DEFENSIVE PLAN AT BATTALION LEVEL

The built-up area defensive plan at battalion level depends on the size and location of the area. Many factors must be considered before instituting such a plan.

4-19. DEFENSE OF A VILLAGE

A battalion TF assigned a defensive sector that includes a village may incorporate the village as a strongpoint in its defense. This use of a built-up area is most common where the village stands astride a highspeed avenue of approach or where it lies between two difficult obstacles. To incorporate such an area into its defense, the battalion TF must control the high ground on either side of the village to prevent the enemy from firing from those areas into the village.

a. The majority of the TF tanks and BFVs should be employed where the maneuver room is the greatest (on the key terrain to the flanks of the village). This is also where the TF antiarmor vehicles (BFVs and or ITVs) should be employed. As the security force withdraws and companies and or teams assume the fight, BFVs and ITVs can assume support by fire positions in depth.

b. Although the battalion TFs disposition should prevent large enemy forces from threatening the rear and flanks of the village, the danger of small-unit enemy infiltration means that the village must be prepared for all-round defense.

c. Engineers required for team mobility operations should stay with the company or company team in the town to provide continuous engineer support if that company or company team becomes isolated. Engineer support for the rest of the TF should be centrally controlled by the TF commander. Engineer assets may be in DS of the other companies or company teams. The priority of barrier materials, demolitions, and mines should go to the company or company team in the village.

d. The TF commander should use the key terrain on the village's flanks for maneuver to prevent the village's defense from becoming isolated. The strongpoints in the town should provide a firm location where the enemy can be stopped, around which counterattacks can be launched (Figure 4-9).

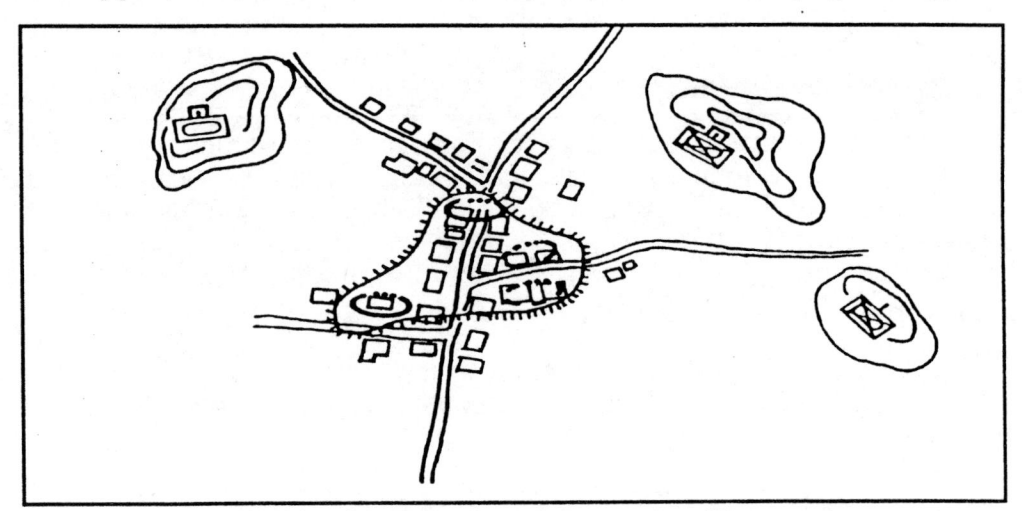

Figure 4-9. Battalion defense of a village.

4-20. DEFENSE IN SECTOR

Along with defending a village, a battalion TF may be given the mission of defending a sector in a city (Figure 4-10). The battalion should take advantage of the outlying structures to provide early warning and to delay the enemy, and of the tougher interior buildings to provide fixed defense. This defense should cover an area about 4 to 12 blocks square.

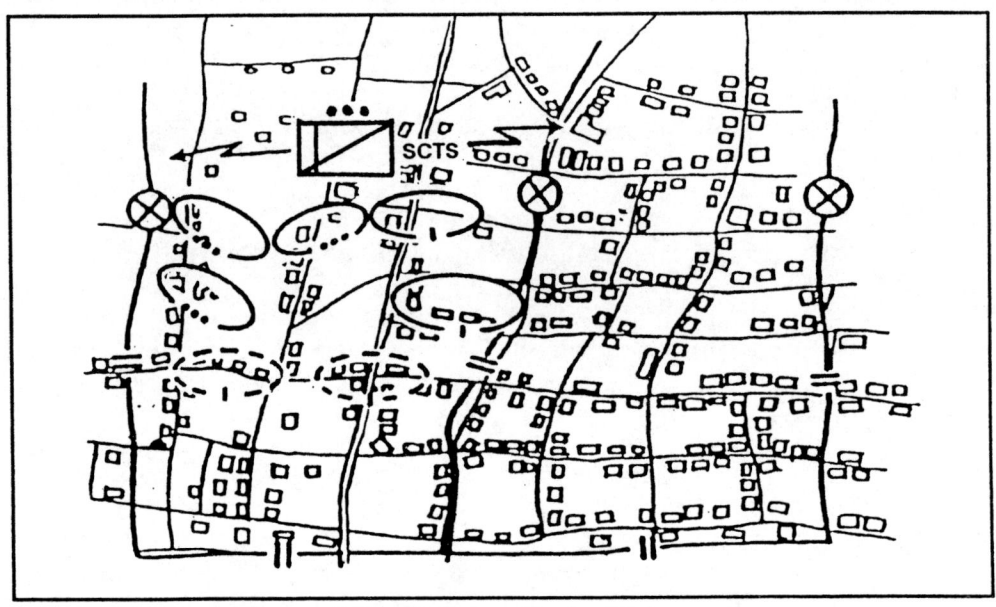

Figure 4-10. Defense of a built-up sector.

a. The battalion TF deployment begins with the reconnaissance platoon reconnoitering the built-up area to provide an area reconnaissance and location of the enemy. At the edge of the area, where fields of fire are the greatest, the battalion TF should deploy BFVs and ITVs and other antiarmor weapon systems to provide long-range antiarmor defense.

b. The FEBA should include the most formidable buildings in the sector. Forward of the FEBA, the battalion TF should organize a guard force, which could be a reinforced company. The guard force should concentrate on causing the enemy to deploy without engaging the enemy in decisive combat. This can be done through maximum use of ambushes and obstacles, and of covered and concealed routes through buildings for disengagement. The guard force inflicts casualties and delays the enemy but avoids decisive engagement since buildings beyond the FEBA do not favor the defense. As the action nears the FEBA, the guard force detects the location of the enemy's main attack. Upon reaching the FEBA, the guard force can be used as a reserve and reinforce other elements of the battalion, or it can counterattack to destroy an enemy strongpoint.

c. The defense along the FEBA consists of a series of positions set up similar to that described in the company defense of the village (see paragraph 4-18). Key terrain features, such as strong buildings, road junctions, and good firing positions, should be the center of the strongpoint defense. Based on METT-T considerations, the defense in sector may consist of either strongpoints or battle positions. Strongpoints located on or covering decisive terrain are extremely effective in the defense. Buildings should be prepared for defense as outlined in Appendix D.

d. BFVs should be used to engage BMPs, BTRs, and BRDMs; to cover obstacles with fire; and to engage in counterattacks with tanks. They can also be used to transport casualties and supplies to and from the fight.

e. The battalion's attached tanks should be used to engage enemy tanks, cover obstacles by fire, and engage in counterattacks. They should be employed in platoons where possible, but in congested areas they may be employed in sections.

f. Artillery and mortar fire should be used to suppress and blind enemy overwatch elements, to engage enemy infantry on the approaches to the door, to provide counterbattery fire, and to support counterattacks using both indirect and direct fire.

g. Engineers should be attached to the delaying force to help in laying mines and constructing obstacles, clearing fields of fire, and preparing routes to the rear. These routes should also have obstacles. Engineers should be in support of the force in the strongpoints to help prepare fighting positions.

4-21. DELAY IN A BUILT-UP AREA

The purpose of a delay is to slow the enemy, cause enemy casualties, and stop the enemy (where possible) without becoming decisively engaged or being outmaneuvered. The delay can either be oriented on the enemy or on specified terrain such as a key building or manufacturing complex.

a. A delay in a built-up area consists of a succession of ambushes and battle positions (Figure 4-11, page 4-22).

(1) Ambushes are planned on overwatching obstacles and are closely coordinated but decentrally executed. The deployment of the battalion TF is realigned at important cross streets. The ambushes can be combined with

limited objective attacks on the enemy's flanks. These are usually effective in the edge of open spaces, parks, wide streets, and so on. These should be executed by tanks and BFVs along with dismounted infantry.

(2) Battle positions should be placed where heavy weapons, such as tanks, BFVs, TOWs, antiarmor weapons, and machine guns, will have the best fields of fire. Such locations are normally found at major street intersections, parks, and at the edge of open residential areas. Battle positions should be carefully and deliberately prepared, reinforced by obstacles and demolished buildings, and supported by artillery and mortars. They should inflict maximum losses on the enemy and cause him to deploy for a deliberate attack.

b. Tanks, BFVs, and antiarmor weapons should have prepared primary and alternate positions to reduce their vulnerability. Coordination must be ongoing with withdrawing ambushes until they are safely within the battle position.

c. The battalion TF is most effective when deployed in two delaying echelons, alternating between conducting ambushes and fighting from battle positions. As the enemy threatens to overrun a battle position, the company disengages and delays back toward the next battle position. As the company passes through the company to the rear, it establishes another battle position. Smoke and demolitions are used to aid in the disengagement. Security elements on the flank can be employed to prevent the enemy from out-flanking the delay. A small reserve can be used to react to unexpected enemy action and to conduct continued attacks on the enemy's flank.

d. The direction of the engineer effort should be centralized to support the preparation of battle positions. It should be decentralized to support the force committed to ambush.

e. The width of the TF zone depends upon the nature of the buildings and obstacles along the street and the time that the enemy must be delayed.

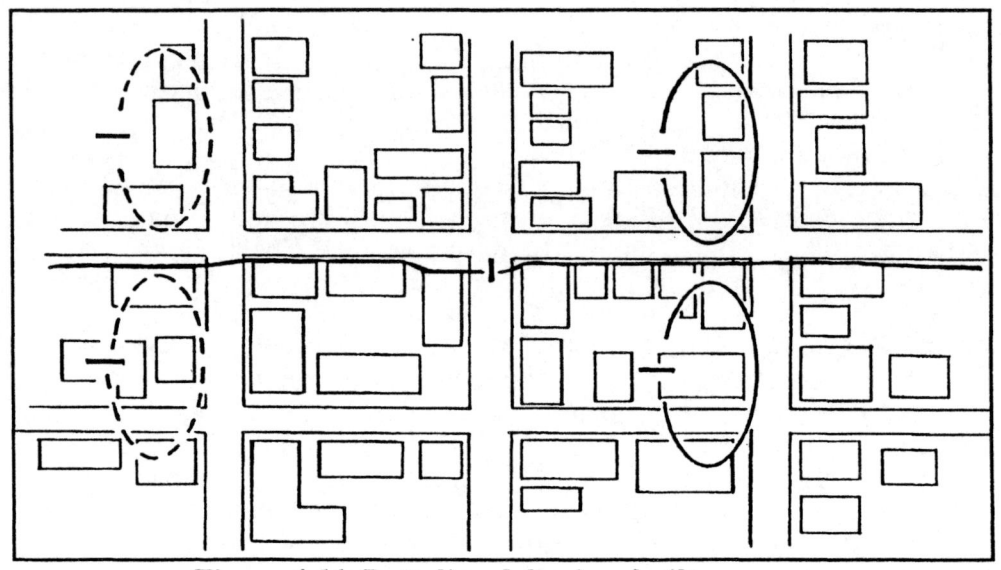

Figure 4-11. Battalion delay in a built-up area.

Section VI. DEFENSIVE PLAN AT COMPANY LEVEL

The defensive plan in built-up areas at company level depends on the size and location of the area. Many factors must be considered before instituting such a plan.

4-22. DEFENSE OF A VILLAGE

Once the company commander has completed his reconnaissance of the village, he scouts the surrounding terrain and, with the information assembled, develops his plan for the defense (Figure 4-12). One of his first decisions is whether to defend with his infantry on the leading edge of the village or farther back within the village.

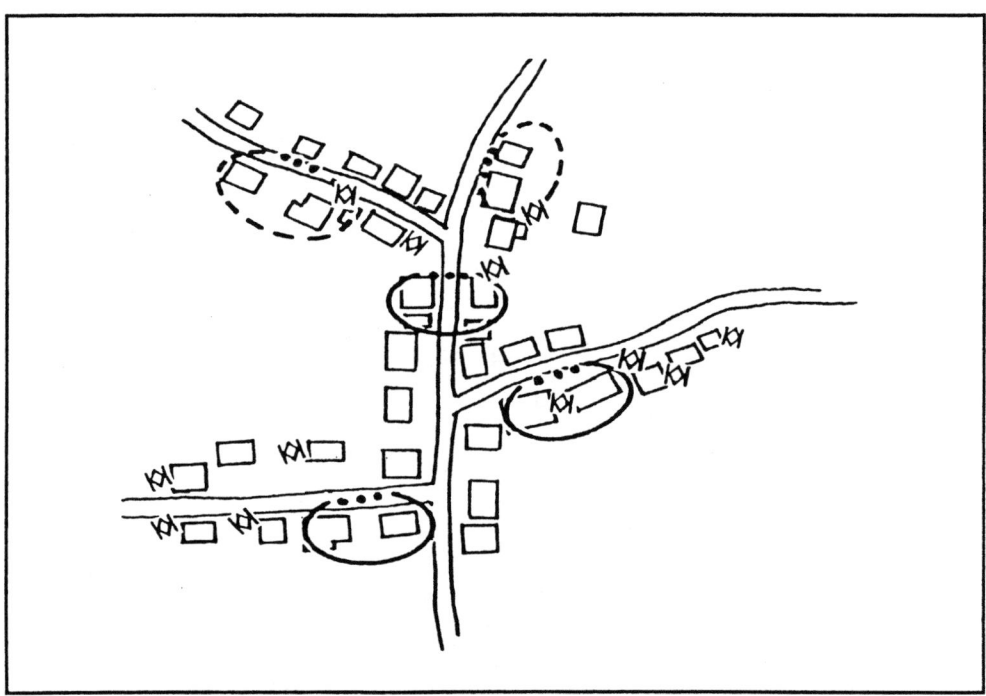

Figure 4-12. Company defense of a village.

   a. Several factors influence the commander's decision. First, he must know the type of enemy that his company will defend against. If the threat is mainly dismounted infantry, the greater danger is allowing them to gain a foothold in the town. If the threat is armor or motorized infantry, the greatest danger is that massive direct fire will destroy the company's defensive positions. The company commander must also consider the terrain forward and to the flanks of the village from which the enemy can direct fires against his positions.

   b. Platoons are given a small group of buildings in which to prepare their defense, permitting the platoon leader to establish mutually supporting squad-sized positions. This increases the area that the platoon can control and hampers the enemy's ability to isolate or bypass a platoon. A platoon is responsible for the road through the village. The rest of the company is positioned to provide all-round security and defense in depth.

c. A position for the company mortars must be chosen that protects mortars from direct fire and allows for overhead clearance. The company's BFVs or APCs are placed in positions to the rear of the buildings and interior courtyards where their weapon systems can provide added rear and flank security. Combat vehicles are assigned primary, alternate, and supplementary positions as well as primary and secondary sectors of fire. They should be positioned in defilade behind rubble and walls, or inside buildings for movement into and out of the area. Control of the platoon's BFVs or APCs by the platoon leader is required for resupply, MEDEVAC, and rapid repositioning during the battle.

d. The company commander locates a forward area where he can position his company trains. A location is chosen near the highway to ease recovery and maintenance operations. A company OP is established where the fields of observation are best.

e. The company commander must also decide which buildings must be rubbled. To defeat the enemy he must have good fields of fire, but rubbling the buildings too soon or rubbling too many may disclose his exact locations and destroy cover from direct fire. The company's TOWs are positioned on high ground in and around the town to attain good fields of fire to the front and flanks.

f. If a tank platoon is available from the TF, the company commander could place the tanks along the leading edge where rapid fire would complement the TOWs and Dragons. The tank platoon leader should select exact firing positions and assign sectors of fire. If faced by enemy infantry, the tanks move to alternate positions with the protection of the infantry. These alternate positions allow the tanks to engage to the front as well as the flanks with as little movement as possible. After they are withdrawn from the leading edge of the town, the tanks could provide a mobile reserve for the team.

g. FPFs are planned to address the biggest threat to the platoon—the enemy's infantry. When firing an FPF inside a built-up area is required, mortars are more effective than artillery. This is due to their higher angle of fall which gives them a greater chance of impacting on the street.

h. Obstacles, mainly antivehicle obstacles, are easily constructed in a built-up area. The company commander must stop enemy vehicles without interfering with his own movement in the village. Therefore, he executes the emplacement of cratering charges at key street locations on order. Mines are laid on the outskirts of the town and along routes the company will not use.

i. The supporting engineers use C4 and other explosives to make firing ports, mouseholes, and demolition obstacles. Based upon his priority of work, the commander tells the engineer squad leader to assist each of the infantry platoons preparing the village for defense and to execute the team's obstacle plan. The squad leader's mission is to tell the infantrymen exactly where to place the demolitions and how much is needed for the desired effect. He also assists in the emplacement and recording of the minefields as well as the preparation of fighting positions.

j. Ammunition expenditure is usually high when fighting in a built-up area. To avoid moving around the village with ammunition resupply during the battle, the commander directs that more ammunition be stockpiled in each occupied platoon and squad position. He also orders the platoons to

stockpile firefighting equipment, drinking water, food, and first-aid supplies at each squad position. Other factors the company commander must consider are:
- Resupply.
- Medical evacuation.
- Communications.
- Firefighting.
- Sleep and alert plans.
- Security.
- Limited visibility.
- Civilian control.

k. To ensure adequate communications, the company installs a wire net and develops a plan for pyrotechnic signals. Backup wire should be laid in case primary lines are cut by vehicles, fires, or the enemy. The commander also plans for the use of messengers throughout the village.

4-23. DEFENSE OF A CITY BLOCK

A company in a built-up area may have to defend a city block in a core periphery or residential area. It conducts this operation according to the defensive scheme of the battalion. The operation should be coordinated with the action of security forces charged with delaying to the front of the company's position. The defense should take advantage of the protection of buildings that dominate the roads.

a. A well-organized company defense—
- Stops the attack of the enemy on the roads by using obstacles and enfilade fire.
- Destroys the enemy by ambush and direct fire from prepared positions.
- Ejects the enemy from footholds or remains in place for a counterattack conducted by battalion.

b. The operation of the company is more effective if it has time to reconnoiter the terrain, and to prepare obstacles and fire lanes. Vehicles not needed for the defense should be grouped in the combat trains at battalion. The OPs should be supplemented by patrols, mainly at night, and communications should be wire. The company should be organized to provide a series of OPs, a defense, and a reserve that is tasked with counterattacks.

c. The defensive forces should ambush on the avenues of approach, cover the obstacles by fire, and prepare a strong defense inside the buildings (Appendix D). The reserve can be tasked—
- To reinforce the fires of the defense.
- To react to a danger on the flank.
- To counterattack to throw the enemy from a foothold.

d. Engineers should be controlled at company level. They construct obstacles, prepare access routes, and assist in preparing defensive positions.

A company or section of tanks attached to the company should provide heavy direct-fire support, engage enemy tanks, and support counterattack.

4-24. COMPANY DELAY
A company delay can be part of a battalion's defense (Figure 4-13). Its operations destroy enemy reconnaisance elements forward of the outskirts of the town, prevent their penetration of the built-up areas, and gain and maintain contact with the enemy to determine the strength and location of the main attack.

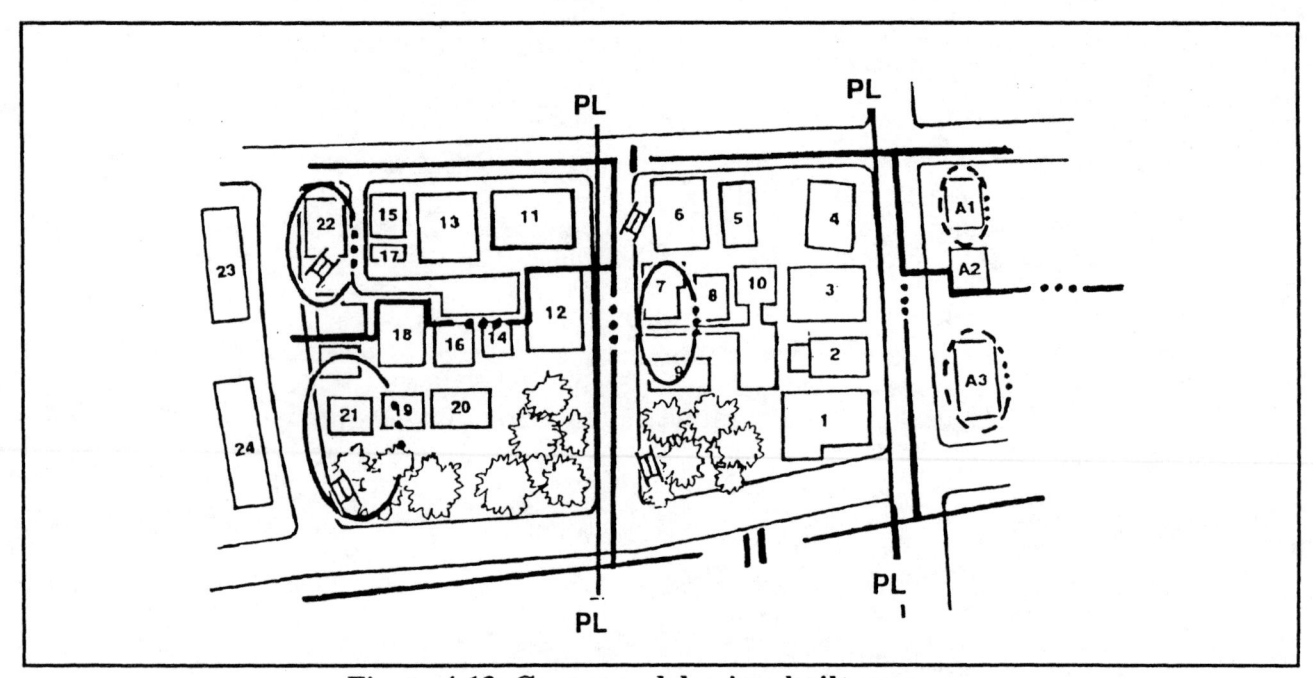

Figure 4-13. Company delay in a built-up area.

   a. The company's sector should be prepared with obstacles to increase the effect of the delay. Engineers prepare obstacles on main routes but avoid some covered and concealed routes that are known by the friendly troops for reinforcement, displacement, and resupply. These routes are destroyed when no longer needed.

   b. Antiarmor weapon systems are positioned on the outskirts of the town to destroy the enemy at maximum range. They should be located in defilade positions or in prepared shelters. They fire at visible targets and then fall back or proceed to alternate positions. Platoons should be assigned sectors from 500 to 700 meters (one to two blocks) wide. They should be reinforced with sensors or GSRs, which can be emplaced on the outskirts or on higher ground. Platoons delay by using patrols, OPs, and ambushes and by taking advantage of all obstacles. Each action is followed by a disengagement and withdrawal. By day, the defense is dispersed; at night, it is more concentrated. Close coordination is vital.

   c. Tanks support the platoon by engaging enemy tanks, providing reinforcing fires, aiding the disengagement of the platoons, and covering obstacles by fire.

d. BFVs support the platoon in the same manner as tanks except they engage BTRs, BMPs, and BRDMs.

4-25. DEFENSE OF A TRAFFIC CIRCLE
A rifle company or company team may be assigned the mission of defending a key traffic circle in a built-up area to prevent the enemy from seizing it (Figure 4-14).

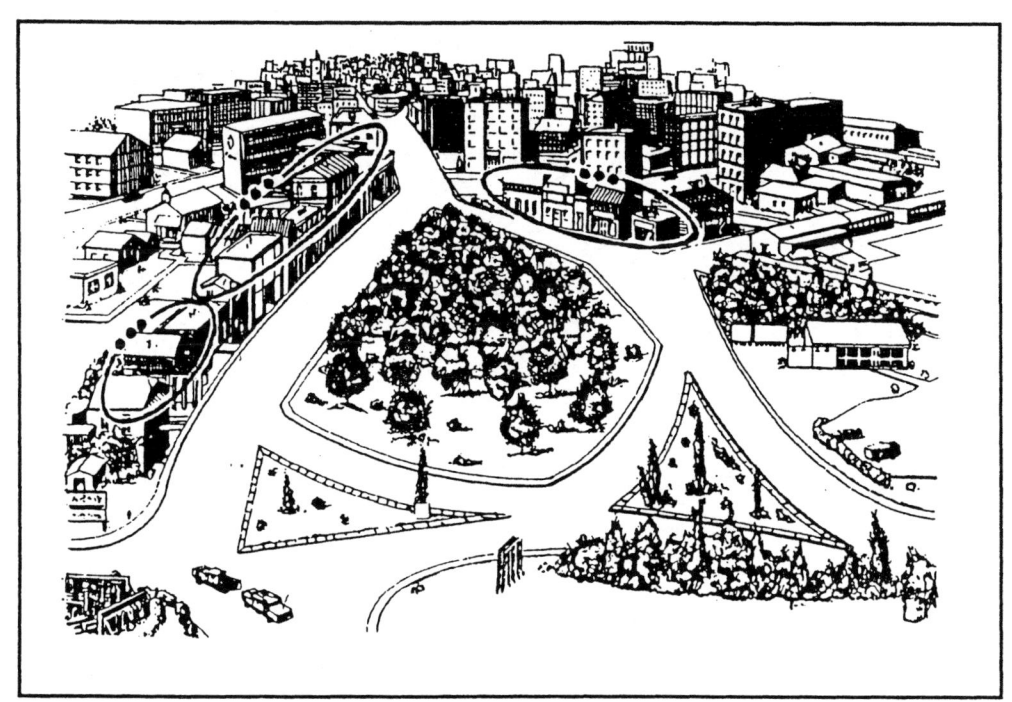

Figure 4-14. Defense of a traffic circle.

a. The company commander with this mission should analyze enemy avenues of approach and buildings that dominate those avenues. He should plan all possible fire power on the traffic circle itself and on the approaches to it. He should also plan for all-round defense of the buildings that dominate the traffic circle to prevent encirclement. The commander should prepare as many covered and concealed routes between these buildings as possible. This makes it easier to mass or shift fires, and to execute counterattacks.

b. Obstacles can also deny the enemy the use of the traffic circle. Obstacle planning in this case must consider if friendly forces are supposed to use the traffic circle. TOWs and Dragons can fire across the traffic circle if fields of fire are long enough. Tanks should engage enemy tanks and provide heavy direct-fire support for counterattacks. BFVs should engage BTRs, BMPs, and BRDMs and provide direct fire to protect obstacles.

Section VII. DEFENSIVE PLAN AT PLATOON LEVEL

The defensive plan in built-up areas at platoon level is METT-T and ROE dependent.

4-26. DEFENSE OF A STRONGPOINT

One of the most common defensive tasks a platoon will be given is the strongpoint defense of a building, part of a building, or a group of small buildings (Figure 4-15). The platoon's defense is normally integrated into the company's mission (defense of a traffic circle, and so forth). The platoon must keep the enemy from gaining a foothold in buildings. It makes the best use of its weapons and supporting fires, organizes all-round defense, and counterattacks or calls for a company counterattack to eject an enemy that has a foothold. The platoon leader analyzes his defensive sector to recommend to the company commander the best use of obstacles and supporting fires.

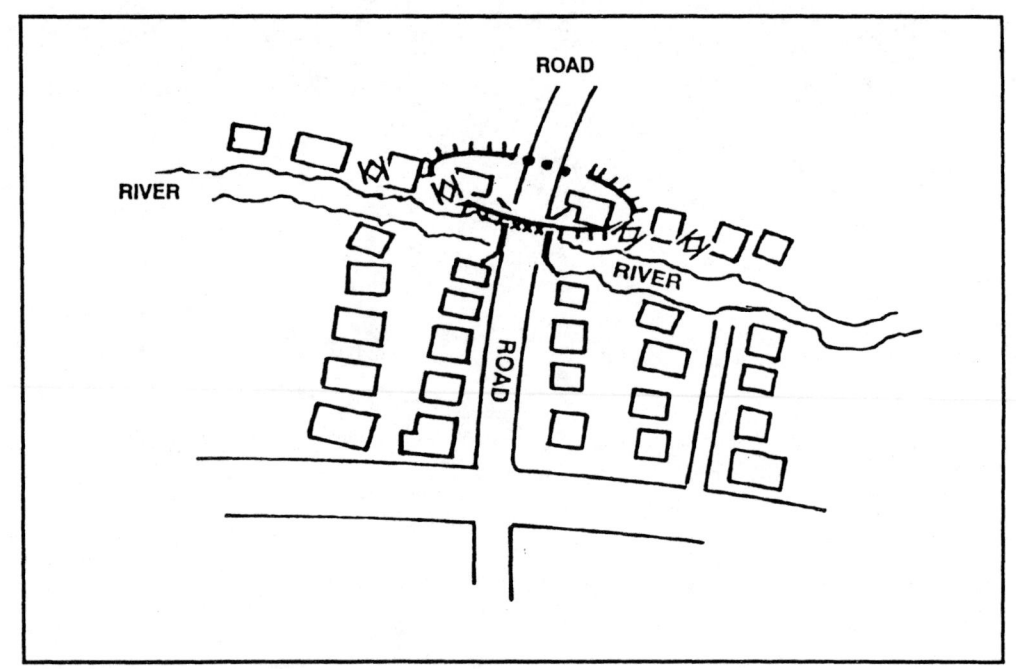

Figure 4-15. Defense of a strongpoint.

    a. The platoon should be organized into a series of firing positions located to cover avenues of approach, to cover obstacles, and to provide mutual support. Snipers may be located on the upper floors of the buildings. Unengaged elements should be ready to counterattack, fight fires, or reinforce other elements of the platoon.

    b. Depending on the length of the mission, the platoon should stockpile the following:

- Pioneer equipment (axes, shovels, hammers, picket pounders).
- Barrier material (barbed wire, sandbags).
- Munitions (especially grenades).
- Food and water.
- Medical supplies.
- Firefighting equipment.

4-27. DEFENSE AGAINST ARMOR

The terrain common to built-up areas is well-suited to an infantry's defense against mechanized infantry and armored forces. Mechanized infantry and armored forces try to avoid built-up areas but may be forced to pass through them. A well-trained infantry can inflict heavy casualties on such forces.

   a. Built-up areas have certain traits that favor infantry antiarmor operations.

     (1) Rubble in the streets can be used to block enemy vehicles, conceal mines, and cover and conceal defending infantry.

     (2) The streets restrict armor maneuver, fields of fire, and communications, thereby reducing the enemy's ability to reinforce.

     (3) Buildings provide cover and concealment for defending infantry.

     (4) Rooftops, alleys, and upper floors provide good firing positions.

     (5) Sewers, storm drains, and subways provide underground routes for infantry forces.

   b. Antiarmor operations in built-up areas involve the following planning steps:

STEP 1: Choose a good engagement area. Enemy tanks should be engaged where most restricted in their ability to support each other. The best way for infantrymen to engage tanks is one at a time, so that they can destroy one tank without being open to the fires of another. Typical locations include narrow streets, turns in the road, "T" intersections, bridges, tunnels, split-level roads, and rubbled areas. Less obvious locations can also be used by using demolitions or mines to create obstacles.

STEP 2: Select good weapons positions. The best weapons positions are places where the tank is weakest and the infantry is most protected. A tank's ability to see and fire are limited, mainly to the rear and flanks, if the tanks are buttoned up. Figure 4-16 shows the weapons and visual dead space of a buttoned-up tank against targets located at ground level. Similar dead space exists against targets located overhead.

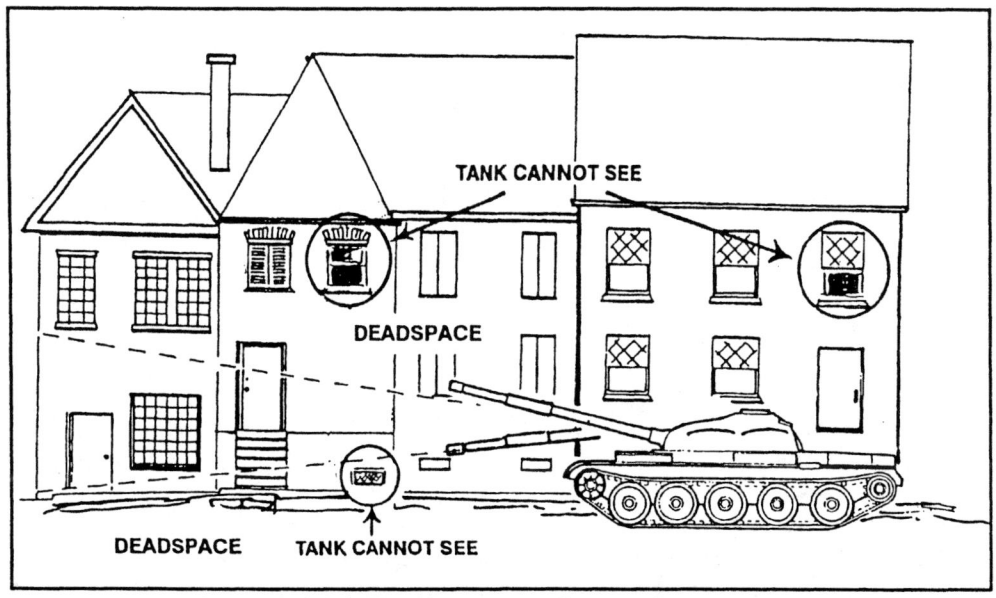

Figure 4-16. Tanks cannot fire at close-range, street-level, and overhead targets.

STEP 3: Assign target reference points and select method of engagement. After selecting the weapons positions, assign target reference points (TRPs) to ensure coverage of the areas and as a tool in controlling fires. The TRPs should be clearly visible through the gunner's sights and should be resistant to battle damage (for example, large buildings or bridge abatements, but not trees or cars). The leader of the antiarmor operation should specify what type of engagement should be used, such as frontal, cross-fire, or depth. Frontal fire is the least preferred since it exposes the gunner to the greatest probability of detection and it is where the armor is the thickest. (For more information on target engagement techniques, see FM 7-91 and or FM 23-1.)

(a) To the infantry force, the best places to fire on tanks are at the flanks and rear at ground level or at the top of tanks if the force is in an elevated position in a building (see Appendix H for minimum arming distance). A suitable antiarmor defense might be set up as shown in Figure 4-17.

Figure 4-17. A platoon's antiarmor defense.

(b) The best place to engage a tank from a flank is over the second road wheel at close range. This can be done using a corner so that the tank cannot traverse the turret to counterattack.

(c) For a safe engagement from an elevated position, infantrymen should allow the tank to approach to a range three times the elevation of the weapons.

(d) To engage at a longer range is to risk counterfire, since the weapon's position will not be in the tank's overhead dead space. However, overhead fire at the rear or flank of the tank is even more effective. Alternate and supplementary positions should be selected to enforce all-round security and to increase flexibility.

STEP 4: Coordinate target engagement. Tanks are most vulnerable when buttoned up. The first task of the tank-killing force is to force the tanks to button up, using all available direct and indirect fire. The proper use of fire control measures and graphics will greatly diminish the probability of fratricide. The next task is to coordinate the fires of the antitank weapons so that if there is more than one target in the engagement area, all targets are engaged at the same time.

c. Armored vehicles are often accompanied by infantry in built-up areas. Antiarmor weapons must be supported by an effective all-round antipersonnel defense (Figure 4-18).

d. At a planned signal (for example, the detonation of a mine) all targets are engaged at the same time. If targets cannot, they are engaged in the order of the most dangerous first. Although tanks present the greatest threat, BMPs are also dangerous because their infantry can dismount and destroy friendly antiarmor positions. If the friendly force is not secured by several infantrymen, priority of engagement might be given to enemy APCs. Rubble and mines should be used to reduce target mobility to present more targets to engage.

Figure 4-18. Coordinated antiarmor ambush.

4-28. CONDUCT OF ARMORED AMBUSH

A rifle company can use an attached tank platoon to conduct an armored ambush in a built-up area (Figure 4-19). To do so, the tank platoon should be reinforced with a BFV or APC and one or two squads from the rifle company. The ambush can be effective against enemy armor if it is conducted in an area cleared and reconnoitered by friendly forces.

Figure 4-19. Armored ambush.

    a. The operation involves maneuver on a road network that is free of obstacles. Obstacles outside the ambush area can be used to canalize and delay the enemy. The ambushing tank platoon must know the area.

    b. The ambushing tanks should be located in a hide position situated about 1,000 meters from the expected enemy avenue of approach. A security post, located at a choke point, observes and reports the approach, speed, security posture, and activity of the enemy. This role is assigned to a scout who uses the BFV, ITV, or APC to move from OP to OP. When the enemy is reported at a trigger point or TRP, the tank platoon leader knows how much he must move his tanks to execute the ambush.

    c. The tanks move quickly from their hide positions to firing positions, taking advantage of all available concealment. They try for flank shots on the approaching enemy—the average range is 300 to 400 meters. Such long ranges do not expose tanks to the enemy infantry. Once the enemy is engaged, tanks break contact and move to a rally point with close security provided by an infantry squad. They then move to a new ambush site.

CHAPTER 5

FUNDAMENTAL COMBAT SKILLS

Successful combat operations in built-up areas depend on the proper employment of the rifle squad. Each member must be skilled in the techniques of combat in built-up areas: moving, entering buildings, clearing buildings, employing hand grenades, selecting and using firing positions, navigating in built-up areas, and camouflaging. Soldiers must remember to remain in buddy teams when moving through a MOUT environment.

Section I. MOVEMENT

Movement in built-up areas is the first fundamental skill the soldier must master. Movement techniques must be practiced until they become habitual. To reduce exposure to enemy fire, the soldier avoids silhouetting himself, avoids open areas, and selects his next covered position before movement.

5-1. CROSSING OF A WALL

Each soldier must learn the correct method of crossing a wall (Figure 5-1). After he has reconnoitered the other side, he quickly rolls over the wall, keeping a low silhouette. The speed of his move and a low silhouette deny the enemy a good target.

Figure 5-1. Soldier crossing a wall.

5-2. MOVEMENT AROUND CORNERS

The area around a corner must be observed before the soldier moves beyond it. The most common mistake a soldier makes at a corner is allowing his weapon to extend beyond the corner, exposing his position. He should show his head below the height an enemy soldier would expect to see it. When using the correct techniques for looking around a corner (Figure 5-2), the soldier lies flat on the ground and does not extend his weapon beyond the corner of the building. He wears his kevlar helmet, and exposes his head (at ground level) only enough to permit observation.

Figure 5-2. Correct technique for looking around a corner.

5-3. MOVEMENT PAST WINDOWS

Windows present another hazard to the soldier and small-unit leader. The most common mistake in passing a window is exposing the head. If the soldier shows his head (Figure 5-3), an enemy gunner inside the building could engage him through the window without exposing himself to friendly covering fires.

Figure 5-3. Soldier moving past windows.

a. When using the correct technique for passing a window, the soldier stays below the window level. He makes sure he does not silhouette himself in the window; he "hugs" the side of the building. An enemy gunner inside the building would have to expose himself to covering fires if he tried to engage the soldier.

b. The same techniques used in passing first-floor windows are used when passing basement windows (Figure 5-4); however, the most common mistake in passing a basement window is not being aware of it. A soldier should not walk or run past a basement window, since he presents a good target to an enemy gunner inside the building. When using the correct procedure for negotiating a basement window, the soldier stays close to the wall of the building and steps or jumps past the window without exposing his legs.

Figure 5-4. Soldier passing basement windows.

5-4. USE OF DOORWAYS

Doorways should not be used as entrances or exits since they are normally covered by enemy fire. If a soldier must use a doorway as an exit, he should move quickly through it to his next position, staying as low as possible to avoid silhouetting himself (Figure 5-5, page 5-4). Preselection of positions, speed, a low silhouette, and the use of covering fires must be emphasized in exiting doorways.

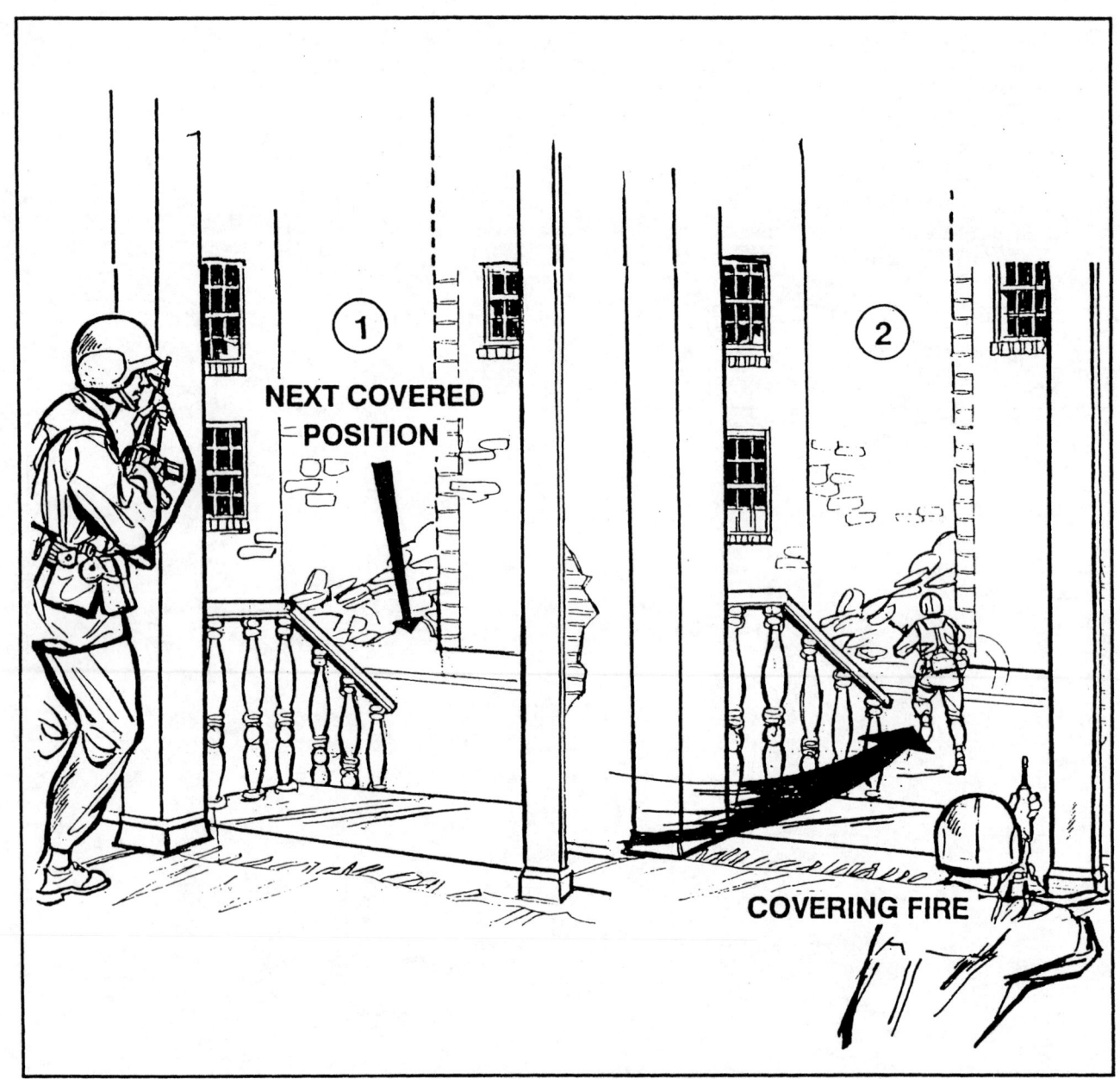

Figure 5-5. Soldier exiting a doorway.

5-5. MOVEMENT PARALLEL TO BUILDINGS

Soldiers and small units may not always be able to use the inside of buildings as a route of advance. Therefore, they must move on the outside of the buildings (Figure 5-6). Smoke and covering fires, and cover and concealment should be used to hide movement. In correctly moving on the outside of a building, the soldier "hugs" the side of the building, stays in the shadow, presents a low silhouette, and moves rapidly to his next position (Figure 5-7). If an enemy gunner inside the building fires on a soldier, he exposes himself to fire from other squad members. Furthermore, an enemy gunner farther down the street would have difficulty detecting and engaging the soldier.

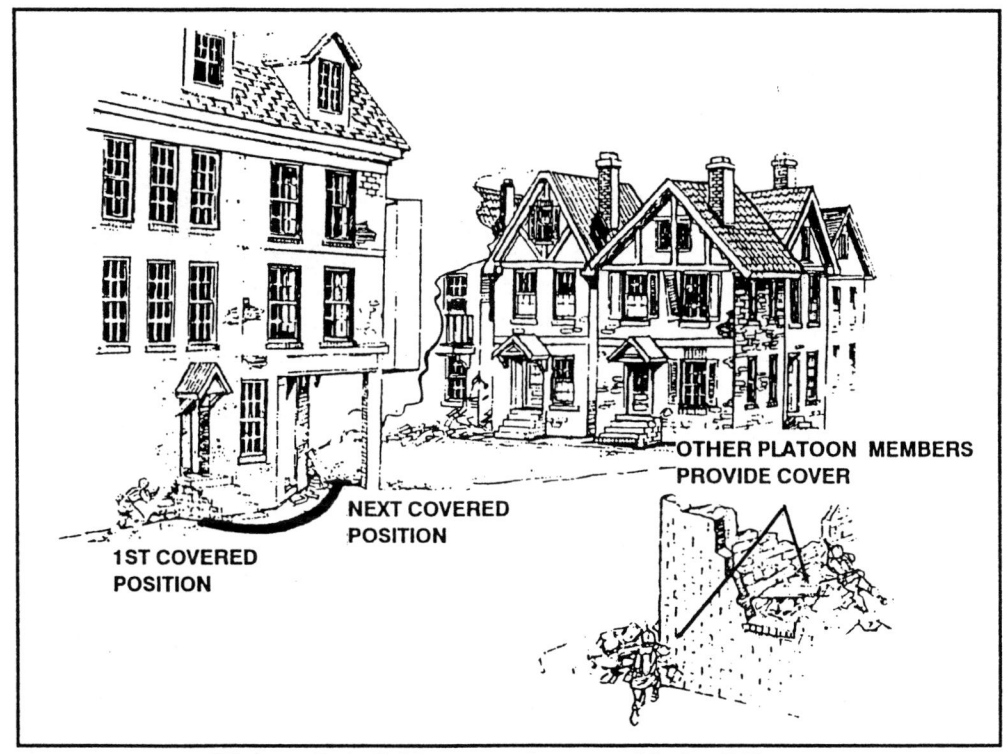

Figure 5-6. Soldier moving outside building.

Figure 5-7. Selection of the next position.

5-6. CROSSING OF OPEN AREAS

Open areas, such as streets, alleys, and parks, should be avoided. They are natural kill zones for enemy crew-served weapons. They can be crossed safely if certain fundamentals are applied by the individual or small-unit leader.

   a. When using the correct procedure for crossing an open area, the soldier develops a plan for his own movement. (Smoke from hand grenades or smoke pots should be used to conceal the movement of all soldiers.) He runs the shortest distance between the buildings and moves along the far building to the next position. By doing so, he reduces the time he is exposed to enemy fire.

   b. Before moving to another position, the soldier should make a visual reconnaissance and select the position for the best cover and concealment. At the same time, he should select the route that he will take to get to that position.

5-7. FIRE TEAM EMPLOYMENT

Moving as a fire team, from building to building or between buildings, presents a problem because a fire team presents a large target to enemy fire (Figure 5-8). When moving from the corner of one building to another, the fire team should move across the open area in a group. Moving from the side of one building to the side of another presents a similar problem, and the technique of movement employed is the same. The fire team should use the building as cover. In moving to the adjacent building (Figure 5-9), team members should keep a distance of 3 to 5 meters between themselves and, using a planned signal, make an abrupt flanking movement (on line) across the open area to the next building.

Figure 5-8. Fire team movement.

Figure 5-9. Movement to adjacent building.

5-8. MOVEMENT BETWEEN POSITIONS

When moving from position to position, each soldier must be careful not to mask his supporting fires. When he reaches his next position, he must be prepared to cover the movement of other members of his fire team or squad. He must use his new position effectively, and fire his weapon from either shoulder.

   a. The most common errors a soldier can make when firing from a position are firing over the top of his cover and silhouetting himself against the building to his rear, providing the enemy with an easy target. The correct technique for firing from a covered position is to fire around the side of the cover, reducing exposure to the enemy (Figure 5-10, page 5-8).

   b. Another common error is for a right-handed firer to try to fire from the right shoulder around the left corner of a building. Firing left-handed around the left corner of a building takes advantage of the cover afforded by the building (Figure 5-11, page 5-8). Right-handed and left-handed soldiers should be trained to adapt cover and concealment to fit their manual orientation. Also, soldiers should be able to fire from their opposite shoulder, if needed.

Figure 5-10. Soldier firing from a covered position.

Figure 5-11. Firing left-handed around the corner of a building.

5-9. MOVEMENT INSIDE A BUILDING

When moving within a building that is under attack (Figure 5-12), the soldier avoids silhouetting himself in doors and windows. If forced to use a hallway (Figure 5-13), he must stay against the wall to avoid presenting a target to the enemy. When operating under precision MOUT conditions, movement techniques may be modified or omitted based on the ROE in effect.

Figure 5-12. Movement within a building under attack.

Figure 5-13. Hallway procedures.

a. The enemy often boobytraps windows and doors. When entering a room, a soldier avoids using the door handle. Instead, he fires a short burst of automatic fire through the door around the latch and then kicks it open. If booby traps are detected, they should be marked, reported, and bypassed.

b. Before entering each room, the first soldier cooks off a concussion hand grenade by removing the grenade's safety pin, releasing the safety lever, counting by thousands (one thousand and one, one thousand and two), and then throwing the grenade into the room. He must be careful of thin walls and floors. Voice alerts should be given while throwing grenades. When friendly forces throw grenades, the command is "Frag Out;" when an enemy grenade has been identified, friendly forces shout "Grenade."

> **WARNING**
> Because fragments from M67 fragmentation grenades may injure soldiers outside the room, they should not be used. Soldiers should use MK3A2 offensive hand grenades instead. Also, cooking off hand grenades can be dangerous unless properly performed.

c. When the hand grenade goes off, the second man immediately enters the room and engages any targets with short bursts of automatic fire (Figure 5-14). He then systematically searches the room. The first man follows the second man and takes a position opposite the side of the doorway from the second man. Meanwhile, the support party, in position outside the room being cleared, provides outside security. (See FM 7-8 for more detailed information on entering a room.)

Figure 5-14. Procedures for the first man entering a room.

d. **The soldier uses voice alerts.** Voice alerts and signals within the assault team are extremely important. The soldier must always let others in the assault team know where he is and what he is doing. Once a room has been cleared, the assault team yells, "Clear," to inform the support party. Before leaving the room and rejoining the support party, the assault team yells "Coming out." The team then marks the room according to unit SOP. When moving up or down a staircase, the assault team yells "Coming up" or "Coming down."

e. **Mouseholes** measure about 2 feet wide and are blown or cut through a wall so soldiers can enter a room (Figure 5-15). These are safer entrances than doors because doors can be easily booby trapped and should be avoided. As with any entry, a hand grenade is thrown in first.

Figure 5-15. Soldiers entering through a mousehole.

Section II. ENTRY TECHNIQUES

When entering a building, a soldier must enter with minimum exposure. He must select the entry point before moving toward the building; avoid windows and doors; use smoke to conceal his advance to the building; use demolitions, tank rounds, combat engineer vehicles (CEVs), and so on to make new entrances; precede his entry with a grenade; enter immediately after the grenade explodes; and be covered by one of his buddies.

5-10. UPPER BUILDING LEVELS

Clearing a building from the top down is the preferred method. Clearing or defending a building is easier from an upper story. Gravity and the building's floor plan become assets when throwing hand grenades and moving from floor to floor.

FM 90-10-1

a. An enemy who is forced to the top of a building may be cornered and fight desperately or escape over the roof. But an enemy who is forced down to the ground level may withdraw from the building, thus exposing himself to friendly fires from the outside.

b. Various means, such as ladders, drainpipes, vines, helicopters, or the roofs and windows of adjoining buildings may be used to reach the top floor or roof of a building. In some cases, one soldier can climb onto the shoulders of another and reach high enough to pull himself up. Another method is to attach a grappling hook to the end of a scaling rope so that a rifleman can scale a wall, spring from one building to another, or gain entrance through an upstairs window.

5-11. USE OF LADDERS
Ladders offer the quickest method to gain access to the upper levels of a building (Figure 5-16). Units can get ladders from local civilians or stores or material to build ladders can be obtained through supply channels. If required, ladders can be built with resources that are available throughout the urban area; for example, lumber can be taken from inside the walls of buildings (Figure 5-17). Although ladders will not permit access to the top of some buildings, they will offer security and safety through speed.

Figure 5-16. Using ladders to get to upper levels.

Figure 5-17. Getting lumber from inside the walls.

5-12. USE OF GRAPPLING HOOK

A suitable grappling hook and rope are selected. The grappling hook should be sturdy, portable, easily thrown, and equipped with hooks that can hold inside a window. The scaling rope should be 5/8 to 1 inch in diameter and long enough to reach the objective window. Knots are tied in the rope at 1-foot intervals to make climbing easier. The soldier should follow the procedures outlined below.

    a. When throwing the grappling hook, stand as close to the building as possible (Figure 5-18, page 5-14). The closer you stand, the less exposure to enemy fires. The closer the range, the less horizontal distance the hook must be thrown.

    b. Making sure there is enough rope to reach the target, hold the hook and a few coils of rope in the throwing hand. The remainder of the rope, in loose coils, should be in the other hand. Allow the rope to play out freely. The throw should be a gentle, even, upward lob of the hook, with the other hand releasing the rope as it plays out.

    c. Once the grappling hook is inside the window (or on the roof), pull on the rope to obtain a good hold before beginning to climb. When using a window, pull the hook to one corner to ensure the chances of a good "bite" and to reduce exposure to lower windows during the climb.

    d. The use of grappling hooks is the least preferred method for gaining entry to upper levels of buildings. It should be used only as a last resort and away from potential enemy positions. This method may potentially be used on adjacent buildings that offer concealed locations and a connecting roof to enemy positions.

Figure 5-18. Grappling hook thrown at close range.

5-13. SCALING OF WALLS

When forced to scale a wall during exposure to enemy fire, all available concealment must be used. Smoke and diversionary measures improve the chances of a successful exposed movement. When using smoke for concealment, soldiers must plan for wind direction and smoke use. They should use fire, shouting, and fake movement to distract the enemy.

    a. A soldier scaling an outside wall is vulnerable to enemy sniper fire. Soldiers who are moving from building to building and climbing buildings should be covered with friendly fire. Areas between buildings offer good fields of fire to the enemy. Properly positioned friendly weapons can suppress and eliminate enemy fire. The M203 grenade launcher is effective in clearing the enemy from rooms inside buildings (Figure 5-19).

    b. The soldier scaling a wall with a rope should avoid silhouetting himself in windows of uncleared rooms and avoid exposing himself to enemy fires from lower windows. He should climb with his weapon slung over the firing shoulder to quickly bring it to a firing position. He should clear the lower room with a hand grenade before going outside the window. The soldier does this by first loosening the safety pin so that he only needs one hand to throw the grenade. The objective upper-story window should not be entered before a hand grenade has been thrown in.

    c. The soldier enters the objective window with a low silhouette (Figure 5-20). Entry can be head first; however, a preferred method is to hook a leg over the window sill and enter sideways, straddling the ledge.

Figure 5-19. Employment of M203 grenade launcher for clearing enemy snipers.

ENTER WINDOW ONLY AFTER GRENADE HAS BEEN THROWN IN

Figure 5-20. Soldier entering the objective window.

5-14. RAPPELLING

Rappelling (Figure 5-21) is an entry technique for upper floors that soldiers can use to descend from the rooftop of a tall building into a window. (See TC 21-24 for more information on rappelling.)

Figure 5-21. Rappelling.

5-15. ENTRY AT LOWER LEVELS

Buildings should be cleared from the top down. However, it may be impossible to enter a building at the top; therefore, entry at the bottom or lower level may be the only course of action. When entering a build at the lower level, soldiers avoid entry through windows and doors since both can be easily boobytrapped and are usually covered by enemy fire.

    a. Ideally, when entering at lower levels, demolitions, artillery, tank fire, antiarmor weapons fire, or similar means are used to create a new entrance to avoid booby traps. Quick entry is required to follow up the effects of the blast and concussion.

    b. When the only entry to a building is through a window or door, supporting fire should be directed at that location. If no supporting fire is available, LAWs can be employed instead.

    c. Before entering, soldiers throw a cooked off hand grenade into the new entrance to reinforce the effects of the original blast. When making a new entrance in a building, they consider the effects of the blast on the building and adjacent buildings. If there is the possibility of a fire in adjacent building, soldiers coordinate with adjacent units and obtain permission before starting the operation. In wooden frame buildings, the blast may cause the building to collapse. In stone, brick, or cement buildings, supporting fires are aimed at the corner of the building or at weak points in the building construction. (Specific lower-level entry techniques are shown in Figure 5-22.)

THE TWO-MAN LIFT, SUPPORTED

(1) Two men stand facing one another, holding a support (a board or bar).

(2) Another soldier steps onto the support.

(3) Once both feet are on the support, the two men raise it, lifting the third man upward and into the entrance.

THE TWO-MAN LIFT WITH HEELS RAISED

(1, 2) One man, standing with palms flat against the building, feet out from the building about 2 feet with heels raised, is lifted by two men.

(3) Two men bend over facing each other. They each grasp a heel of the third man, and with one quick move lift him up and into the entrance.

Figure 5-22. Lower-level entry techniques.

FM 90-10-1

ONE-MAN LIFT

One man, with his back or side against the building and with his hands cupped, allows another man to raise one foot up into his cupped hands, and then lifts him up and into the entrance.

THE TWO-MAN PULL

When the first two soldiers are inside the building and other soldiers seek entrance, the two already inside may assist the others by pulling them up into the building.

Figure 5-22. Lower-level entry techniques (continued).

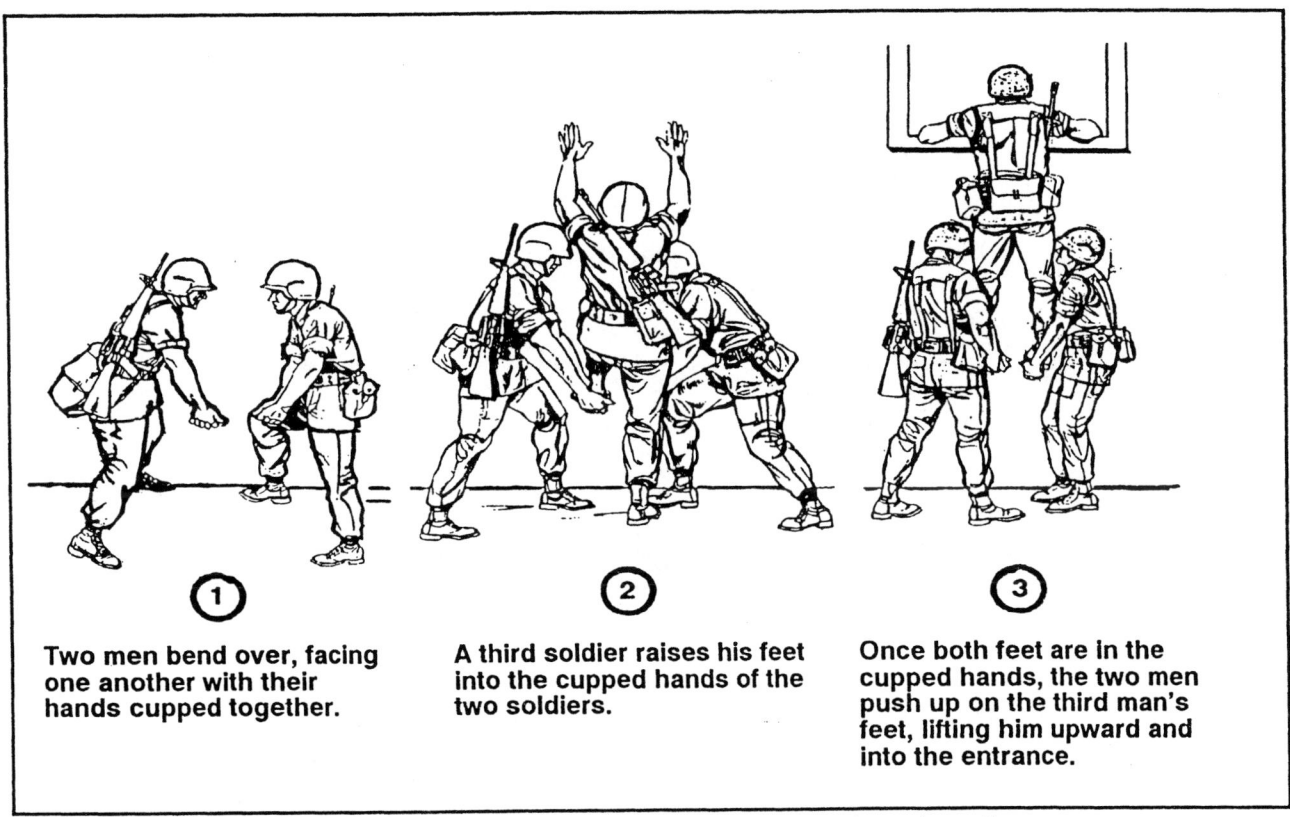

| ① Two men bend over, facing one another with their hands cupped together. | ② A third soldier raises his feet into the cupped hands of the two soldiers. | ③ Once both feet are in the cupped hands, the two men push up on the third man's feet, lifting him upward and into the entrance. |

Figure 5-22. Lower-level entry techniques (continued).

5-16. HAND GRENADES

Combat in built-up areas (mainly during the attack) requires extensive use of hand grenades. The soldier should throw a grenade before negotiating staircases, mouseholes, and so on. This usually requires the use of both hands and the overhand and underhand methods of throwing. The grenade should be allowed to cook off for two seconds to prevent the enemy from grabbing the grenade and tossing it back.

   a. The construction material used in the building being cleared influences the use of grenades. Concussion or offensive grenades are preferred over fragmentary grenades during offensive operations or when defending from hasty defensive positions. If the walls of a building are made of thin material, such as sheetrock or thin plyboard, the soldier must either lie flat on the floor with his helmet pointing towards the area of detonation, or move away from any wall that might be penetrated by grenade fragments.

   b. Soldiers should throw grenades in an opening before entering a building to eliminate enemy that might be near the entrance (Figure 5-23, page 5-20). The M203 greande launcher is the best method for putting a grenade in an upper story window.

   c. When a hand grenade must be used, the soldier throwing the grenade should stand close to the building, using it for cover. At the same time, the individual and the rest of the element should have a planned area to move to for safety if the grenade does not go through the window but falls back to the ground.

FM 90-10-1

d. The soldier throwing the grenade should allow the grenade to cook off for at least two seconds, and then step out far enough to lob the grenade in the upper-story opening. The weapon should be kept in the nonthrowing hand so it can be used if needed. The weapon should never be laid outside or inside the building. Once the grenade has been thrown into the opening (Figure 5-23), assaulting troops must move swiftly to enter the building. This technique should only be employed when the window has been broken. Otherwise, the chances are high that the thrown grenade will fall back onto the ground without going into the room.

e. If soldiers must enter the building by the stairs, they first look for booby traps. Then they throw a grenade through the stairwell door, let it detonate, and quickly move inside. They can use the staircase for cover.

Figure 5-23. Hand grenade thrown through window.

WARNING
After throwing the grenade, the soldier must immediately announce "frag out" to indicate that a grenade has been thrown. He then takes cover since the grenade may bounce back or be thrown back, or the enemy may fire at him.

f. The best way to enter a building is to breach the exterior wall. Again, a grenade must be thrown through the hole using all available cover, such as the lower corner of the building (Figure 5-24).

Figure 5-24. Hand grenade being thrown into a loophole.

g. When a door is the only means of entering a room, soldiers must beware of fire from enemy soldiers within the room and beware of booby traps. Doors can be opened by using the hand, by kicking, by firing, or by using pioneer tools such as an ax. When opening a door, soldiers should not expose themselves to firers through the door. A two-man team should be used when doors are opened by using the hand. Each soldier should stay close to one side of the doorway so as not to expose himself in the open doorframe. However, it is better to open the door by kicking or firing (Figure 5-25, page 5-22). When kicking, one man stands to the side while the other kicks.

h. Soldiers force the door open using short bursts of automatic fire aimed at the door locking mechanism. Other techniques are to use an ax or demolitions, if they are available. As a last resort, soldiers can resort to kicking the door open. This is the least favored technique since it is difficult and tiring to the soldier. It rarely works the first time, thereby giving any enemy soldiers within the room ample warning (it also gives the enemy time to shoot through the door). Once the door is open, a hand grenade is tossed in. After the grenade explodes, the first soldier entering the room positions himself to the right (left) of the entrance, up against the wall; engages targets with rapid, short bursts of automatic fire; and scans the room. The rest of the team provide immediate security. The first man in the room decides where the next man should position himself and gives the command **NEXT MAN IN, LEFT (RIGHT)**. The next man shouts **COMING IN, LEFT (RIGHT)**, enters the room, and positions himself up against the wall left (right) of the entrance and scans the room. Once in position, the senior soldier can call in additional team members with the **NEXT MAN IN** command, as the situation dictates. It is critical that all assault team members tell each other where they are to avoid fratricide.

Figure 5-25. Soldier shooting the door open.

 i. Another way to enter a room is to blast mouseholes with demolitions. In moving from room to room through mouseholes, soldiers should use grenades as in moving through doorways. As they enter the mousehole, they should stay low and use all available cover.

 j. Although buildings are best cleared from the top down, this is not always possible. While clearing the bottom floor of a building, soldiers may encounter stairs, which must also be cleared. Once again, grenades play an important role. To climb the stairs, soldiers should first inspect for booby traps, then toss a cooked-off grenade to the head of the stairs (Figure 5-26). Soldiers must use voice alerts when throwing grenades. Once the first grenade has detonated, another grenade should be thrown over and behind the staircase banister and into the hallway, destroying any enemy hiding to the rear. Using the staircase for cover, soldiers throw the grenade underhand to reduce the risk of it bouncing back and rolling down the stairs.

 k. After the stairs have been cleared, assaulting forces move to the top floor and clear it, using the methods already described. Upon clearing the top floor, forces move downstairs to clear the center and bottom floors, and to continue with the mission.

> NOTE: Since large quantities of hand grenades are used when clearing buildings, a continuous supply must be available to forces having this mission within a built-up area.

Figure 5-26. Soldier tossing grenade up stairway.

Section III. FIRING POSITIONS

Whether a unit is attacking, defending, or conducting retrograde operations, its success or failure depends on the ability of the individual soldier to place accurate fire on the enemy with the least exposure to return fire. Consequently, the soldier must immediately seek and properly use firing positions.

5-17. HASTY FIRING POSITION

A hasty firing position is normally occupied in the attack or the early stages of the defense. It is a position from which the soldier can place fire upon the enemy while using available cover for protection from return fire. The soldier may occupy it voluntarily, or he may be forced to occupy it due to enemy fire. In either case, the position lacks preparation before occupation. Some of the more common hasty firing positions in a built-up area and techniques for occupying them are: corners of buildings, firing from behind walls, firing from windows, firing from unprepared loopholes, and firing from the peak of a roof.

   a. **Corners of Buildings.** The corner of a building provides cover for a hasty firing position if used properly.

   (1) The firer must be capable of firing his weapon both right- and left-handed to be effective around corners. A common error made in firing around corners is firing from the wrong shoulder. This exposes more of the firer's body to return fire than necessary. By firing from the proper shoulder, the firer can reduce the target exposed to enemy fire.

   (2) Another common mistake when firing around corners is firing from the standing position. The firer exposes himself at the height the enemy would expect a target to appear, and risks exposing the entire length of his body as a target for the enemy.

b. **Walls.** When firing behind walls, the soldier must fire around cover—not over it (Figure 5-27).

Figure 5-27. Soldier firing around cover.

c. **Windows.** In a built-up area, windows provide convenient firing ports. The soldier must avoid firing from the standing position since it exposes most of his body to return fire from the enemy and could silhouette him against a light-colored interior beyond the window. This is an obvious sign of the firer's position, especially at night when the muzzle flash can easily be observed. In using the proper method of firing from a window (Figure 5-28), the soldier is well back into the room to prevent the muzzle flash from being seen, and he is kneeling to limit exposure and avoid silhouetting himself.

Figure 5-28. Soldier firing from window.

d. **Loopholes.** The soldier may fire through a hole torn in the wall and avoid windows (Figure 5-29). He stays well back from the loophole so the muzzle of the weapon does not protrude beyond the wall, and the muzzle flash is concealed.

Figure 5-29. Soldier firing from loophole.

e. **Roof.** The peak of a roof provides a vantage point for snipers that increases their field of vision and the ranges at which they can engage targets (Figure 5-30). A chimney, a smokestack, or any other object protruding from the roof of a building can reduce the size of the target exposed and should be used.

Figure 5-30. Soldier firing from peak of a roof.

f. **No Position Available.** When the soldier is subjected to enemy fire and none of the positions mentioned above are available, he must try to expose as little of himself as possible. When a soldier in an open area between buildings (a street or alley) is fired upon by enemy in one of the buildings to his front and no cover is available, he should lie prone as close as possible to a building on the same side of the open area as the enemy. To engage the soldier, the enemy must then lean out the window and expose himself to return fire.

g. **No Cover Available.** When no cover is available, target exposure can be reduced by firing from the prone position, by firing from shadows, and by presenting no silhouette against buildings.

5-18. PREPARED FIRING POSITION

A prepared firing position is one built or improved to allow the firer to engage a particular area, avenue of approach, or enemy position, reducing his exposure to return fire. Examples of prepared positions include: barricaded windows, fortified loopholes, sniper positions, antiarmor positions, and machine gun positions.

a. The natural firing port provided by windows can be improved by barricading the window, leaving a small hole for the firer's use (Figure 5-31). The barricading may be accomplished with materials torn from the interior walls of the building or any other available material. When barricading windows, avoid—

(1) Barricading only the windows that will be used as firing ports. The enemy will soon determine that the barricaded windows are firing positions.

(2) Neat, square, or rectangular holes that are easily identified by the enemy. A barricaded window should not have a neat, regular firing port. The window should keep its original shape so that the position of the firer is hard to detect. Firing from the bottom of the window gives the firer the advantage of the wall because the firing port is less obvious to the enemy. Sandbags are used to reinforce the wall below the window and to increase protection for the firer. All glass must be removed from the window to prevent injury to the firer. Lace curtains permit the firer to see out and prevent the enemy from seeing in. Wet blankets should be placed under weapons to reduce dust. Wire mesh over the window keeps the enemy from throwing in hand grenades.

b. Although windows usually are good firing positions, they do not always allow the firer to engage targets in his sector.

(1) To avoid establishing a pattern of always firing from windows, an alternate position is required such as the prepared loophole (Figure 5-32). This involves cutting or blowing a small hole into the wall to allow the firer to observe and engage targets in his sector.

(2) Sandbags are used to reinforce the walls below, around, and above the loophole. Two layers of sandbags are placed on the floor under the firer to protect him from an explosion on a lower floor (if the position is on the second floor or higher). A wall of sandbags, rubble, furniture, and so on should be constructed to the rear of the position to protect the firer from explosions in the room.

(3) A table, bedstead, or other available material provides overhead cover for the position. This prevents injury to the firer from falling debris or explosions above his position.

(4) The position should be camouflaged by knocking other holes in the wall, making it difficult for the enemy to determine which hole the fire is coming from. Siding material should be removed from the building in several places to make loopholes less noticeable.

Figure 5-31. Window firing position.

Figure 5-32. Prepared loopholes.

c. A chimney or other protruding structure provides a base from which a sniper position can be prepared. Part of the roofing material is removed to allow the sniper to fire around the chimney. He should stand inside the building on the beams or on a platform with only his head and shoulders above the roof (behind the chimney). Sandbags placed on the sides of the position protect the sniper's flanks.

d. When the roof has no protruding structure to provide protection (Figure 5-33), the sniper position should be prepared from underneath on the enemy side of the roof. The position is reinforced with sandbags, and a small piece of roofing material should be removed to allow the sniper to engage targets in his sector. The missing piece of roofing material should be the only sign that a position exists. Other pieces of roofing should be removed to deceive the enemy as to the true sniper position. The sniper should be invisible from outside the building, and the muzzle flash must be hidden from view.

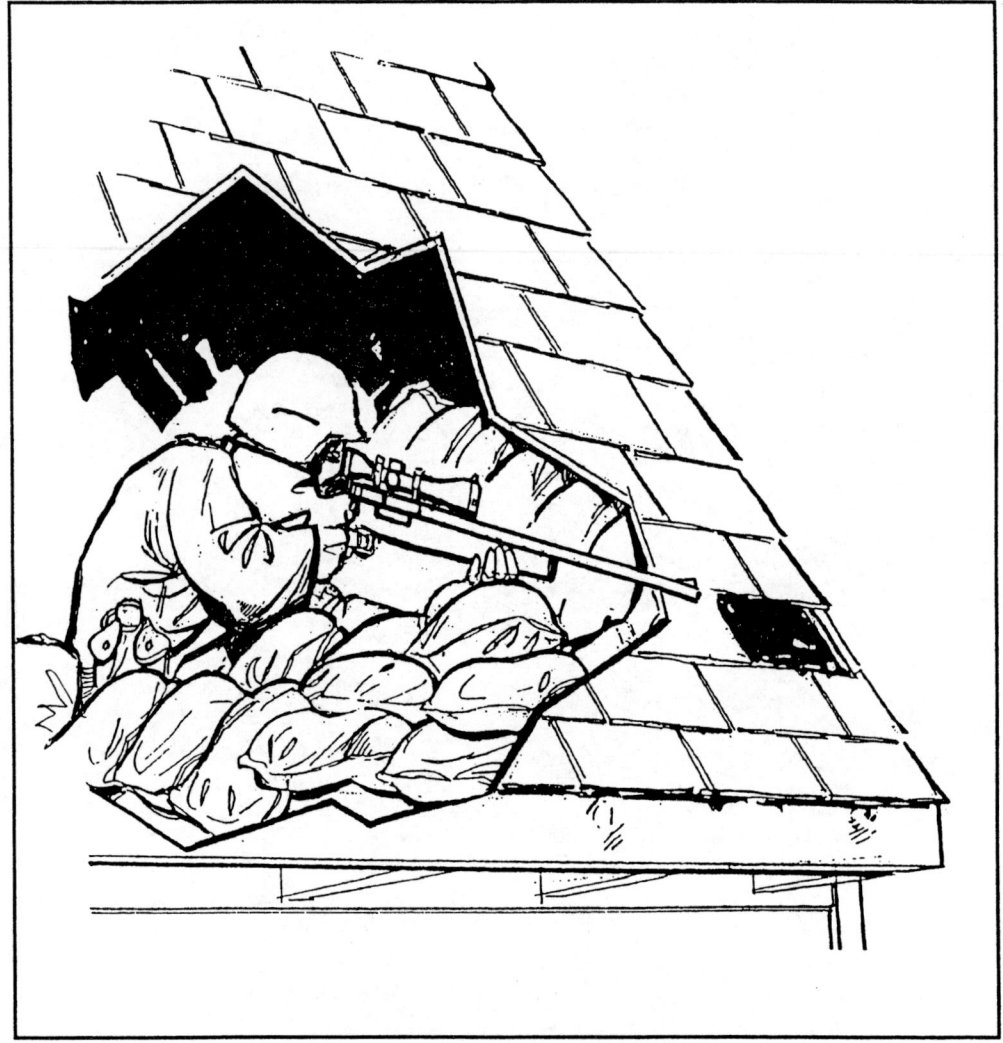

Figure 5-33. Sniper position.

e. Some rules and considerations for selecting and occupying individual firing positions are:
  (1) Make maximum use of available cover and concealment.
  (2) Avoid firing over cover; when possible, fire around it.
  (3) Avoid silhouetting against light-colored buildings, the skyline, and so on.
  (4) Carefully select a new firing position before leaving an old one.
  (5) Avoid setting a pattern; fire from both barricaded and unbarricaded windows.
  (6) Keep exposure time to a minimum.
  (7) Begin improving a hasty position immediately after occupation.
  (8) Use construction material for prepared positions that is readily available in a built-up area.
  (9) Remember that positions that provide cover at ground level may not provide cover on higher floors.

f. In attacking a built-up area, the recoilless weapon and ATGM crews are severely hampered in choosing firing positions due to the backblast of their weapons. They may not have enough time to knock out walls in buildings and clear backblast areas. They should select positions that allow the backblast to escape such as corner windows where the round fired goes out one window and the backblast escapes from another. The corner of a building can be improved with sandbags to create a firing position (Figure 5-34).

Figure 5-34. Corner firing position.

g. The rifle squad during an attack on and in defense of a built-up area is often reinforced with attached antitank weapons. Therefore, the rifle squad leader must be able to choose good firing positions for the antitank weapons under his control.

h. Various principles of employing antitank weapons have universal applications such as: making maximum use of available cover; trying to achieve mutual support; and allowing for the backblast when positioning recoilless weapons, TOWs, Dragons, and LAWs or AT4s.

i. Operating in a built-up area presents new considerations. Soldiers must select numerous alternate positions, particularly when the structure does not provide cover from small-arms fire. They must position their weapons in the shadows and within the building.

j. Recoilless weapons and ATGMs firing from the top of a building can use the chimney for cover (Figure 5-35). The rear of this position should be reinforced with sandbags.

Figure 5-35. A recoilless weapon crew firing from a rooftop.

k. When selecting firing positions for recoilless weapons and ATGMs, make maximum use of rubble, corners of buildings, and destroyed vehicles to provide cover for the crew. Recoilless weapons and ATGMs can also be moved along rooftops to obtain a better angle in which to engage enemy armor. When buildings are elevated, positions can be prepared using a building for overhead cover (Figure 5-36). The backblast under the building must not damage or collapse the building or injure the crew.

NOTE: When firing from a slope, ensure that the angle of the launcher relative to the ground or firing platform is not greater than 20 degrees. When firing within a building, ensure the enclosure is at least 10 feet by 15 feet, is clear of debris and loose objects, and has windows, doors, or holes in the walls for the backblast to escape.

Figure 5-36. Prepared positions using a building for overhead cover.

FM 90-10-1

l. The machine gun has no backblast, so it can be emplaced almost anywhere. In the attack, windows and doors offer ready-made firing ports (Figure 5-37). For this reason, the enemy normally has windows and doors under observation and fire, which should be avoided. Any opening in walls that was created during the fighting may be used. When other holes are not present, small explosive charges can create loopholes (Figure 5-38). Regardless of what openings are used, machine guns should be within the building and in the shadows.

Figure 5-37. Emplacement of machine gun in a doorway.

Figure 5-38. Use of a loophole with a machine gun.

m. Upon occupying a building, soldiers board up all windows and doors. By leaving small gaps between the slots, soldiers can use windows and doors as good alternative firing positions.

n. Loopholes should be used extensively in the defense. They should not be constructed in any logical pattern, nor should they all be at floor or table-top level. Varying their height and location makes them hard to pinpoint and identify. Dummy loopholes, shingles knocked off, or holes cut that are not intended to be used as firing positions aid in the deception. Loopholes located behind shrubbery, under doorjams, and under the eaves of a building are hard to detect. In the defense, as in the offense, a firing position can be constructed using the building for overhead cover.

o. Increased fields of fire can be obtained by locating the machine gun in the corner of the building or sandbagged under a building (Figure 5-39). Available materials, such as desks, overstuffed chairs, couches, and other items of furniture, should be integrated into the construction of bunkers to add cover and concealment (Figure 5-40).

Figure 5-39. Sandbagged machine gun emplacement under a building.

Figure 5-40. Corner machine gun bunker.

FM 90-10-1

p. Although grazing fire is desirable when employing the machine gun, it may not always be practical or possible. Where destroyed vehicles, rubble, and other obstructions restrict the fields of grazing fire, the gun can be elevated to where it can fire over obstacles. Therefore, firing from loopholes on the second or third story may be necessary. A firing platform can be built under the roof (Figure 5-41) and a loophole constructed. Again, the exact location of the position must be concealed by knocking off shingles in isolated patches over the entire roof.

Figure 5-41. Firing platform built under roof.

5-19. TARGET ACQUISITION
Built-up areas provide unique challenges to units. Buildings mask movement and the effects of direct and indirect fires. The rubble from destroyed buildings, along with the buildings themselves, provide concealment and protection for attackers and defenders, making target acquisition difficult. A city offers definite avenues of approach that can easily be divided into sectors.

a. The techniques of patrolling and using observation posts apply in the city as well as in wooded terrain. These techniques enable units to locate the enemy, to develop targets for direct and indirect fires in the defense, and to find uncovered avenues of approach in the offense.

b. Most weapons and vehicles have distinguishing signatures. These come from design features or from the environment in which the equipment is used. For example, firing a tank main gun in dry, dusty, and debris-covered streets raises a dust cloud; a tank being driven in built-up areas produces more noise than one moving through an open field; soldiers moving through rubble on a street or in the halls of a damaged building create more noise than in a wooded area. Soldiers must recognize signatures so they can locate and identify targets. Seeing, hearing, and smelling assist in detecting and identifying signatures that lead to target location, identification, and rapid engagement. Soldiers must look for targets in areas where they are most likely to be employed.

c. Target acquisition must be continuous, whether halted or moving. Built-up areas provide both the attacker and defender with good cover and concealment, but it usually favors the defender because of the advantages achieved. This makes target acquisition extremely important since the side that fires first may win the engagement.

d. When a unit is moving and enemy contact is likely, the unit must have an overwatching element. This principle applies in built-up areas as it does in other kinds of terrain except that the overwatching element must observe both the upper floors of buildings and street level.

e. Stealth should be used when moving in built-up areas since little distance separates attackers and defenders. Only arm-and-hand signals should be used until contact is made. The unit should stop periodically to listen and watch, ensuring it is not being followed or that the enemy is not moving parallel to the unit's flank for an ambush. Routes should be carefully chosen so that buildings and piles of rubble can be used to mask the unit's movement.

f. Observation duties must be clearly given to squad members to ensure all-round security as they move. This security continues at the halt. All the senses must be used to acquire targets, especially hearing and smelling. Soldiers soon recognize the sounds of vehicles and people moving through streets that are littered with rubble. The smell of fuel, cologne, and food cooking can disclose enemy positions.

g. Observation posts are positions from which soldiers can watch and listen to enemy activity in a specific sector. They warn the unit of an enemy approach and are ideally suited for built-up areas. OPs can be positioned in the upper floors of buildings, giving soldiers a better vantage point than at street level.

h. In the defense, a platoon leader positions OPs for local security as ordered by the company commander. The platoon leader selects the general location but the squad leader sets up the OP (Figure 5-42, page 5-36). Normally, there is at least one OP for each platoon. An OP consists of two to four men and is within small-arms supporting range of the platoon. Leaders look for positions that have good observation of the target sector. Ideally, an OP has a field of observation that overlays those of adjacent OPs. The position selected for the OP should have cover and concealment for units moving to and from the OP. The upper floors of houses or other buildings should be used. The squad leader should not select obvious positions, such as water towers or church steeples, that attract the enemy's attention.

> A squad leader is given the general location of the OP by the platoon leader. The squad leader selects the exact position.
>
> Leaders look for positions that—
>
> - Have good observation of the desired area or sector (Ideally, an OP has a field of observation which overlaps those adjacent OPs).
>
> - Have cover and concealment (good observation of the sector may require the OP to accept less cover and concealment and require troops to selectively clear fields of observation).

Figure 5-42. Selection of OP location.

i. The soldier should be taught how to scan a target area from OPs or from his fighting positions. Use of proper scanning techniques enable squad members to quickly locate and identify targets. Without optics, the soldier searches quickly for obvious targets, using all his senses to detect target signatures. If no targets are found and time permits, he makes a more detailed search (using binoculars, if available) of the terrain in the assigned sector using the 50-meter method. First, he searches a strip 50 meters deep from right to left; then he searches a strip from left to right that is farther out, overlapping the first strip. This process is continued until the entire sector is searched. In the city core or core periphery where the observer is faced with multistory buildings, the overlapping sectors may be going up rather than out.

j. Soldiers who man OPs and other positions should employ target acquisition devices. These devices include binoculars, image intensification devices, thermal sights, ground surveillance radar (GSR), remote sensors (REMs) and platoon early warning systems (PEWS). All of these devices can enhance the units ability to detect and engage targets. Several types of devices should be used since no single device can meet every need of a unit. A mix might include PEWS sensors to cover out-of-sight areas and dead space, image intensification devices for close range, thermal sights for camouflage, and smoke penetration for low light conditions. A mix of devices is best because several devices permit overlapping sectors and more coverage, and the capabilities of one device can compensate for limitations of another.

k. Target acquisition techniques used at night are similar to those used during the day. At night, whether using daylight optics or the unaided eye, a soldier does not look directly at an object but a few degrees off to the side. The side of the eye is more sensitive to dim light. When scanning with off-center vision, he moves his eyes in short, abrupt, irregular moves. At each likely target area, he pauses a few seconds to detect any motion.

l. Sounds and smells can aid in acquiring targets at night since they transmit better in the cooler, damper, night air. Running engines, vehicles, and soldiers moving through rubble-covered streets can be heard for great distances. Odors from diesel fuel, gasoline, cooking food, burning tobacco, after-shave lotion, and so on reveal enemy and friendly locations.

5-20. FLAME OPERATIONS

Incendiary ammunition, special weapons, and the ease with which incendiary devices can be constructed from gasoline and other flammables make fire a true threat in built-up area operations. During defensive operations, firefighting should be a primary concern. The proper steps must be taken to reduce the risk of a fire that could make a chosen position indefensible.

a. Soldiers choose or create positions that do not have large openings. These positions provide as much built-in cover as possible to prevent penetration by incendiary ammunition. All unnecessary flammable materials are removed, including ammunition boxes, furniture, rugs, newspapers, curtains, and so on. The electricity and gas coming into the building must be shut off.

b. A building of concrete block construction, with concrete floors and a tin roof, is an ideal place for a position. However, most buildings have wooden floors or subfloors, wooden rafters, and wooden inner walls, which require improvement. Inner walls are removed and replaced with blankets to resemble walls from the outside. Sand is spread 2 inches deep on floors and in attics to retard fire.

c. All available firefighting gear is pre-positioned so it can be used during actual combat. For the individual soldier, such gear includes entrenching tools, helmets, sand, and blankets. These items are supplemented with fire extinguishers that are not in use.

d. Fire is so destructive that it can easily overwhelm personnel regardless of extraordinary precautions. Soldiers plan routes of withdrawal so that a priority of evacuation can be sent from fighting positions. This allows soldiers to exit through areas that are free from combustible material and provide cover from enemy direct fire.

e. The confined space and large amounts of combustible material in built-up areas can influence the enemy to use incendiary devices. Two major first-aid problems that are more urgent than in the open battlefield are: burns, and smoke and flame inhalation, which creates a lack of oxygen. These can easily occur in buildings and render the victim combat ineffective. Although there is little defense against flame inhalation and lack of oxygen, smoke inhalation can be greatly reduced by wearing the individual protective mask. Regardless of the fire hazard, defensive planning for combat in built-up areas must include aidmen. Aidmen must reach victims and their equipment, and must have extra supplies for the treatment of burns and inhalation injuries.

f. Offensive operations also require plans for firefighting since the success of the mission can easily be threatened by fire. Poorly planned use of incendiary munitions can make fires so extensive that they become obstacles to offensive operations. The enemy may use fire to cover his withdrawal, and to create obstacles and barriers to the attacker.

g. When planning offensive operations, the attacker must consider all available weapons. The best two weapons for creating fires are the M202 FLASH and the flamethrower, which is currently out of the Army inventory but can be obtained by special request through logistics channels. The flamethrower is the better training weapon, since water can be substituted for the flame, and the effect of the weapon can be measured by the penetration of the water. There is currently no training round for the M202. When using fire in an operation, firefighting support must be available to avoid using soldiers to fight fires. Soldiers chose targets during the initial planning to avoid accidentally destroying critical facilities within the built-up area. When using flame operations in a built-up area, soldiers set priorities to determine which critical installations (hospitals, power stations, radio stations, and historical landmarks) should have primary firefighting support.

h. Every soldier participating in the attack must be ready to deal with fire. The normal firefighting equipment available includes the entrenching tool, helmet (for carrying sand or water), and blankets (for snuffing out small fires). Fire extinguishers are available on each of the vehicles supporting the attack.

5-21. EMPLOYMENT OF SNIPERS
The value of the sniper to a unit operating in a built-up area depends on several factors. These factors include the type of operation, the level of conflict, and the rules of engagement. Where ROE allow destruction, the snipers may not be needed since other weapons systems available to a mechanized force have greater destructive effect. However, they can contribute to the fight. Where the ROE prohibit collateral damage, snipers may be the most valuable tool the commander has. (See FM 7-20; FM 71-2, C1; and TC 23-14 for more information.)

a. Sniper effectiveness depends in part on the terrain. Control is degraded by the characteristics of an urban area. To provide timely and effective support, the sniper must have a clear picture of the commander's concept of operation and intent.

b. Snipers should be positioned in buildings of masonry construction. These buildings should also offer long-range fields of fire and all-round observation. The sniper has an advantage because he does not have to move with, or be positioned with, lead elements. He may occupy a higher position to the rear or flanks and some distance away from the element he is supporting. By operating far from the other elements, a sniper avoids decisive engagement but remains close enough to kill distant targets that threaten the unit. Snipers should not be placed in obvious positions, such as church steeples and roof tops, since the enemy often observes these and targets them for destruction. Indirect fires can generally penetrate rooftops and cause casualties in top floors of buildings. Also, snipers should not be positioned where there is heavy traffic; these areas invite enemy observation as well.

c. Snipers should operate throughout the area of operations, moving with and supporting the companies as necessary. Some teams may operate independent of other forces. They search for targets of opportunity, especially for enemy snipers. The team may occupy multiple positions. A single position may not afford adequate observation for the entire team without increasing the risk of detection by the enemy. Separate positions must maintain mutual support. Alternate and supplementary positions should also be established in urban areas.

d. Snipers may be assigned tasks such as the following:

(1) Killing enemy snipers (countersniper fire).

(2) Killing targets of opportunity. These targets may be prioritized by the commander. Types of targets might include enemy snipers, leaders, vehicle commanders, radio men, sappers, and machine gun crews.

(3) Denying enemy access to certain areas or avenues of approach (controlling key terrain).

(4) Providing fire support for barricades and other obstacles.

(5) Maintaining surveillance of flank and rear avenues of approach (screening).

(6) Supporting local counterattacks with precision fire.

Section IV. NAVIGATION IN BUILT-UP AREAS

Built-up areas present a different set of challenges involving navigation. Deep in the city core, the normal terrain features depicted on maps may not apply—buildings become the major terrain features and units become tied to streets. Fighting in the city destroys buildings whose rubble blocks streets. Street and road signs are destroyed during the fighting if they are not removed by the defenders. Operations in subways and sewers present other unique challenges. However, maps and photographs are available to help the unit overcome these problems. The global positioning system can provide navigation abilities in built-up areas.

5-22. MILITARY MAPS

The military city map is a topographical map of a city that is usually a 1:12,500 scale, delineating streets and showing street names, important buildings, and other urban elements. The scale of a city map can vary from 1:25,000 to 1:50,000, depending on the importance and size of the city, density of detail, and intelligence information.

a. Special maps, prepared by supporting topographic engineers, can assist units in navigating in built-up areas. These maps have been designed or modified to give information not covered on a standard map, which includes maps of road and bridge networks, railroads, built-up areas, and electric power fields. They can be used to supplement military city maps and topographical maps.

b. Once in the built-up area, soldiers use street intersections as reference points much as hills and streams in rural terrain. City maps supplement or replace topographic maps as the basis of navigation. These maps enable units moving in the built-up area to know where they are and to move to new locations even though streets have been blocked or a key building destroyed.

c. The old techniques of compass reading and pace counting can still be used, especially in a blacked-out city where street signs and buildings are not visible. The presence of steel and iron in the MOUT environment may cause inaccurate compass readings. Sewers must be navigated much the same way. Maps providing the basic layout of the sewer system are maintained by city sewer departments. This information includes directions the sewer lines run and distances between manhole covers. Along with basic compass and pace count techniques, such information enables a unit to move through the city sewers.

   d. Operations in a built-up area adversely affect the performance of sophisticated electronic devices such as GPS and data distribution systems. These systems function the same as communications equipment—by line-of-sight. They cannot determine underground locations or positions within a building. These systems must be employed on the tops of buildings, in open areas, and down streets where obstacles will not affect line-of-sight readings.

   e. City utility workers are assets to units fighting in built-up areas. They can provide maps of sewers and electrical fields, and information about the city. This is important especially with regard to the use of the sewers. Sewers can contain pockets of methane gas that are highly toxic to humans. City sewer workers know the locations of these danger areas and can advise a unit on how to avoid them.

5-23. GLOBAL POSITIONING SYSTEMS
Most global positioning systems use a triangular technique using satellites to calculate their position. Preliminary tests have shown that GPS are not affected by small built-up areas, such as villages. However, large built-up areas with a mixture of tall and short buildings cause some degradation of most GPS. This affect may increase as the system is moved into an interior of a large building or taken into subterranean areas.

5-24. AERIAL PHOTOGRAPHS
Current aerial photographs are also excellent supplements to military city maps and can be substituted for a map. A topographic map or military city map could be obsolete if compiled many years ago. A recent aerial photograph shows changes that have taken place since the map was made. This could include destroyed buildings and streets that have been blocked by rubble as well as enemy defensive preparations. More information can be gained by using aerial photographs and maps together than using either one alone.

Section V. CAMOUFLAGE

To survive and win in combat in built-up areas, a unit must supplement cover and concealment with camouflage. To properly camouflage men, carriers, and equipment, soldiers must study the surrounding area and make positions look like the local terrain.

5-25. APPLICATION
Only the material needed for camouflaging a position should be used since excess material could reveal the position. Material must be obtained from a wide area. For example, if defending a cinder block building, do not strip the front, sides, or rear of the building to camouflage a position.

   a. Buildings provide numerous concealed positions. Armored vehicles can often find isolated positions under archways or inside small industrial or commercial structures. Thick masonry, stone, or brick walls offer excellent protection from direct fire and provide concealed routes.

b. After camouflage is completed, the soldier inspects positions from the enemy's viewpoint. He makes routine checks to see if the camouflage remains natural looking and actually conceals the position. If it does not look natural, the soldier must rearrange or replace it.

c. Positions must be progressively camouflaged as they are prepared. Work should continue until all camouflage is complete. When the enemy has air superiority, work may be possible only at night. Shiny or light-colored objects that attract attention from the air must be hidden.

d. Shirts should be worn since exposed skin reflects light and attracts the enemy. Even dark skin reflects light because of its natural oils.

e. Camouflage face paint is issued in three standard, two-tone sticks. When issue-type face-paint sticks are not available, burnt cork, charcoal, or lampblack can be used to tone down exposed skin. Mud may be used as a last resort since it dries and may peel off, leaving the skin exposed; it may also contain harmful bacteria.

5-26. USE OF SHADOWS

Buildings in built-up areas throw sharp shadows, which can be used to conceal vehicles and equipment (Figure 5-43). Soldiers should avoid areas that are not in shadows. Vehicles may have to be moved periodically as shadows shift during the day. Emplacements inside buildings provide better concealment.

Figure 5-43. Use of shadows for concealment.

FM 90-10-1

a. Soldiers should avoid the lighted areas around windows and loopholes. They will be better concealed if they fire from the shadowed interior of a room (Figure 5-44).

b. A lace curtain or piece of cheesecloth provides additional concealment to soldiers in the interior of rooms if curtains are common to the area. Interior lights are prohibited.

Figure 5-44. Concealment inside a building.

5-42

5-27. COLOR AND TEXTURE

Standard camouflage pattern painting of equipment is not as effective in built-up areas as a solid, dull, dark color hidden in shadows. Since repainting vehicles before entering a built-up area is not always practical, the lighter sand-colored patterns should be subdued with mud or dirt.

  a. The need to break up the silhouette of helmets and individual equipment exists in built-up areas the same as it does elsewhere. However, burlap or canvas strips are a more effective camouflage than foliage (Figure 5-45). Predominant colors are normally browns, tans, and sometimes grays rather than greens, but each camouflage location should be evaluated.

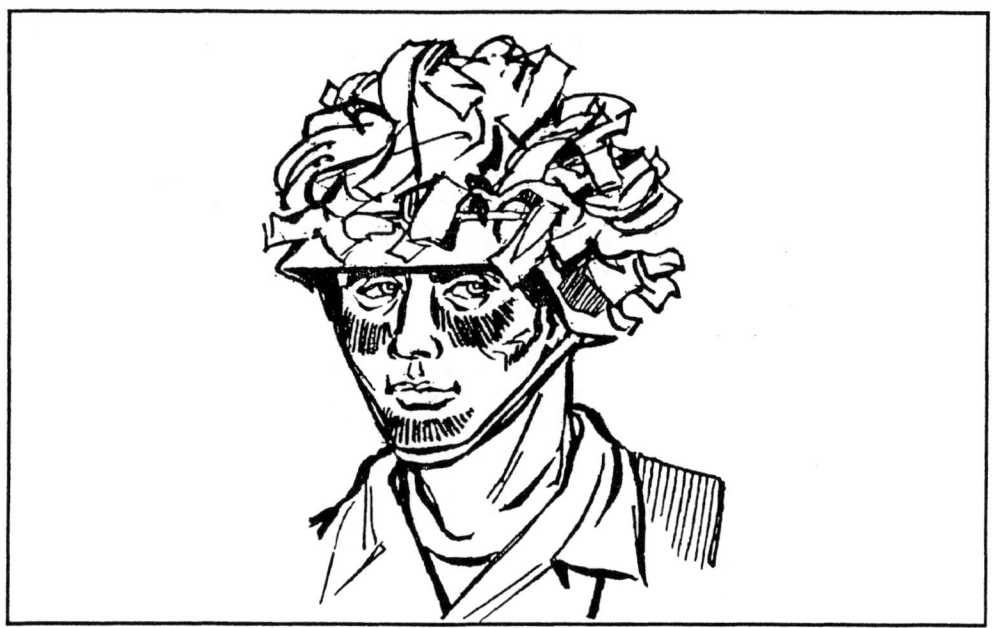

Figure 5-45. Helmet camouflaged with burlap strips.

  b. Weapons emplacements should use a wet blanket (Figure 5-46, page 5-44), canvas, or cloth to keep dust from rising when the weapon is fired.

  c. Command posts and logistical emplacements are easier to camouflage and better protected if located underground. Antennas can be remoted to upper stories or to higher buildings based on remote capabilities. Field telephone wire should be laid in conduits, in sewers, or through buildings.

  d. Soldiers should consider the background to ensure that they are not silhouetted or skylined, but rather blend into their surroundings. To defeat enemy urban camouflage, soldiers should be alert for common camouflage errors such as the following:

- Tracks or other evidence of activity.
- Shine or shadows.
- An unnatural color or texture.
- Muzzle flash, smoke, or dust.
- Unnatural sounds and smells.
- Movement.

Figure 5-46. Wet blanket used to keep dust down.

   e. Dummy positions can be used effectively to distract the enemy and make him reveal his position by firing.

   f. Built-up areas afford cover, resources for camouflage, and locations for concealment. The following basic rules of cover, camouflage, and concealment should be adhered to:

   (1) Use the terrain and alter camouflage habits to suit your surroundings.
   (2) Employ deceptive camouflage of buildings.
   (3) Continue to improve positions. Reinforce fighting positions with sandbags or other fragment- and blast-absorbent material.
   (4) Maintain the natural look of the area.
   (5) Keep positions hidden by clearing away minimal debris for fields of fire.
   (6) Choose firing ports in inconspicuous spots when available.

   NOTE: Remember that a force that COVERS and CONCEALS itself has a significant advantage over a force that does not.

CHAPTER 6
COMBAT SUPPORT

Combat support is fire support and other assistance provided to combat elements. It normally includes field artillery, air defense, aviation (less air cavalry), engineers, military police, communications, electronic warfare, and NBC.

6-1. MORTARS

Mortars are the most responsive indirect fires available to battalion and company commanders. Their mission is to provide close and immediate fire support to the maneuver units. Mortars are well suited for combat in built-up areas because of their high rate of fire, steep angle of fall, and short minimum range. Battalion and company commanders must plan mortar support with the FSO as part of the total fire support system. (See FM 7-90 for detailed information on the tactical employment of mortars.)

a. **Role of Mortar Units.** The role of mortar units is to deliver suppressive fires to support maneuver, especially against dismounted infantry. Mortars can be used to obscure, neutralize, suppress, or illuminate during MOUT. Mortar fires inhibit enemy fires and movement, allowing friendly forces to maneuver to a position of advantage. Effectively integrating mortar fires with dismounted maneuver is key to successful combat in a built-up area at the rifle company and battalion level.

b. **Position Selection.** The selection of mortar positions depends on the size of buildings, the size of the urban area, and the mission. Also, rubble can be used to construct a parapet for firing positions.

(1) The use of existing structures (for example, garages, office buildings, or highway overpasses) for hide positions is recommended to afford maximum protection and minimize the camouflage effort. By proper use of mask, survivability can be enhanced. If the mortar has to fire in excess of 885 mils to clear a frontal mask, the enemy counterbattery threat is reduced. These principles can be used in both the offense and the defense.

(2) Mortars should not be mounted directly on concrete; however, sandbags may be used as a buffer. Sandbags should consist of two or three layers; be butted against a curb or wall; and extend at least one sandbag width beyond the baseplate.

(3) Mortars are usually not placed on top of buildings because lack of cover and mask makes them vulnerable. They should not be placed inside buildings with damaged roofs unless the structure's stability has been checked. Overpressure can injure personnel, and the shock on the floor can weaken or collapse the structure.

c. **Communications.** An increased use of wire, messenger, and visual signals will be required. However, wire should be the primary means of communication between the forward observers, fire support team, fire direction center, and mortars since elements are close to each other. Also, FM radio transmissions in built-up areas are likely to be erratic. Structures reduce radio ranges; however, remoting of antennas to upper floors or roofs may improve communications and enhance operator survivability. Another technique that applies is the use of radio retransmissions. A practical solution is to use existing civilian systems to supplement the unit's capability.

d. **Magnetic Interference.** In an urban environment, all magnetic instruments are affected by surrounding structural steel, electrical cables, and automobiles. Minimum distance guidelines for the use of the M2 aiming circle (FM 23-90) will be difficult to apply. To overcome this problem, an azimuth is obtained to a distant aiming point. From this azimuth, the back azimuth of the direction of fire is subtracted. The difference is indexed on the red scale and the gun manipulated until the vertical cross hair of the sight is on the aiming point. Such features as the direction of a street may be used instead of a distant aiming point.

e. **High-Explosive Ammunition.** During MOUT, mortar HE fires are used more than any other type of indirect fire weapon. The most common and valuable use for mortars is often harassment and interdiction fires. One of their greatest contributions is interdicting supplies, evacuation efforts, and reinforcement in the enemy rear just behind his forward defensive positions. Although mortar fires are often targeted against roads and other open areas, the natural dispersion of indirect fires will result in many hits on buildings. Leaders must use care when planning mortar fires during MOUT to minimize collateral damage.

(1) High-explosive ammunition, especially the 120-mm projectile, gives good results when used on lightly built structures within cities. However, it does not perform well against reinforced concrete found in larger urban areas.

(2) When using HE ammunition in urban fighting, only point detonating fuzes should be used. The use of proximity fuzes should be avoided, because the nature of built-up areas causes proximity fuzes to function prematurely. Proximity fuzes, however, are useful in attacking targets such as OPs on tops of buildings.

(3) During both World War II and recent Middle East conflicts, light mortar HE fires have been used extensively during MOUT to deny the use of streets, parks, and plazas to enemy personnel.

f. **Illumination.** In the offense, illuminating rounds are planned to burst above the objective to put enemy troops in the light. If the illumination is behind the objective, the enemy troops would be in the shadows rather than in the light. In the defense, illumination is planned to burst behind friendly troops to put them in the shadows and place the enemy troops in the light. Buildings reduce the effectiveness of the illumination by creating shadows. Continuous illumination requires close coordination between the FO and FDC to produce the proper effect by bringing the illumination over the defensive positions as the enemy troops approach the buildings.

g. **Special Considerations.** When planning the use of mortars, commanders must consider the following:

(1) FOs should be positioned on tops of buildings so target acquisition and adjustments in fire can best be accomplished.

(2) Commanders must understand ammunition effects to correctly estimate the number of volleys needed for the specific target coverage. Also, the effects of using WP or LP may create unwanted smoke screens or limited visibility conditions that could interfere with the tactical plan.

(3) FOs must be able to determine dead space. Dead space is the area in which indirect fires cannot reach the street level because of buildings. This area is a safe haven for the enemy. For mortars, the dead space is about one-half the height of the building.

(4) Mortar crews should plan to provide their own security.

(5) Commanders must give special consideration to where and when mortars are to displace while providing immediate indirect fires to support the overall tactical plan. Combat in built-up areas adversely affects the ability of mortars to displace because of rubbling and the close nature of MOUT.

6-2. FIELD ARTILLERY

A field artillery battalion is normally assigned the tactical mission of direct support (DS) to a maneuver brigade. A battery may not be placed in DS of a battalion task force, but may be attached.

   a. Appropriate fire support coordination measures should be carefully considered since fighting in built-up areas results in opposing forces fighting in close combat. When planning for fire support in a built-up area, the battalion commander, in coordination with his FSO, considers the following.

   (1) Target acquisition may be more difficult because of the increased cover and concealment afforded by the terrain. Ground observation is limited in built-up areas, therefore FOs should be placed on tops of buildings. Adjusting fires is difficult since buildings block the view of adjusting rounds; therefore, the lateral method of adjustment should be used.

   (2) Initial rounds are adjusted laterally until a round impacts on the street perpendicular to the FEBA. Airburst rounds are best for this adjustment. The adjustments must be made by sound. When rounds impact on the perpendicular street, they are adjusted for range. When the range is correct, a lateral shift is made onto the target and the gunner fires for effect.

   (3) Special consideration must be given to shell and fuze combinations when effects of munitions are limited by buildings.
      (a) Careful use of VT is required to avoid premature arming.
      (b) Indirect fires may create unwanted rubble.
      (c) The close proximity of enemy and unfriendly troops requires careful coordination.
      (d) WP may create unwanted fires and smoke.
      (e) Fuze delay should be used to penetrate fortifications.
      (f) Illumination rounds can be effective; however, friendly positions should remain in shadows and enemy positions should be highlighted. Tall buildings may mask the effects of illumination rounds.
      (g) VT, TI, and ICM are effective for clearing enemy positions, observers, and antennas off rooftops.
      (h) Swirling winds may degrade smoke operations.
      (i) FASCAM may be used to impede enemy movements. FASCAM effectiveness is reduced when delivered on a hard surface.

   (4) Targeting is difficult in urban terrain because the enemy has many covered and concealed positions and movement lanes. The enemy may be on rooftops and in buildings, and may use sewer and subway systems. Aerial observers are extremely valuable for targeting because they can see deep to detect movements, positions on rooftops, and fortifications. Targets should be planned on rooftops to clear away enemy FOs as well as communications and radar equipment. Targets should also be planned on major roads, at road intersections, and on known or likely enemy fortifications. Employing artillery in the direct fire mode to destroy fortifications should be considered. Also, restrictive fire support coordination measures (such as a restrictive fire area or no-fire area) may be imposed to protect civilians and critical installations.

(5) The 155-mm and 8-inch self-propelled howitzers are effective in neutralizing concrete targets with direct fire. Concrete-piercing 155-mm and 8-inch rounds can penetrate 36 inches and 56 inches of concrete, respectively, at ranges up to 2,200 meters. These howitzers must be closely protected when used in the direct-fire mode since none of them have any significant protection for their crews. Restrictions may be placed on types of artillery ammunition used to reduce rubbling on avenues of movement that may be used by friendly forces.

(6) Forward observers must be able to determine where and how large the dead space is. Dead space is the area in which indirect fires cannot reach the street level because of buildings. This area is a safe haven for the enemy because he is protected from indirect fires. For low-angle artillery, the dead space is about five times the height of the building. For mortars and high-angle artillery, the dead space is about one-half the height of the building.

(7) Aerial observers are effective for seeing behind buildings immediately to the front of friendly forces. They are extremely helpful when using the ladder method of adjustment because they may actually see the adjusting rounds impact behind buildings. Aerial observers can also relay calls for fire when communications are degraded due to power lines or building mask.

(8) Radar can locate many artillery and mortar targets in an urban environment because of the high percentage of high-angle fires. If radars are sited too close behind tall buildings, some effectiveness will be lost.

b. The use of airburst fires is an effective means of clearing snipers from rooftops. HE shells with delay fuzes may be effective against enemy troops in the upper floors of buildings, but, due to the overhead cover provided by the building, such shells have little effect on the enemy in the lower floors. (The planning and use of field artillery in offensive and defensive operations are also addressed in Chapters 3 and 4.)

6-3. NAVAL GUNFIRE

When a unit is operating with gunfire support within range, naval gunfire can provide effective fire support. If naval gunfire is used, a supporting arms liaison team (SALT) of a US Marine air naval gunfire liaison company (ANGLICO) may be attached to the battalion. The SALT consists of one liaison section that operates at the battalion main CP. It also has two firepower control teams at the company level, providing ship-to-shore communications and coordination for naval gunfire support. The SALT collocates and coordinates all naval gunfire support with battalion FSE.

6-4. TACTICAL AIR

A battalion may be supported by USAF, USN, USMC, or allied fighters and attack aircraft while fighting in built-up areas.

a. The employment of CAS depends on the following.

(1) *Shock and concussion.* Heavy air bombardment provides tactical advantages to an attacker. The shock and concussion of the bombardment reduce the efficiency of defending troops and destroy defensive positions.

(2) *Rubble and debris.* The rubble and debris resulting from air attacks may increase the defender's cover while creating major obstacles to the movement of attacking forces.

(3) *Proximity of friendly troops.* The proximity of opposing forces to friendly troops may require the use of precision-guided munitions and may

require the temporary disengagement of friendly forces in contact. The AC-130 is the air weapons platform of choice for precision MOUT as the proximity of friendly troops precludes other tactical air use.

(4) *Indigenous civilians or key facilities.* The use of air weapons may be restricted by the presence of civilians or the requirement to preserve key facilities within a city.

(5) *Limited ground observation.* Limited ground observation may require the use of airborne FAC.

b. CAS may be employed during offensive operations—

- To support the isolation of the city by interdicting entry and exit routes.
- To support attacking units by reducing enemy strongpoints with precision-guided munitions.
- To conduct tactical air reconnaissance and to provide detailed intelligence of enemy dispositions, equipment, and strengths.

c. CAS may be employed during defensive operations—

- To strike enemy attack formations and concentrations outside the built-up area.
- To provide precision-guided munitions support to counterattacks for recovering fallen friendly strongpoints.

6-5. AIR DEFENSE

Basic air defense doctrine does not change when units operate in urbanized terrain. The fundamental principles of mix, mass, mobility, and integration all apply to the employment of air defense assets.

a. The ground commander must consider the following when developing his air defense plan.

(1) Enemy air targets, such as principal lines of communications, road and rail networks, and bridges, are often found in and around built-up areas.

(2) Good firing positions may be difficult to find and occupy for long-range air defense missile systems in the built-up areas. Therefore, the number of weapons the commander can employ may be limited.

(3) Movement between positions is normally restricted in built-up areas.

(4) Long-range systems can provide air defense cover from positions on or outside of the edge of the city.

(5) Radar masking and degraded communications reduce air defense warning time for all units. Air defense control measures must be adjusted to permit responsive air defense within this reduced warning environment.

b. Positioning of Vulcan weapons in built-up areas is often limited to more open areas without masking such as parks, fields, and rail yards. Towed Vulcans (separated from their prime movers) may be emplaced by helicopter onto rooftops in dense built-up areas to provide protection against air attacks from all directions. This should be accomplished only when justified by the expected length of occupation of the area and of the enemy air threat.

c. Stingers provide protection for battalions the same as in any operation. When employed within the built-up area, rooftops normally offer the best firing positions.

d. Heavy machine guns emplaced on rooftops can also provide additional air defense.

6-6. ARMY AVIATION

Army aviation support of urban operations includes attack, observation, utility, and cargo helicopters for air movement or air assault operations, command and control, observation, reconnaissance, operations of sensory devices, attack, radio transmissions, and medical evacuation. When using Army aviation, the commander considers the enemy air situation, enemy air defenses, terrain in or adjacent to the city, and the availability of Army or Air Force suppression means.

   a. **Offensive Missions.** Missions for Army aviation in support of urban offensive operations include:

   (1) Air assault operations to secure key terrain adjacent to or in the urban area and key objectives when the area is lightly defended or enemy fires have been suppressed.

   (2) Employment of attack helicopters with aerial weapons to support the commander's scheme of maneuver in or adjacent to the built-up area.

   (3) Air movement and medical evacuation.

   (4) Command and control by providing rapid displacement of command elements to critical areas and an airborne command platform.

   (5) Aerial retransmission.

   (6) Intelligence-gathering operations.

   (7) Long-range antiarmor fire.

   b. **Defensive Missions.** Missions for Army aviation during urban defensive operations include:

   (1) Long-range antiarmor fire.

   (2) Rapid insertion or relocation of personnel (antiarmor teams and reserves).

   (3) Rapid concentration of forces and fires.

   (4) Retrograde movement of friendly forces.

   (5) Combat service support operations.

   (6) Command and control.

   (7) Communications.

   (8) Intelligence-gathering operations.

6-7. HELICOPTERS

An advantage can be gained by air assaulting onto rooftops. Before a mission, an inspection should be made of rooftops to ensure that no obstacles exist, such as electrical wires, telephone poles, antennas, or enemy-emplaced mines and wire, that could damage helicopters or troops. In many modern cities, office buildings often have helipads on their roofs, which are ideal for landing helicopters. Other buildings, such as parking garages, are usually strong enough to support the weight of a helicopter. The delivery of troops onto a building can also be accomplished by rappelling from the helicopter or jumping out of the helicopter while it hovers just above the roof.

   a. **Small-Scale Assaults.** Small units may have to be landed onto the rooftop of a key building. Success depends on minimum exposure and the suppression of all enemy positions that could fire on the helicopter. Depending on the construction of the roof, rappelling troops from the helicopter may be more of an advantage than landing them on the rooftop. The rappel is often more reliable and safer for the troops than a jump from a low hover.

With practice, soldiers can accomplish a rappel insertion with a minimum of exposure.

b. **Large-Scale Assaults.** For large-scale air assaults, rooftop landings are not practical. Therefore, open spaces (parks, parking lots, sports arenas) within the built-up area must be used. Several spaces large enough for helicopter operations normally can be found within 2 kilometers of a city's center.

c. **Air Movement of Troops and Supplies.** In battle in a built-up area, heliborne troop movement may become a major requirement. Units engaged in house-to-house fighting normally suffer more casualties than units fighting in open terrain. The casualties must be evacuated and replaced quickly with new troops. At the same time, roads are likely to be crowded with resupply and evacuation vehicles, and may also be blocked with craters or rubble. Helicopters provide a responsive means to move troops by flying nap-of-the-earth flight techniques down selected streets already secured and cleared of obstacles. Aircraft deliver the troops at the last covered position short of the fighting and then return without exposure to enemy direct fire. Similar flight techniques can be used for air movement of supplies and medical evacuation missions.

d. **Air Assaults.** Air assaults into enemy-held territory are extremely difficult (Figure 6-1). One technique is to fly nap-of-the-earth down a broad street or commercial ribbon while attack helicopters and door gunners from utility helicopters suppress buildings on either side of the street. Scheduled artillery preparations can be incorporated into the air assault plan through the H-hour sequence. Feints and demonstrations in the form of false insertions can confuse the enemy as to the real assault landings.

Figure 6-1. Air assualt of a built-up area.

6-8. ENGINEERS

The engineer terrain team supports the division commander and staff with specialized terrain analyses, products, and information for combat in built-up areas. During fighting in built-up areas, divisional engineers should be attached to the dispersed maneuver elements; for example, one engineer company to each committed brigade, one platoon to each battalion or battalion task force, and a squad to each company or company team. Most engineer manual-labor tasks, however, will have to be completed by infantry units, with reinforcing engineer heavy-equipment support and technical supervision.

   a. **Offensive Missions.** Engineers may perform the following missions during offensive operations.

   (1) Conduct a technical reconnaissance to determine the location and type of enemy obstacles and minefields, and to make breaching recommendations.

   (2) Clear barricades and heavy rubble with earth-moving equipment to assist forward movement.

   (3) Use the fires from the CEV or use hand-emplaced demolitions to destroy fortifications and strongpoints that cannot be reduced with the maneuver unit's organic assets.

   (4) Use the CEV to destroy structures or to clear rubble.
   (5) Lay mines to protect flanks and rear areas.
   (6) Conduct mobility operations (gap crossing).

   b. **Defensive Missions.** Engineers may perform the following missions during the defense of a built-up area.

   (1) Construct complex obstacle systems.
   (2) Provide technical advice to maneuver commanders.
   (3) Rubble buildings.
   (4) Lay mines.
   (5) Assist in the preparation of defensive strongpoints.
   (6) Maintain counterattack, communications, and resupply routes.
   (7) Enhance movement between buildings, catwalks, bridges, and so on.
   (8) Fight as infantry, when needed.

   c. **Defense Against Armor.** In defensive situations, when opposed by an armor-heavy enemy, priority should be given to the construction of antiarmor obstacles throughout the built-up area. Use of local materials, where possible, makes obstacle construction easier and reduces logistics requirements. Streets should be barricaded in front of defensive positions at the effective range of antitank weapons. These weapons are used to increase the destruction by antiarmor fires, to separate dismounted enemy infantry from their supporting tanks, and to assist in the delay and destruction of the attacker. Antitank mines with antihandling devices, integrated with antipersonnel mines in and around obstacles and covered by fires, help stop an enemy attack.

6-9. MILITARY POLICE

Military police operations play a significant role by assisting the tactical commander in meeting the challenges associated with combat in built-up areas. Through their four battlefield missions (battlefield circulation control, area security, EPW operations, and law and order) MP provide a wide range of diverse support in urban terrain. MP operations require continuous coordination with host nation civilian police to maintain control of the civilian population and to enforce law and order.

   a. MP units take measures to support area damage control operations that are frequently found in built-up areas. With the increased possibility of rubbling, MP units report, block off affected areas, and reroute movement to alternate road networks.

   b. MP units also secure critical activities, such as communications centers and water and electrical supply sources. They are responsible for securing critical cells within the corps and TAACOM main CPs, which often use existing "hardstand" structures located in built-up areas.

   c. MP units are tasked with EPW operations and collect them as far forward as possible. They operate collecting points and holding areas to briefly retain EPWs and civilian internees (CIs). EPW operations are of great importance in built-up areas because the rate of capture can be higher than normal.

   d. Commanders must realize that MP support may not be available and that infantry soldiers may have to assume certain MP missions. The following are some of those missions:

   (1) Route reconnaissance, selection of routes and alternate routes, convoy escort, and security of lines of communication.

   (2) Control of roads, waterways, and railroad terminals, which are critical chokepoints in the main supply routes.

   (3) Security of critical sites and facilities to include communication centers, government buildings, water and electrical supply sources, C4 nodes, nuclear or chemical delivery means and storage facilities, and other mission essential areas.

   (4) Refugee control in close cooperation with host nation civil authorities. (See Chapter 7 for more information.)

   (5) Collection and escort of EPWs.

6-10. COMMUNICATIONS

Buildings and electrical power lines reduce the range of FM radios. To overcome this problem, battalions set up retransmission stations or radio relays, which are most effective when placed in high areas. Antennas should be camouflaged by placing them near tall structures. Remoting radio sets or placing antennas on rooftops can also solve the range problem.

   a. **Wire.** Wire is a more secure and effective means of communications in built-up areas. Wires should be laid overhead on existing poles or underground to prevent vehicles from cutting them.

   b. **Messengers and Visual Signals.** Messengers and visual signals can also be used in built-up areas. Messengers must plan routes that avoid pockets of resistance. Routes and time schedules should be varied to avoid establishing a pattern. Visual signals must be planned so they can be seen from the buildings.

   c. **Sound.** Sound signals are normally not effective in built-up areas due to too much surrounding noise.

   d. **Existing Systems.** If existing civil or military communications facilities can be captured intact, they can also be used by the infantry battalion. A civilian phone system, for instance, can provide a reliable, secure means of communication if codes and authentication tables are used. Other civilian media can also be used to broadcast messages to the public.

(1) Evacuation notices, evacuation routes, and other emergency notices designed to warn or advise the civilian population must be coordinated through the civil affairs officer. Such notices should be issued by the local civil government through printed or electronic news media.

(2) Use of news media channels in the immediate area of combat operations for other than emergency communications must also be coordinated through the civil affairs officer. A record copy of such communications will be sent to the first public affairs office in the chain of command.

CHAPTER 7

COMBAT SERVICE SUPPORT AND LEGAL ASPECTS OF COMBAT

During combat in built-up areas, the terrain and the nature of operations create unique demands on the battalion CSS system. Increased ammunition consumption, high casualty rates, transportation difficulties resulting from rubble, and the decentralized nature of operations all challenge the battalion CSS operators and planners. The solutions to these problems require innovative techniques and in-depth planning.

Section I. COMBAT SERVICE SUPPORT

Combat in built-up areas presents a different set of problems, but the supply and movement operations of the support platoon change minimally. The guidelines and principal functions of CSS are explained in this section.

7-1. GUIDELINES

Guidelines for providing effective CSS to units fighting in built-up areas are explained in this paragraph.

　　a. Provide supplies to using units in the required quantities as close as possible to the location where those supplies are needed.

　　b. Protect supplies and CSS elements from the effects of enemy fire by both seeking cover and avoiding detection.

　　c. Disperse and decentralize CSS elements with proper emphasis on communication, command and control, security, and proximity of MSR for resupply.

　　d. Plan for the use of carrying parties and litter bearers.

　　e. Plan for and use host country support and civil resources when authorized and practical.

　　f. Position support units as far forward as the tactical situation permits.

　　g. Plan for requesting and arranging special equipment such as the M202 FLASH, toggle ropes with grappling hooks, ladders, and so on.

　　h. Position support units near drop or landing zones for resupply from corps to forward units to reduce surface movement.

7-2. PRINCIPAL FUNCTIONS

The principal functions of CSS in built-up areas are to arm, fuel, fix, and man the combat systems.

　　a. **Arm.** Combat in built-up areas is characterized by extremely high ammunition expenditure rates. Not only do individual soldiers fire more, but they also use more munitions such as smoke, concussion, and fragmentation grenades; LAWs; AT4s; Claymore mines; and demolitions. The ammunition consumption rate for the first day of combat in a built-up area can be up to four times the normal rate. Even though it decreases during succeeding days, consumption remains high. Commanders and S4s must plan to meet these high consumption rates. The plan must include how ammunition and demolitions are to be moved forward to the companies. BFVs and M113 APCs may have to be allocated for the movement of ammunition if rubble or glass prevents wheeled-vehicle traffic. Carrying parties may also have to be used if streets are blocked by rubble.

b. **Fuel.** The amount of bulk fuel needed by a battalion during combat in built-up areas is greatly reduced. Combat vehicles normally use less fuel in built-up areas, because they travel shorter distances and perform less cross-country traveling. Engineer equipment and power generation equipment may use more fuel but requirements are small. A company may not use much fuel daily, but when it does need fuel, a problem exists in delivering bulk fuel to the vehicle. In open terrain, a vehicle that has run out of fuel can be recovered later. But in built-up areas, the same vehicle is probably going to be lost quickly. Commanders and S4s must plan and provide the means of moving limited amounts of bulk fuel forward to combat units.

c. **Fix.** Maintenance teams must operate well forward to support units fighting in built-up areas. Although some maintenance operations may be consolidated in civilian facilities, many vehicles will have to be fixed near the fighting positions. Battle damage assessment and repair (BDAR) procedures allow mechanics to be inventive and make maximum use of battlefield damage, analysis, and repair techniques to return damaged vehicles to a serviceable condition (see applicable TMs).

(1) Combat in built-up areas generates a high demand for tires.

(2) The dust and rough handling characteristic of combat in built-up areas also places great strains on communication and night observation devices.

(3) The unit armorers and their small-arms repair kits provide only limited maintenance. S4s should plan for increased weapon maintenance demands and coordinate maintenance support from higher headquarters. Based on recommendations from the staff (S3, S4, motor officer), the commander may choose to consolidate and cross-level major items of equipment and weapons.

d. **Man.** Units conducting combat in built-up areas must expect high casualty rates. According to the factors outlined in FM 101-10-1, Volume 2, units may experience 6.6 percent casualites on the first day of an attack and 3.5 percent each succeeding day. In the defense, the planning percentage is a casualty rate of 3.5 percent on the first day and 1.9 percent each successive day. Units attacking a defended built-up area experience casualties of more than 6 percent. Casualty feeder reports must be prepared scrupulously and forwarded to the battalion personnel and administration center (PAC).

(1) The S1 with the medical platoon leader must plan to expedite the evacuation of wounded out of the built-up area. Forward aid station locations and evacuation routes must be planned and disseminated to the lowest level. Higher casualty rates should be expected and may require the stockpiling of medical supplies and augmentation of medical personnel from higher headquarters.

(2) The battalion PAC should process replacements quickly and transport them to their new unit. The battalion PAC is responsible for reviewing assignment orders, welcoming soldiers to the battalion, assigning soldiers IAW commanders priorities, obtaining personal information, and collecting medical records and forwarding them to the aid station. It is also responsible for adding names to the battle roster, preparing SIDPERS input for each one, and processing the names into the servicing postal activity. The S1 and

PAC should brief the new soldiers on the tactical situation, provide mess and medical support as needed, inspect for combat critical equipment shortages, and coordinate transportation to units. Replacements should be brought forward from the field trains with the LOGPAC and linked up with their new unit's first sergeant. If replacements are brought forward at unscheduled times, the LRP should still be used as the linkup point.

(3) Proper accountability of platoon personnel and accurate strength reporting are essential to support decision making by platoon leaders, company commanders, and, the battalion commander. Using battle rosters, leaders in the platoon maintain accurate, up-to-date records of their personnel. At periodic intervals, they provide strength figures to the company CP. During combat, they provide hasty strength reports on request or when significant changes in strength occur.

(4) By-name casualty information is reported by wire or by messenger to company headquarters during lulls in the tactical situation. This information should not be transmitted by radio since it could adversely affect unit morale, and the enemy could gain valuable information. Soldiers having direct knowledge of an incident complete a DA Form 1155 to report missing or captured soldiers, or casualties no longer under US control. (See AR 600-8-1 for instructions on how to complete this form.) DA Form 1156 is used to report soldiers who are killed or wounded. (See AR 600-10 for instructions on how to complete this form.) After being collected and reviewed for accuracy by the platoon leader or platoon sergeant, these forms are forwarded to the company headquarters. These forms provide important casualty information and are also used to determine the platoon's replacement requirements.

(5) The S1 must coordinate with the S3 or S4 for the transport of replacements over long distances, and for the issue of missing individual combat equipment. At night, replacements may need to be sent forward with guides to their new unit. These groups may be used to carry critical supplies and ammunition forward.

(6) The S1 must be prepared to deal with not only physical wounds but also psychological wounds.

(a) Prolonged combat in built-up areas generates incredible stress. Some soldiers show signs of inability to cope with such stress. Stress management is the responsibility of commanders at all levels. The S1 coordinates trained personnel, such as medical personnel and unit ministry team personnel, to support units when the situation dictates.

(b) The more intense the combat, the higher the casualties; the more extreme the weather, the longer the battle lasts; the more combat exhaustion and stress, the more casualties. The battalion PA, brigade surgeon, or other qualified medical personnel should be brought forward to screen stress casualties.

(c) The S1 should plan to provide the soldier with a short rest period in a protected section of the battalion rear area, along with warm food and hot liquids. He should take this opportunity to give the soldier command information products (obtained through public affairs channels). These inform the soldier of the larger picture of the battle, the theater of operations, the Army, and the welfare of the nation as a whole. As a result of treating stress problems in the battalion area, a higher percentage of stress casualties can be returned to duty than if they had been evacuated farther to the rear. When recovered, they should be returned to their original units the same as all hospital returnees.

7-3. SUPPLY AND MOVEMENT FUNCTIONS
The S4, support platoon leader, and battalion motor officer share the responsibility for coordinating all supply and movement functions within the battalion. The use of preconfigured LOGPACs that are pushed forward to the elements in contact will be the key to successful resupply operations. The support platoon contains the trucks and trained drivers needed to move supplies forward. Some classes of supply, and how they are moved, may assume greater importance than during combat outside the city or village.

    a. **Class I (Rations).** The process of ordering and moving rations to the battalion's forward positions is complicated by the dispersed nature of combat in built-up areas, and its increased caloric demands on soldiers. The battalion mess section must try to provide a hot meal.

    (1) Combat in built-up areas not only causes great stress on soldiers but also requires great physical exertion. This combination of stress and exertion quickly causes dehydration. Unless potable water is continuously provided, soldiers seek local sources, which are usually contaminated by POL runoff, sewage, bacteria, or unburied corpses. Soldiers who are not provided sufficient quantities of potable water become casualties due to drinking from contaminated sources or from dehydration. Waterborne contaminates can quickly render entire units combat-ineffective.

    (2) Water and other liquid supplements, such as coffee, tea, or soup, that must be forwarded to exposed positions may need to be backpacked at night.

    b. **Class II (General Supplies).** Combat in built-up areas places a great strain on combat uniforms and footgear. The battalion S4 should increase his on-hand stocks of uniforms, boots, and individual combat equipment such as protective masks and armored vests. NBC protective suits either tear or wear out quickly when worn in the rubble, which is typical of combat in built-up areas. Extra stocks of these and protective mask filters should be kept on hand. Limited amounts of other Class II and IV items may be available locally. These should be gathered and used if authorized and practical. Local shops may provide such items as hand tools, nails, bolts, chains, and light construction equipment, which are useful in preparing a defense or reducing enemy-held positions. The unit's organic wire communications net may be augmented with locally obtained telephone wire and electrical wire.

    c. **Class III (POL).** Bulk fuel may have to be brought forward from fuel tankers by use of 5-gallon cans. One man can carry a fuel can long distances, even over rubble, if it is lashed to a pack frame. Supplies of bulk Class III items and some prepackaged POL may be available at local gas stations and garages. These may be contaminated or of poor quality. The S4 should coordinate with the brigade S4 to have a fuel test performed by a qualified member of the supporting FSB or FAST.

    d. **Class IV (Barrier Materials).** If a unit is defending a built-up area, the required Class IV materials are less than in other areas. This class of supply is probably the most available locally. After coordinating the effort with higher headquarters, the S4, support platoon leader, and supporting engineer officer can gather materials for use in strengthening a defense. Cargo trucks from the support platoon, wreckers or recovery vehicles from the maintenance platoon, and engineer construction equipment can be used to load and move barrier materials. Normally, division- or corps-level assets bring Class IV materials forward. Defense of a built-up area may require concertina wire and or barbed wire to restrict the enemy infantry's movements. Barriers can be built of abandoned cars and buses, which are dragged into position, turned on their sides, and chained together through the axles.

e. **Class V (Munitions).** Combat in built-up areas causes ammunition to be expended at extremely high rates. Commanders should plan for early resupply of explosives, grenades, and ammunition for small arms, direct fire, and indirect fire.

(1) In the defense, the S4 should prestock as much ammunition as practical in dispersed storage areas. These storage areas should be protected and be of easy access from the forward defensive positions. In the offense, attacking troops should not be overburdened with excessive ammunition. Mobile distribution points may be set up as low as company level.

(2) Commanders and S4s must plan to continuously deliver ammunition to the leading elements as they advance. This may be carried by armored vehicles close behind the advancing troops or by designated carrying parties. Modern ammunition, particularly missiles, is characterized by extensive amounts of packing material. The S4 must plan to have an element remove the ammunition depot overpack before it is transported forward. Resupply by helicopter (prepackaged slingloads) may be feasible.

(3) Removing the overpack from large amounts of ammunition can be a time-consuming process. It may require the efforts of the entire support platoon, augmented by available soldiers. If carrying parties are used to move ammunition forward, an individual can carry about 75 to 90 pounds using a pack frame or rucksack. Bulky and heavier loads can be carried by lashing them to litters and using teams of two to four men. Loads up to 400 pounds can be carried moderate distances using four-man teams.

NOTE: **DO NOT** use aidmen to carry ammunition forward as described above—it is a violation of the Geneva Accords.

f. **Class VIII (Medical Supplies).** Due to the decentralized nature of combat in built-up areas, medical supplies should be dispersed throughout the battalion, not just consolidated with the aid station and the individual aidmen. Individual soldiers, especially trained combat lifesavers, should carry additional bandages, cravats, and intravenous sets. Companies should request additional splints and stretchers.

7-4. MEDICAL

The battalion S1, battalion surgeon, physician's assistant, and medical platoon leader are responsible for planning and executing medical functions within the battalion. The most critical functions during combat in built-up areas include preventive medicine, trauma treatment, and evacuation. In addition, there should be a plan for the treatment and evacuation of NBC-related casualties that would occur in combat in built-up areas.

a. Combat in built-up areas exposes soldiers not only to combat wounds but also to the diseases endemic to the area of operations. Commanders must enforce prevention measures against the spread of infectious diseases.

The medical platoon advises the commander on how best to implement the use of prophylactics.

b. Although the aidman normally attached to each rifle platoon is the soldier best trained in the treatment of traumatic injury, he can quickly become overwhelmed by the number of casualties needing care. The commander must train selected soldiers within the platoons to perform basic trauma treatment. The work of these combat lifesavers, plus the buddy-aid efforts of individual soldier, eases the burden of the aidman and allows him to concentrate on the seriously wounded. The medical platoon should plan to care for the mass casualties inherent in combat in built-up areas. The incidence of crushing injuries, eye injuries, burns, and fractures increases.

c. The difficulties encountered when evacuating casualties from urban terrain are many and require innovative techniques and procedures. The planning for medical evacuation in urban terrain must include special equipment requirements, use of litter teams, use of air ambulances and the rescue hoist, use of the ambulance shuttle system, and communications requirements and techniques for locating casualties.

(1) Special equipment requirements include ropes, pulleys, sked litters, axes, crowbars, and other tools used to break through barriers.

(2) Although litter teams are labor intensive, they are required for evacuation from buildings, where casualties can occur on any level. Also, rubble in the streets, barricades, and demolition of roads impede the use of ground ambulances, requiring a heavy reliance on litter teams. The medical personnel assigned to the unit must dismount from the ambulance, and search for and rescue casualties. However, there are not sufficient medical assets to accomplish the evacuation mission, requiring assistance from the supported units.

(3) Air ambulances equipped with the rescue hoist may be able to evacuate casualties from the roofs of buildings or to insert medical personnel where they are needed. The vulnerability to sniper fire must be considered and weighed against probable success of the evacuation mission. Also, pilots must be familiar with overflying built-up areas and the atmospheric conditions they may encounter. Air ambulances can also be used at secured ambulance exchange points to hasten evacuation time.

(4) An ambulance shuttle system with collecting points, ambulance exchange points (AXPs), and relay points must be established. The battalion aid station may be located in a park or sports arena within the city's boundaries, or outside the built-up area. In either case, the existence of rubble and other obstructions hamper the mobility and accessibility of the treatment element. By establishing an ambulance shuttle system, the distance required to carry casualties by litter teams is shortened. This also allows personnel familiar with the area to remain in that area and to continue their search, rescue, recovery, and evacuation mission. By predesignating collecting points, soldiers who are wounded but still ambulatory can walk to these points, hastening the evacuation effort.

(5) The area of communications presents one of the biggest obstacles to casualty evacuation. Due to the terrain, line of sight radios are not effective. Also, individual soldiers normally do not have access to radios. Therefore, when wounded within a building, a soldier may be difficult to find and evacuate. The unit SOP should contain alternate forms of communications such as colored panels or other forms of markers that can be displayed

to hasten rescue when the battle is over. Also, a systematic search of the area after the battle may be required to recover casualties.

   d. The use of local medical facilities, hospitals, professional medical help, and medical supplies may be available during combat in large built-up areas. The commander must adhere to the guidelines established within the theater as to when and how these facilities can be used. If civilians are wounded in the battalion area, the commander is responsible for providing them aid and protection without disrupting military operations. A commander cannot confiscate civilian medical supplies unless he makes provisions to provide adequate replacements if civilians are wounded.

   e. The commander is responsible for the evacuation of deceased personnel to the nearest mortuary affairs collection point, whether they are US, allied, enemy, or civilian. (See FMs 10-63 and 10-497 for specific information on the handling of deceased personnel.) Some general considerations for the handling of deceased personnel include:

   (1) The Theater commander is the approval authority for hasty burial.
   (2) The deceased person's personal effects must remain with the body to assist in the identification of the body and to facilitate shipment of personal effects to the next of kin. Retention of personal items is considered looting and is, therefore, punishable by UCMJ.
   (3) When operating under NBC conditions, the bodies of deceased personnel should be decontaminated before removal from contaminated areas to prevent further contamination and casualties.
   (4) Care must be exercised when handling deceased personnel. Improper handling of deceased personnel can result in a significant decrease in unit and civilian morale.

7-5. PERSONNEL SERVICES

Timely and accurate personnel services are just as important during combat in built-up areas as in any other operation. The close, intense, isolated fighting places great stress on the soldier.

   a. The S1 plans for all personnel services that support and sustain the morale and fighting spirit of the battalion. Among the most important of these services are:

- Religious support.
- Postal services
- Awards and decorations.
- Rest and recuperation.
- Replacement operations.
- Strength accounting.
- Casualty reporting.
- Finance support.
- Legal support and services.
- Public Affairs activities.

   b. A unit may lose a battle if it allows civilians to steal or destroy its equipment. Even friendly civilians may steal supplies or furnish intelligence to the enemy. Civilians should be evacuated, if possible, to prevent pilferage, sabotage, and espionage. Control of the civilian population is normally provided by military police and civil affairs units. Collection points for noncombatants are established in rear areas. The S1 is the battalion's link to the population control programs of the higher command.

Section II. LEGAL ASPECTS OF COMBAT

Commanders must be well educated in the legal aspects of combat in built-up areas that include the control of large groups of civilians, the protection of key facilities, and civil affairs operations.

7-6. CIVILIAN IMPACT IN THE BATTLE AREA

The presence of large concentrations of civilians can greatly impede tactical operations. Civilians attempting to escape from the battle area may have the following impact on military operations.

a. **Mobility.** Fleeing civilians, attempting to escape over roads, can block military movement. Commanders should plan routes to be used by civilians and should seek the assistance of the civil police in traffic control.

b. **Firepower.** The presence of civilians can restrict the use of potential firepower available to a commander. Areas may be designated no-fire areas to prevent civilian casualties. Other areas may be limited to small-arms fire and grenades with prohibitions on air strikes, artillery, mortars, and flame. Target acquisition and the direction of fire missions are complicated by the requirement for positive target identification. Detailed guidance on the use of firepower in the presence of civilians is published by the division G3. In the absence of guidance, the general rules of the law of the land warfare apply.

c. **Security.** Security should be increased to preclude:
- Civilians being used as cover by enemy forces or agents.
- Civilians wandering around defensive areas.
- Pilferage of equipment.
- Sabotage.

d. **Obstacle Employment.** The presence of local civilians and movement of refugees influence the location and type of obstacles that may be employed. Minefields may not be allowed on designated refugee routes or, if allowed, must be guarded until the passage of refugees is completed. Booby traps and flame obstacles cannot be emplaced until civilians are evacuated.

7-7. COMMAND AUTHORITY

The limits of authority of commanders at all levels over civilian government officials and the civilian populace must be established and understood. A commander must have that degree of authority necessary to accomplish his mission. However, the host government's responsibility for its populace and territory can affect the commander's authority in civil-military matters. In less secure areas, where the host government may be only partly effective, the commander may be called upon to assume greater responsibility for the safety and well being of the civilian populace.

7-8. SOURCE UTILIZATION
Operations in highly populated areas require the diversion of men, time, equipment, and supplies to accomplish humanitarian tasks. If host government agencies collapse, the impact on military resources could be substantial.

7-9. HEALTH AND WELFARE
The disruption of civilian health and sanitary services sharply increases the risk of disease among both civilian and military personnel.

7-10. LAW AND ORDER
The host government may not be able to control mobs. US forces may have to augment civilian forces to protect life and property and to restore order. US forces may also have to secure vital government facilities for the host nation. (For more information on how to control civilians violating civil law, see FM 9-15.)

7-11. PUBLIC AFFAIRS OFFICER AND MEDIA RELATIONS
The best way to relate the Army's story is through the media. While free access to units in the field is desirable, operational security, existing guidelines, and or rules of engagement considerations take the first priority. All members of the media visiting the field should have an escort officer. This officer may be detailed from line units due to the shortage of trained public affairs personnel. Ensuring the media follows the established guidelines or rules of engagement will help prevent negative publicity that could jeopardize the operation or US national objectives. If operations permit, the battalion should also appoint a representative to serve as a point of contact with the local population to deal with their concerns (usually maneuver damage).

7-12. CIVIL AFFAIRS UNITS AND PSYCHOLOGICAL OPERATIONS
Civil affairs units and psychological operations have prominent and essential roles in MOUT. They are critical force multipliers that can save lives. The battle in urban terrain is won through effective military operations, but PSYOP and CA can make that victory more easily attained. In an ideal setting, PSYOP and CA offer the possibility of victory in an urban setting without the destruction, suffering, and horror of battle. They should be included in any study of MOUT. Civil affairs units are normally placed in support of units to assist and conduct CA operations.

    a. The primary responsibility of the S5 (Civil-Military Operations) in MOUT is to coordinate activities necessary for the evacuation of civilians from the battle area. This is accomplished in two separate but supporting actions.

    (1) CA personnel coordinate with the military police and local police officials for evacuation planning. They plan for establishing evacuation routes and thoroughfare crossing control, and for removing civilians from the military supply routes (MSRs).

    (2) CA personnel coordinate with US Army PSYOP assets, local government officials, radio and television stations, newspapers, and so on, to publicize the evacuation plan.

    b. The civil military operations officer also has the responsibility to advise the commander concerning his legal and moral obligations to the civilian population. This requirement can be fulfilled by CA assets conducting coordination for the health and well being of civilians. It can include the reestablishment of water systems; distribution of available food stocks, clothing, and medical supplies; and establishment of displaced persons, refugee, and evacuee (DPRE) camps.

    c. If the civil government is not functioning because of battlefield devastation, it is the commander's responsibility to conduct evacuation planning and to provide for the well being of the civilian population. He must do this with only those internal assets available. Because of foreign sovereignty and the utilization of all available host nation assets, this should only be used as a last resort.

    d. Tactical PSYOP in support of MOUT operations are planned and conducted in combat areas to achieve immediate and short-term objectives. PSYOP are an integral and coordinated part of the overall tactical plan. They provide the tactical commander with a system that can weaken the enemy soldier's will to fight, thereby reducing his combat effectiveness. They can also help prevent civilian interference with military operations. PSYOP are designed to exploit individual and group weaknesses.

    e. Psychological operations units provide support in MOUT using television, radio, posters, leaflets, and loudspeakers to disseminate propaganda and information. Television, including video tapes, is one of the most effective media for persuasion. It offers many advantages for PSYOP and is appropriate for use in a limited, general, or cold war. In areas where television is not common, receivers may be distributed to public facilities and selected individuals.

    NOTE: See FM 51-5 and FM 41-10 for further discussion on civil affairs.

7-13. PROVOST MARSHAL

The provost marshal recommends measures required to control civilians and directs military police activities in support of refugee control operations. The provost marshal coordinates his activities with the staff sections and supporting units in the area. Refugee control operations are the responsibility of the G5 or S5, host nation authorities, or both. MPs assist, direct, or deny the movement of civilians whose location, direction of movement, or actions may hinder operations. The host nation government is responsible for identifying routes for the safe movement of refugees out of an area of operations.

    NOTE: Other military police responsibilities, regarding civil affairs and civilian control, are contained in FM 19-1.

7-14. COMMANDER'S LEGAL AUTHORITY AND RESPONSIBILITIES

Commanders and leaders at all levels are responsible for protecting civilians and their property to the maximum extent allowed by military operations. Looting, vandalism, and brutal treatment of civilians are strictly prohibited, and individuals who commit such acts should be severely punished. Civilians, and their religions and customs, must be treated with respect. Women must be especially protected against any form of abuse. In urban fighting, however, some situations are not quite so explicit as the above rules imply. Discussed herein are those civilian-military confrontations most common in built-up areas and how an infantry commander might manage them to legally accomplish his mission.

a. **Control Measures.** Commanders may enforce control measures to conduct operations, maintain security, or ensure the safety and well-being of the civilians.

(1) *Curfew.* A commander with the mission of defending a town could establish a curfew to maintain security or to aid in control of military traffic. However, a curfew would not be legal if imposed strictly as punishment.

(2) *Evacuation.* A commander can require civilians to evacuate towns or buildings if the purpose of the evacuation is to use the town or building for imperative military purposes, to enhance security, or to safeguard those civilians being evacuated. If a commander takes this action, he must specify and safeguard the evacuation route. Food, clothing, and sanitary facilities should be provided at the destination until the evacuees can care for themselves.

(3) *Forced labor.* The Geneva Accords prohibit the use of civilians in combat. However, they may be used before the battle reaches the city. Guidelines for use of civilian labor should be published by the division G5. The commander may force civilians over 18 years of age to work if the work does not oblige them to take part in military operations. Permitted jobs include maintenance of public utilities as long as those utilities are not used in the general conduct of the war. Jobs can also include services to local population such as care of the wounded and burial. Civilians can also be forced to help evacuate and care for military wounded, as long as doing so does not involve any physical danger. Prohibited jobs include digging entrenchments, constructing fortifications, transporting supplies or ammunition, or acting as guards. Volunteer civilians can be employed in such work.

b. **Civilian Resistance Groups.** Another situation that commanders might encounter is combat with a civilian resistance group.

(1) Civilians accompanying their armed forces with an identity card authorizing them to do so, are treated as PWs when captured—for example, civilian members of military aircraft crews, war correspondents, supply contractors, and members of labor units or of service organizations responsible for the welfare of the armed forces.

(2) Civilians of a nonoccupying territory who take up arms against an invading enemy without time to form regular armed forces; wear a fixed, distinctive insignia that can be seen at a distance; carry their weapons openly; and operate according to the rules and customs of warfare are treated as PWs when captured. Other civilians who provide assistance to such groups may not be entitled to status as combatants, depending upon whether they are actually members of the resistance group. They are normally best treated as combatants until a higher authority determines their status.

(3) Armed civilian groups that do not meet the criteria of a legal resistance (civilians accompanying their armed forces and levee en masse) or individuals caught in the act of sabotage, terrorism, or espionage are not legal combatants. If captured, they must be considered criminals under the provisions of the law of land warfare. They should be detained in a facility separate from EPWs and should be quickly transferred to the military police. Reprisals, mass punishments, taking of hostages, corporal punishment, pillage, or destruction of property are prohibited punishments.

(4) The law of land warfare lets a commander control the civil population under the conditions already described using his own resources. However, language and cultural differences between US and foreign personnel make it good practice to use native authorities, such as the police, for such purposes. Use of the police does not relieve a commander of his responsibility to safeguard civilians in his area.

c. **Protection of Property.** Like civilian personnel, civilian buildings and towns normally have a protected status—for example, they are not legal targets. Buildings and towns lose their protected status if authorities determine that the enemy is using them for military purposes. If doubt exists as to whether a town or building is defended, that doubt should be settled by reconnaissance—not by fire.

(1) If the enemy is using a building or a portion of the town for military purposes—for example, as a supply point or a strongpoint—that building or that portion of the town is a legal target. Before engaging the target, the commander must decide if the bombardment of the target is necessary. Only such destruction as is required for military purposes is justified.

(2) Normally, religious, historical, and cultural objects and buildings are not legal targets. They are sometimes marked with symbols to signify cultural objects. Medical facilities are protected under the internationally recognized Red Cross, Red Crescent, Red Lion, or Red Star of David symbols. The fact that such symbols are absent does not relieve a commander of his responsibility to protect objects he recognizes as having religious, cultural, medical, or historical value.

(3) The misuse of such objects by the enemy is grounds to disregard their protected status. Whenever possible, a demand should be made for the enemy to stop his misuse of the protected object within a reasonable time. If an enemy forward observer uses a church for an OP, for example, a commander would be justified in destroying it immediately, because a delay would allow the enemy to continue the misuse of the church. If a religious shrine was used as a telephone switchboard, a warning would be appropriate, since it would take some time to dismantle the wires. Once the decision to call fires on those objects is reached, destruction should be limited to the least necessary to neutralize the enemy installations.

(4) The destruction, demolition, or military use of other buildings is permitted under the law of land warfare, if required by clear military necessity. Thus, destroying a house to obtain a better field of fire would be a legal act—destroying it as a reprisal would not be. Likewise, firing on any houses that are occupied or defended by an enemy force is legal.

CHAPTER 8

EMPLOYMENT AND EFFECTS OF WEAPONS

This chapter supplements the technical manuals and field manuals that describe weapons capabilities and effects against generic targets. It focuses on specific employment considerations pertaining to combat in built-up areas, and it addresses both organic infantry weapons and combat support weapons.

8-1. EFFECTIVENESS OF WEAPONS AND DEMOLITIONS

The characteristics and nature of combat in built-up areas affect the results and employment of weapons. Leaders at all levels must consider the following factors in various combinations when choosing their weapons.

   a. Hard, smooth, flat surfaces are characteristic of urban targets. Rarely do rounds impact perpendicular to these flat surfaces but at some angle of obliquity. This reduces the effect of a round and increases the threat of ricochets. The tendency of rounds to strike glancing blows against hard surfaces means that up to 25 percent of impact-fuzed explosive rounds may not detonate when fired onto rubbled areas.

   b. Engagement ranges are close. Studies and historical analyses have shown that only 5 percent of all targets are more than 100 meters away. About 90 percent of all targets are located 50 meters or less from the identifying soldier. Few personnel targets will be visible beyond 50 meters and usually occur at 35 meters or less. Minimum arming ranges and troop safety from backblast or fragmentation effects must be considered.

   c. Engagement times are short. Enemy personnel present only fleeting targets. Enemy-held buildings or structures are normally covered by fire and often cannot be engaged with deliberate, well-aimed shots.

   d. Depression and elevation limits for some weapons create dead space. Tall buildings form deep canyons that are often safe from indirect fires. Some weapons can fire rounds to ricochet behind cover and inflict casualties. Target engagement from oblique angles, both horizontal and vertical, demands superior marksmanship skills.

   e. Smoke from burning buildings, dust from explosions, shadows from tall buildings, and the lack of light penetrating inner rooms all combine to reduce visibility and to increase a sense of isolation. Added to this is the masking of fires caused by rubble and man-made structures. Targets, even those at close range, tend to be indistinct.

   f. Urban fighting often becomes confused melees with several small units attacking on converging axes. The risks from friendly fires, ricochets, and fratracide must be considered during the planning phase of operations and control measures continually adjusted to lower these risks. Soldiers and leaders must maintain a sense of situational awareness and clearly mark their progress IAW with unit SOP to avoid fratricide.

   g. Both the firer and target may be inside or outside buildings, or they may both be inside the same or separate buildings. The enclosed nature of combat in built-up areas means that the weapon's effect, such as muzzle blast and backblast, must be considered as well as the round's impact on the target.

   h. Usually the man-made structure must be attacked before enemy personnel inside are attacked. Therefore, weapons and demolitions can be chosen for employment based on their effects against masonry and concrete rather than against enemy personnel.

i. Modern engineering and design improvements mean that most large buildings constructed since World War II are resilient to the blast effects of bomb and artillery attack. Even though modern buildings may burn easily, they often retain their structual integrity and remain standing. Once high-rise buildings burn out, they are still useful to the military and are almost impossible to damage further. A large structure can take 24 to 48 hours to burn out and get cool enough for soldiers to enter.

j. The most common worldwide building type is the 12- to 24-inch brick building. Table 8-1 lists the frequency of occurrence of building types worldwide.

TYPE OF BUILDING	FREQUENCY OF OCCURRENCE (Percentage)
30-Inch Stone	1
8- to 10-Inch Reinforced concrete	6.9
12-to 24-Inch Brick	63
6-inch wood	16
14-inch steel and concrete (heavy clad)	2
7-inch steel and concrete (light clad)	12

Table 8-1. Types of buildings and frequency of occurrence.

8-2. M16 RIFLE AND M249 SQUAD AUTOMATIC WEAPON/MACHINE GUN

The M16A1/M16A2 rifle is the most common weapon fired in built-up areas. The M16A1/M16A2 rifle and the M249 are used to kill enemy personnel, to suppress enemy fire and observation, and to penetrate light cover. Leaders can use 5.56-mm tracer fire to designate targets for other weapons.

a. **Employment.** Close combat is the predominant characteristic of urban engagements. Riflemen must be able to hit small, fleeting targets from bunker apertures, windows, and loopholes. This requires pinpoint accuracy with weapons fired in the semiautomatic mode. Killing an enemy through an 8-inch loophole at a range of 50 meters is a challenge, but one that may be common in combat in built-up areas.

(1) When fighting inside buildings, three-round bursts or rapid semiautomatic fire should be used. To suppress defenders while entering a room, a series of rapid three-round bursts should be fired at all identified targets and likely enemy positions. This is more effective than long bursts or spraying the room with automatic fire. Soldiers should fire from an underarm or shoulder position; not from the hip.

(2) When targets reveal themselves in buildings, the most effective engagement is the quick-fire technique with the weapon up and both eyes

open. (See FM 23-9 for more detailed information on this technique.) Accurate quick fire not only kills enemy soldiers but also gives the attacker fire superiority.

(3) Within built-up areas, burning debris, reduced ambient light, strong shadow patterns of varying density, and smoke all limit the effect of night vision and sighting devices. The use of aiming stakes in the defense and of the pointing technique in the offense, both using three-round bursts, are night firing skills required of all infantrymen. The individual laser aiming light can sometimes be used effectively with night vision goggles. Any soldier using NVG should be teamed with at least one soldier not wearing them.

b. Weapon Penetration. The penetration that can be achieved with a 5.56-mm round depends on the range to the target and the type of material being fired against. The M16A2 and M249 achieve greater penetration than the older M16A1, but only at longer ranges. At close range, both weapons perform the same. Single 5.56-mm rounds are not effective against structural materials (as opposed to partitions) when fired at close range—the closer the range, the less the penetration.

(1) For the 5.56-mm round, maximum penetration occurs at 200 meters. At ranges less then 25 meters, penetration is greatly reduced. At 10 meters, penetration by the M16 round is poor due to the tremendous stress placed on this high-speed round, which causes it to yaw upon striking a target. Stress causes the projectile to break up, and the resulting fragments are often too small to penetrate.

(2) Even with reduced penetration at short ranges, interior walls made of thin wood paneling, sheetrock, or plaster are no protection against 5.56-mm rounds. Common office furniture such as desks and chairs cannot stop these rounds, but a layer of books 18 to 24 inches thick can.

(3) Wooden frame buildings and single cinder block walls offer little protection from 5.56-mm rounds. When clearing such structures, soldiers must ensure that friendly casualties do not result from rounds passing through walls, floors, or ceilings.

(4) Armor-piercing rounds are slightly more effective than ball ammunition in penetrating urban targets at all ranges. They are more likely to ricochet than ball ammunition, especially when the target presents a high degree of obliquity.

c. Protection. The following common barriers in built-up areas stop a 5.56-mm round fired at less than 50 meters:

- One thickness of sandbags.
- A 2-inch concrete wall (unreinforced).
- A 55-gallon drum filled with water or sand.
- A small ammunition can filled with sand.
- A cinder block filled with sand (block will probably shatter).
- A plate glass windowpane at a 45-degree angle (glass fragments will be thrown behind the glass).
- A brick veneer.
- A car body (an M16A1/M16A2 rifle penetrates but normally will not exit).

d. **Wall Penetration.** Although most structural materials repel single 5.56-mm rounds, continued and concentrated firing can breach some typical urban structures (see Table 8-2).

(1) The best method for breaching a masonary wall is by firing short bursts (three to five rounds) in a U-shaped pattern. The distance from the gunner to the wall should be minimized for best results—ranges as close as 25 meters are relatively safe from ricochet. Ballistic eye protection, protective vest, and helmet should be worn.

(2) Ball ammunition and armor-piercing rounds produce almost the same results, but armor-piercing rounds are more likely to fly back at the firer. The 5.56-mm round can be used to create either a loophole (about 7 inches in diameter) or a breach hole (large enough for a man to enter). When used against reinforced concrete, the M16 rifle and M249 cannot cut the reinforcing bars.

TYPE	PENETRATION	ROUNDS (REQUIRED)
8-inch reinforced concrete	Initial loophole	35 250
14-inch triple brick	Initial loophole	90 160
12-inch cinder block with single-brick veneer	Loophole Breach hole	60 250
9-inch double brick	Initial Loophole	70 120
16-inch tree trunk or log wall	Initial*	1 to 3
12-inch cinder block (filled with sand)	Loophole	35
24-inch double sandbag wall	Initial*	220
3/8-inch mild steel door	Initial*	1
*Penetration only, no loophole.		

Table 8-2. Structure penetration capabilities of the 5.56-mm round against typical urban targets (range 25 to 100 meters).

8-3. MEDIUM AND HEAVY MACHINE GUNS
(7.62-mm and .50-caliber)

In the urban environment, the Browning .50-caliber machine gun and the 7.62-mm M60 machine gun provide high-volume, long-range, automatic fires for the suppression or destruction of targets. They provide final protective fire along fixed lines and can be used to penetrate light structures—the .50-caliber machine gun is most effective in this role. Tracers from both machine guns are likely to start fires, but the .50-caliber tracer is more apt to do so.

a. **Employment.** The primary consideration impacting on the employment of machine guns within built-up areas is the limited availability of long-range fields of fire. Although machine guns should be emplaced at the lowest level possible, grazing fire at ground level is often obstructed by rubble.

(1) The .50-caliber machine gun is often employed on its vehicular mount during both offensive and defensive operations. If necessary, it can be mounted on the M3 tripod mount for use in the ground role or in the upper level of buildings. When mounted on a tripod, the .50-caliber machine gun can be used as an accurate, long-range weapon and can supplement sniper fires.

(2) The M60 machine gun is cumbersome, making it difficult to use inside while clearing a building. However, it is useful outside to suppress and isolate enemy defenders. The M60 can be fired from either the shoulder or the hip to provide a high volume of assault and suppressive fires. The use of the long sling to support the weapon and ammunition is preferred.

(3) Because of their reduced penetration power, M60 machine guns are less effective against masonary targets than .50-caliber machine guns. However, their availability and light weight make them well suited to augment heavy machine gun fire or to be used in areas where .50-caliber machine guns cannot be positioned, or as a substitute when heavy machine guns are not available. The M60 machine gun can be employed on its tripod to deliver accurate fire along fixed lines and then can quickly be converted to bipod fire to cover alternate fields of fire.

b. **Weapon Penetration.** The ability of the 7.62-mm and .50-caliber rounds to penetrate are also affected by the range to the target and type of material fired against. The 7.62-mm round is affected less by close ranges than the 5.56-mm; the .50-caliber's penetration is reduced least of all.

(1) At 50 meters, the 7.62-mm ball round cannot penetrate a single layer of sandbags. It can penetrate a single layer at 200 meters, but not a double layer. The armor-piercing round does only slightly better against sandbags. It cannot penetrate a double layer but can penetrate up to 10 inches at 600 meters.

(2) The penetration of the 7.62-mm round is best at 600 meters but most urban targets are closer. The longest effective range is usually 200 meters or less. Table 8-3 explains the penetration capabilities of a single 7.62-mm (ball) round at closer ranges.

RANGE (meters)	PINE BOARD (inches)	DRY LOOSE SAND (inches)	CINDER BLOCK (inches)	CONCRETE (inches)
25	13	5	8	2
100	18	4.5	10	2
200	41	7	8	2

Table 8-3. Penetration capabilities of a single 7.62-mm (ball) round.

FM 90-10-1

(3) The .50-caliber round is also optimized for penetration at long ranges (about 800 meters). For hard targets, .50-caliber penetration is affected by obliquity and range. Both armor-piercing and ball ammunition penetrate 14 inches of sand or 28 inches of packed earth at 200 meters, if the rounds impact perpendicular to the flat face of the target. Table 8-4 explains the effect of a 25-degree obliquity on a .50-caliber penetration.

THICKNESS (meters)	100 METERS (rounds)	200 METERS (rounds)
2	300	1,200
3	450	1,800
4	600	2,400

Table 8-4. Number of rounds needed to penetrate a reinforced concrete wall at a 25-degree obliquity.

c. **Protection.** Barriers that offer protection against 5.56-mm rounds are also effective against 7.62-mm rounds with some exceptions. The 7.62-mm round can penetrate a windowpane at a 45-degree obliquity, a hollow cinder block, or both sides of a car body. It can also easily penetrate wooden frame buildings. The .50-caliber round can penetrate all of the commonly found urban barriers except a sand-filled 55-gallon drum.

d. **Wall Penetration.** Continued and concentrated machine gun fire can breach most typical urban walls. Such fire cannot breach thick reinforced concrete structures or dense natural stone walls. Internal walls, partitions, plaster, floors, ceilings, common office furniture, home appliances, and bedding can be easily penetrated by both 7.62-mm and .50-caliber rounds (Tables 8-5 and 8-6).

TYPE	THICKNESS (inches)	HOLE DIAMETER (inches)	ROUNDS REQUIRED
Reinforced concrete	8	7	100
Triple brick wall	14	7	170
Concrete block with single brick veneer	12	6 and 24	30 and 200
Cinder block (filled)	12	*	18
Double brick wall	9	*	45
Double sandbag wall	24	*	110
Log wall	16	*	1
Mild steel door	3/8	*	1
*Penetration only, no loophole.			

Table 8-5. Structure penetrating capabilities of 7.62-mm round (NATO ball) against typical urban targets (range 25 meters).

(1) The M60 machine gun can be hard to hold steady to repeatedly hit the same point on a wall. The dust created by the bullet strikes also makes precise aiming difficult. Firing from a tripod is usually more effective than without, especially if sandbags are used to steady the weapon. Short bursts of three to five rounds fired in a U-type pattern are best.

(2) Breaching a brick veneer presents a special problem for the M60 machine gun. Rounds penetrate the cinder block but leave a net-like structure of unbroken block. Excessive ammunition is required to destroy a net since most rounds only pass through a previously eroded hole. One or two minutes work with an E-tool, crowbar, or axe can remove this web and allow entry through the breach hole.

(3) The .50-caliber machine gun can be fired accurately from the tripod using the single-shot mode. This is the most efficient method for producing a loop hole. Automatic fire in three- to five-round bursts, in a U-type pattern, is more effective in producing a breach.

TYPE	THICKNESS (inches)	HOLE DIAMETER (inches)	ROUNDS REQUIRED
Reinforced concrete	10 18	12 24 7	50 100 140
Triple brick wall	12	8 26	15 50
Concrete block with single brick veneer	12	10 33	25 45
Armor plate	1	*	1
Double sandbag wall	24	*	5
Log wall	16	*	1

*Penetration only, no loophole.

Table 8-6. Structure penetrating capabilities of .50-caliber ball against typical urban targrets (range 35 meters).

8-4. GRENADE LAUNCHERS, 40-MM (M203 AND MK 19)

Both the M203 dual-purpose weapon and the MK 19 grenade machine gun fire 40-mm HE and HEDP ammunition. Ammunition for these weapons is not interchangable, but the grenade and fuze assembly that actually hits the target are identical. Both weapons provide point and area destructive fires as well as suppression. The MK 19 has a much higher rate of fire and a longer range; the M203 is much lighter and more maneuverable.

a. **Employment.** The main consideration affecting the employment of 40-mm grenades within built-up areas is the typically short engagement range. The 40-mm grenade has a minimum arming range of 14 to 28 meters. If the round strikes an object before it is armed, it will not detonate. Both the HE and HEDP rounds have 5-meter burst radii against exposed troops, which means that the minimum safe firing range for combat is 31 meters. The 40-mm grenades can be used to suppress the enemy in a building, or inflict casualties by firing through apertures or windows. The MK 19 can use its high rate of fire to concentrate rounds against light structures. This concentrated fire can create extensive damage. The 40-mm HEDP round can penetrate the armor on the flank, rear, and top of Soviet-made BMPs and BTRs. Troops can use the M203 from upper stories to deliver accurate fire against the top decks of armored vehicles. Multiple hits are normally required to achieve a kill.

b. **Weapon Penetration.** The 40-mm HEDP grenade has a small shaped charge that penetrates better than the HE round. It also has a thin wire wrapping that bursts into a dense fragmentation pattern, creating casualties out to 5 meters. Because they explode on contact, 40-mm rounds achieve the same penetration regardless of range. Table 8-7 explains the penetration capabilities of the HEDP round.

TARGET	PENETRATION (Inches)
Sandbags	20 (double layer)
Sand-filled cinder block	16
Pine logs	12
Armor plate	2

Table 8-7. Penetration capabilities of the HEDP round.

(1) If projected into an interior room, the 40-mm HEDP can penetrate all interior partition-type walls. It splinters plywood and plaster walls, making a hole large enough to fire a rifle through. It is better to have HEDP rounds pass into a room and explode on a far wall, even though much of the round's energy is wasted penetrating the back wall (see Figure 8-1). The fragmentation produced in the room causes more casualties then the high-explosive jet formed by the shaped charge.

(2) The fragments from the HEDP round do not reliably penetrate interior walls. They are also stopped by office furniture, sandbags, helmets, and protective vests (flak jackets). The M203 dual-purpose weapon has the inherent accuracy to place grenades into windows at 125 meters and bunker apertures at 50 meters. These ranges are significantly reduced as the angle of obliquity increases. Combat experience shows that M203 gunners cannot consistently hit windows at 50 meters when forced to aim and fire quickly.

c. **Wall Penetration.** The M203 cannot reasonably deliver the rounds needed to breach a typical exterior wall. The MK 19 can concentrate its fire and achieve wall penetration. Firing from a tripod, using a locked down traversing and elevating mechanism, is best for this role. Brick, cinder block, and concrete can be breached using the MK 19; individual HEDP rounds can penetrate 6 to 8 inches of brick. The only material that has proven resistant to concentrated 40-mm fire is dense stone such as that used in some European building construction. No precise data exist as to the number of rounds required to produce loopholes or breach holes with the MK 19. However, the rounds' explosive effects are dramatic and should exceed the performance of the .50 caliber machine gun.

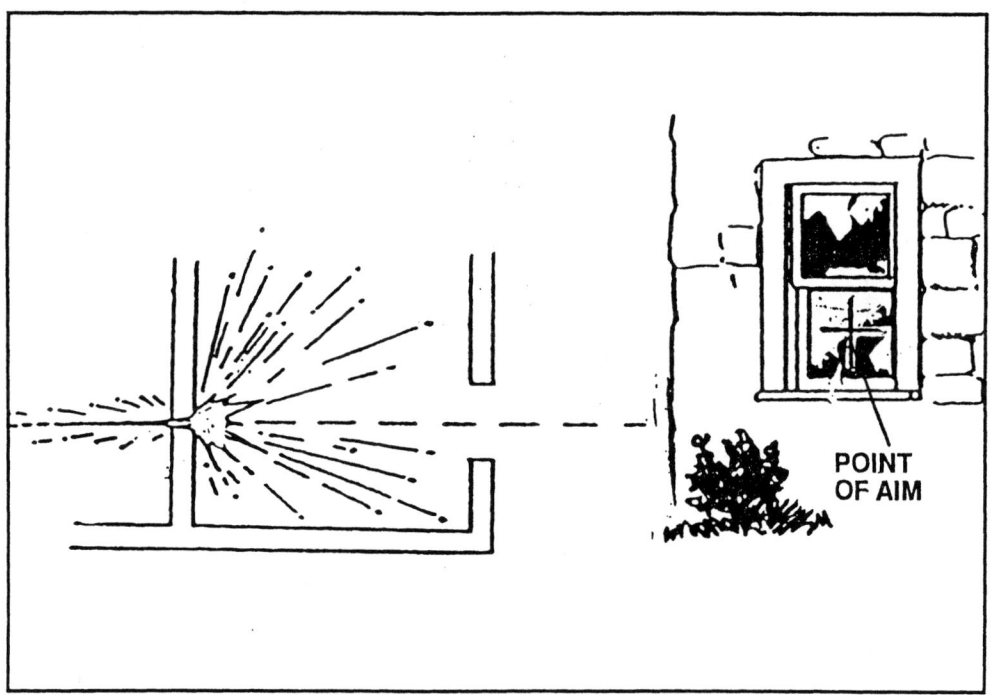

Figure 8-1. Aim point for 40-mm HEDP.

8-5. LIGHT AND MEDIUM RECOILLESS WEAPONS

Light and medium recoilless weapons are used to attack enemy personnel, field fortifications, and light armored vehicles. They have limited capability against main battle tanks, especially those equipped with reactive armor, except when attacking from the top, flanks, or rear. This category of weapons includes the M72 LAW; the AT4 or AT8; the M47 Dragon; the 90-mm and 84-mm recoilless rifles; the shoulder-launched, multipurpose, assault weapon (SMAW); and available foreign weapons such as the RPG-7.

   a. **Employment.** Other then defeating light armored vehicles, the most common task for which light recoilless weapons are used is to neutralize fortified firing positions. Due to the design of the warhead and the narrow blast effect, these weapons are not as effective in this role as heavier weapons such as a tank main gun round. Their light weight allows soldiers to carry several LAWs or AT4s. Light recoilless weapons can be fired from the tops of buildings or from areas with extensive ventilation.

   (1) Light and medium recoilless weapons, with the exception of the SMAW and AT8, employ shaped-charge warheads. As a result, the hole they punch in walls is often too small to use as a loophole. The fragmentation and spall these weapons produce are limited. Normally, shaped-charge warheads do not neutralize enemy soldiers behind walls unless they are located directly in line with the point of impact.

(2) Against structures, shaped-charge weapons should be aimed about 6 inches below or to the side of a firing aperture (see Figure 8-2). This enhances the probability of killing the enemy behind the wall. A round that passes through a window wastes much of its energy on the back wall. Since these shaped-charge rounds lack the wire wrapping of the 40-mm HEDP, they burst into few fragments and are often ineffective casualty producers.

Figure 8-2. Point of aim for a shaped-charge weapon against a masonry structure.

(3) Sandbagged emplacements present a different problem (see Figure 8-3). Because sandbags absorb much of the energy from a shaped-charge, the rounds should be aimed at the center of the firing aperture. Even if the round misses the aperture, the bunker wall area near it is usually easiest to penetrate.

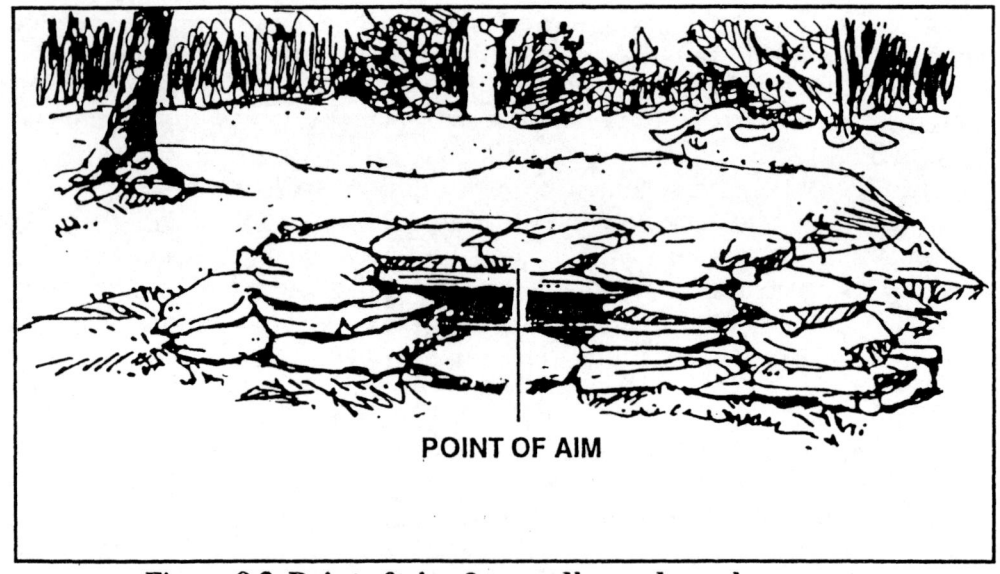

Figure 8-3. Point of aim for sandbagged emplacement.

(4) Light and medium recoilless weapons obtain their most effective short-range antiarmor shots by firing from upper stories, or from the flanks and rear. When firing at main battle tanks, these weapons should always be employed against these weaker areas in volley or paired firing. They normally require multiple hits to achieve a kill on a tank. Flanks, top, and rear shots hit the most vulnerable parts of armored vehicles. Firing from upper stories protects the firer from tank main gun and coaxial machine gun fire since tanks cannot sharply elevate their cannons. The BMP-2 can elevate its 30-mm cannon to engage targets in upper stories. The BTR-series armored vehicles can also fire into upper stories with their heavy machine gun.

(5) Modern infantry fighting vehicles, such as the BMP-2 and the BTR-80, have significantly improved frontal protection against shaped-charge weapons. Many main battle tanks have some form of reactive armor in addition to their thick armor plate. Head-on, ground-level shots against these vehicles have little probability of obtaining a kill. Even without reactive armor, modern main battle tanks are hard to destroy with a light antiarmor weapon.

(6) The easiest technique to use that will improve the probability of hitting and killing an armored vehicle is to increase the firing depression angle. A 45-degree downward firing angle doubles the probability of a first-round hit as compared to a ground-level shot (see Figure 8-4).

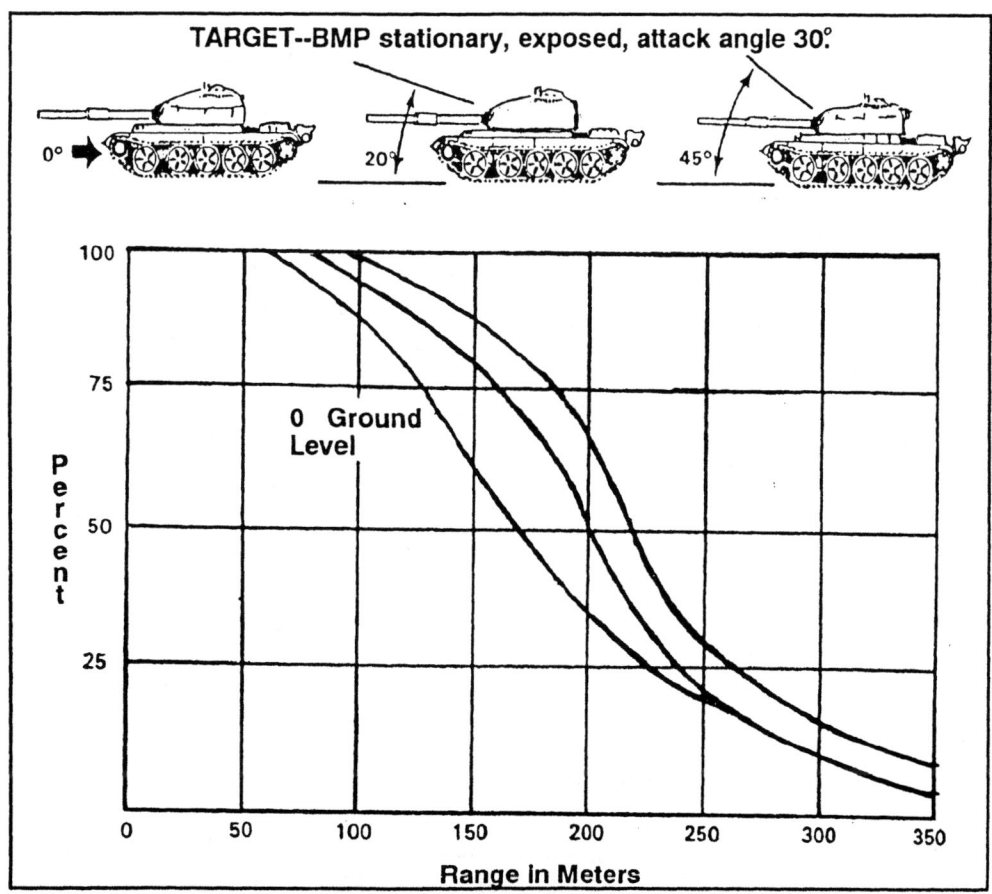

Figure 8-4. Probability of achieving a hit at different angles using an M72A2 LAW.

b. **Backblast.** Backblast characteristics must be considered when employing all recoilless weapons. During combat in built-up areas, the backblast area in the open is more hazardous due to all the loose rubble, and the channeling effect of the narrow streets and alleys. Figure 8-5 shows the backblast areas of United States light and medium recoilless weapons in the open.

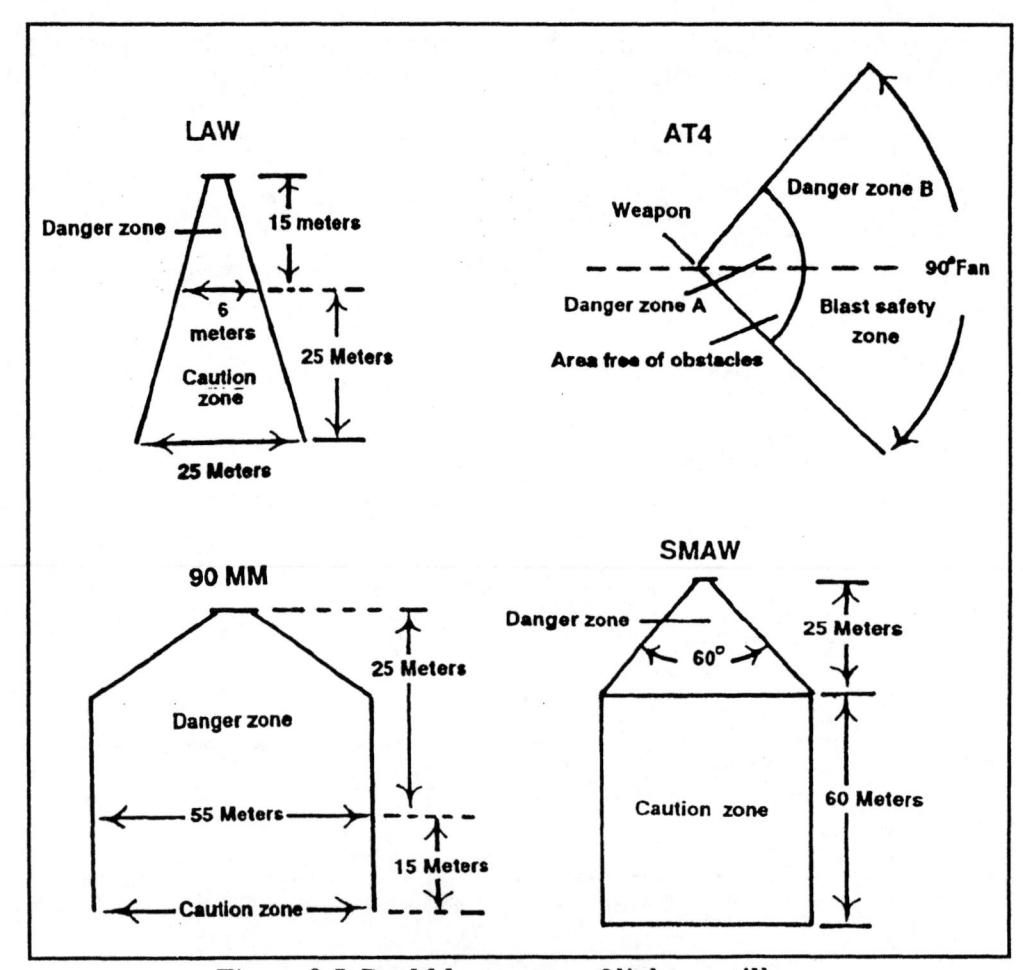

Figure 8-5. Backblast areas of light recoilless weapons in the open.

(1) When firing recoilless weapons in the open, soldiers should protect themselves from blast and burn injuries caused by the backblast. All personnel should be out of the danger zone. Anyone not able to vacate the caution zone should be behind cover. Soldiers in the caution zone should wear helmets, protective vests, and eye protection. The firer and all soldiers in the area should wear earplugs.

(2) Since the end of World War II, the US Army has conducted extensive testing on the effects of firing recoilless weapons from within enclosures. Beginning as early as 1948, tests have been conducted on every type of recoilless weapon available. In 1975, the US Army Human Engineering Laboratory at Aberdeen Proving Grounds, Maryland, conducted extensive firing of LAW, Dragon, 90-mm RCLR, and TOW from masonry and frame

buildings, and from sandbag bunkers. These tests showed that firing these weapons from enclosures presented no serious hazards, even when the overpressure was enough to produce structural damage to the building. The following were other findings of this test.

(a) Little hazard exists to the gunnery or crew from any type of flying debris. Loose items were not hurled around the room.

(b) No substantial degradation occurs to the operator's tracking performance as a result of obscuration or blast overpressure.

(c) The most serious hazard that can be expected is hearing loss. This must be evaluated against the advantage gained in combat from firing from cover. To place this hazard in perspective, a gunner wearing earplugs and firing the loudest combination (the Dragon from within a masonary building) is exposed to less noise hazard than if he fired a LAW in the open without earplugs.

(d) The safest place for other soldiers in the room with the firer is against the wall from which the weapon is fired. Plastic ignition plugs are a hazard to anyone standing directly behind a LAW or TOW when it is fired.

(e) Firers should take advantage of all available sources of ventilation by opening doors and windows. Ventilation does not reduce the noise hazard, but it helps clear the room of smoke and dust, and reduces the effective duration of the overpressure.

(f) The only difference between firing these weapons from enclosures and firing them in the open is the duration of the pressure fluctuation.

(g) Frame buildings, especially small ones, can suffer structural damage to the rear walls, windows, and doors. Large rooms suffer slight damage, if any.

(3) Recoilless weapons fired from within enclosures create some obscuration inside the room, but almost none from the gunner's position looking out. Inside the room, obscuration can be intense, but the room remains inhabitable. Table 8-8 shows the effects of smoke and obscuration.

BUILDING	WEAPON	FROM GUNNER'S POSITION LOOKING OUT	INSIDE THE ROOM	FROM OUTSIDE AT A DISTANCE
Masonry	LAW Dragon	None Slight	Moderate Moderate	Slight smoke Small flash
Bunker	Dragon TOW	None None	Slight Slight	Moderate flash Moderate smoke
Small frame	LAW Dragon	None None	Moderate Severe	Moderate Moderate
Medium frame	LAW Dragon	None None	Slight Severe	Moderate Slight flash
Large frame	LAW Dragon TOW	None Slight None	Slight Severe Severe	None Slight flash Slight smoke

Table 8-8. Smoke and obscuration.

(4) The Dragon causes the most structural damage but only in frame buildings. There does not seem to be any threat of injury to the gunner, since the damage is usually to the walls away from the gunner. The most damage and debris is from flying plaster chips and pieces of wood trim. Large chunks of plasterboard can be dislodged from ceilings. The backblast from LAW, Dragon, or TOW rarely displaces furniture. Table 8-9 shows the test results of structural damage and debris.

BUILDING	WEAPON	DAMAGE STRUCTURE	WALL COVERING	DEBRIS MOVEMENT
Masonry	LAW Dragon	None None	Slight Slight	Slight Slight
Bunker	Dragon TOW	None None	None None	None Leaves and dust disturbed
Small frame	LAW Dragon	None Severe	Slight Severe	None None
Medium frame	LAW Dragon	None Slight	None Slight	Slight Lamp and chair overturned
Large frame	LAW Dragon TOW	None Slight Slight	Slight Moderate Severe	Slight None None

Table 8-9. Structural damage and debris movement.

(5) To fire a LAW from inside a room, the following safety precautions must be taken (see Figure 8-6).

Figure 8-6. Firing a LAW from inside a room.

(a) At least 4 feet of clearance should exist between the rear of the LAW and the nearest wall.

(b) At least 20 square feet of ventilation (an open 7- by 3-foot door is sufficient) should exist to reduce or prevent structural damage to the building—the more ventilation, the better.

(c) All glass should be removed from windows.

(d) All personnel in the room should be forward of the rear of the weapon and should wear helmets, protective vests, ballistic eye protection, and earplugs.

(e) All combustible material should be removed from the rear of the weapon.

(f) Ceiling height should be at least 7 feet.

(6) To fire a 90-mm RCLR, AT4 or AT8, or SMAW from inside a room, the following safety precautions must be taken (see Figure 8-7).

Figure 8-7. Firing a 90-mm RCLR, AT4, AT8 or SMAW from inside a building.

(a) The building should be of a sturdy construction.

(b) The ceiling should be at least 7 feet high with loose plaster or ceiling boards removed.

(c) The floor size should be at least 15 feet by 12 feet. (The larger the room, the better.)

(d) At least 20 square feet of ventilation (room openings) should exist to the rear or side of the weapon. An open 7- by 3-foot door would provide minimum ventilation.

(e) All glass should be removed from windows and small, loose objects removed from the room.

(f) Floors should be wet to prevent dust and dirt from blowing around and obscuring the gunner's vision.

(g) All personnel in the room should be forward of the rear of the weapon.

(h) All personnel in the room should wear helmets, protective vests, ballistic eye protection, and earplugs.

(i) If the gunner is firing from the prone position, his lower body must be perpendicular to the bore of the weapon or the blast could cause injury to his legs.

c. **Weapon Penetration.** The most important tasks to be performed against structures are the neutralization of fortified firing positions, personnel, and weapons behind barriers. Recoilless weapons can be used in this role; none, however, is as effective as heavy direct-fire weapons or standard demolitions. Each recoilless weapon has different penetrating ability against various targets. Penetration does not always mean the destruction of the integrity of a position. Usually, only those enemy soldiers directly in the path of the spall from a HEAT round become casualties. Other soldiers inside a fortification could be deafened, dazed, or shocked but eventually return to action.

(1) *M72 LAW.* The LAW, although light and easy to use, has a small explosive charge and limited penetration. It can be defeated by a double-layer brick wall backed by 4 feet of sandbags since it cannot produce a loophole in this type construction. The LAW requires at least 10 meters to arm. If it hits a target before it arms, it usually does not detonate. (The LAW is being replaced by the AT4 in the US Army inventory of munitions.) The LAW can penetrate—

- 2 feet of reinforced concrete, leaving a dime-sized hole and creating little spall.
- 6 feet of earth, leaving a quarter-sized hole with no spall.
- 12 inches of steel (flanks, rear, and top armor of most armored vehicles), leaving a dime-sized hole.

(2) *M136 84-mm Launcher (AT4).* The AT4 is heavier than the LAW with a diameter of 84 millimeters, which gives the warhead much greater penetration. The AT4 can penetrate more than 17.5 inches (450 mm) of armor plate. Its warhead produces highly destructive results behind the armor. Tests against typical urban targets are still ongoing, but the AT4 should penetrate at least as well as the 90-mm recoilless rifle if not better. The AT4 has a minimum arming distance of 10 meters, which allows it to be fired successfully against close targets. Firers should be well covered by protective equipment when firing at close targets.

(3) *84-mm Launcher (AT8).* The AT8 is a lightweight disposable, multi-purpose, direct fire weapon designed especially for MOUT. Externally, the AT8 is almost identical to the AT4, and it is fired in the same manner. The AT8 was procured in limited amounts and issued to selected US Army and USMC units during the Persian Gulf War. Its fuze has the ability to distinguish between armor and soft earth, maximizing its capabilities against buildings, bunkers, or light armor. The warhead detonates immediately against hard targets, but delays detonation against soft targets and burrows in to explode inside. The AT8 destroys earth and timber bunkers, blows large holes in light-armored vehicles, and breaches 8-inch reinforced concrete walls and 12-inch triple brick walls.

(4) *Recoilless rifles.* The 90-mm recoilless rifle is being phased out of the US Army inventory of weapons, but it is still used in engineer battalions. The 84-mm Ranger antiarmor weapon system (RAAWS) is issued to some light forces. The recoilless rifles' light weight and maneuverability, combined with great penetrating power, make them useful weapons during combat in built-up areas.

(a) The 90-mm RCLR has an antipersonnel round that is effective against exposed enemy. The flechette projectiles fired by this antipersonnel round cannot penetrate structural walls but can pierce partitions and wooden-framed buildings. The antipersonnel round has no minimum range, but the HEAT round is not armed until it has traveled 35 to 50 feet. The 90-mm HEAT round can penetrate—

- 3 1/2 feet of packed earth, leaving a 2-inch hole with no spall.
- 2 1/2 feet of reinforced concrete, creating a small loophole (less than 3 inches wide) with little spall.
- 10 inches of armor plate, leaving a quarter-sized hole.

(b) The RAAWS has a HEAT round for use against armored targets and an HE and HEDP round for use against other targets. The HE round can be set for either air burst or impact burst. It contains 800 steel balls that are distributed in a lethal pattern upon detonation. The HE round is effective against troops in the open or behind vertical cover such as a low wall. The HEDP round is probably the most useful during MOUT. It is effective against light-armored vehicles, thick concrete and brick walls, thin wood walls and field fortifications, and also unprotected troops. The RAAWS also fires illumination and smoke rounds. The smoke round is useful to cover friendly units as they cross small open areas. The HEAT round arms at 5 to 8 meters and may throw fragments back as far as 50 meters. The HE round arms at 20 to 70 meters and may throw its steel balls back as far as 250 meters. The HEDP round arms at 15 to 40 meters and produces only slight fragmentation out to 50 meters.

(5) *Shoulder-launched, multipurpose, assault weapon (SMAW).* The SMAW is being issued to US Marine Corps units. It has been type-classified and in time of war Army units could find it available. The SMAW is a lightweight, man-portable, assault weapon that is easily carried and placed into action by one man. It is used against fortified positions, but it is also effective against light-armored vehicles. The SMAW has a 9-mm spotting rifle and a 3.8-power telescope, which ensure accuracy over ranges common to combat in built-up areas. The SMAW has excellent incapacitating effects behind walls and inside bunkers, and can arm within 10 meters. It fires the same dual-mode fuzed round as the AT8, and it has another round designed for even greater effect against armored vehicles. The SMAW has the same penetration ability as the AT8—it can destroy most bunkers with a single hit. Multiple shots can create breach holes even in reinforced concrete.

(6) *RPG-7.* The RPG-7 is a common threat weapon worldwide. It is lightweight and maneuverable, and is accurate over ranges common to combat in built-up areas. In a conflict almost anywhere in the world, US forces must protect themselves against RPGs. The RPG warhead is moderately effective against armored vehicles particularly M113 armored personnel carriers. It is less effective against common urban hard targets. It has a

limited effect against reinforced concrete or stone. Typically, the round produces a small hole with little spall. The RPG produces a small hole in earth berms with little blast effect and no spall. A triple layer of sandbags is usually protection against RPG rounds. Because of its fuze design, the RPG can often be defeated by a chain-link fence erected about 4 meters in front of a position. Even without such a barrier, a high percentage of RPG rounds fired against urban targets are duds due to glancing blows.

   d. **Wall Breaching.** Wall breaching is a common combat task in built-up areas for which light recoilless weapons can be used. Breaching operations improve mobility by providing access to building interiors without using existing doors or windows. Breaching techniques can also be used to create loopholes for weapons positions or to allow hand grenades to be thrown into defended structures. Breach holes for troop mobility should be about 24 inches (60 centimeters) in diameter. Loopholes should be about 8 inches (20 centimeters) in diameter (see Figure 8-8). None of the light recoilless weapons organic to maneuver battalions (with the possible exception of the AT8 and SMAW) provide a one-shot wall-breaching ability. To breach walls, a number of shots should be planned.

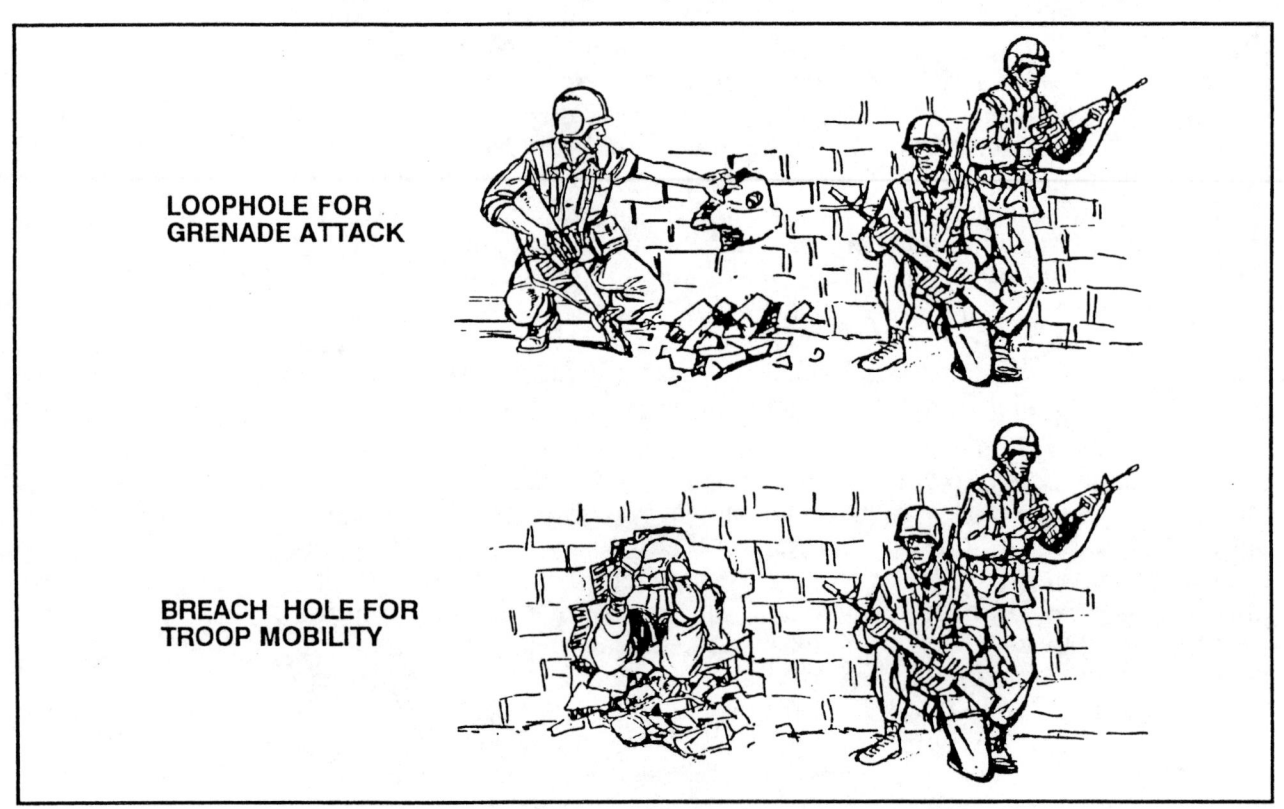

Figure 8-8. Tactical use of holes in masonry walls.

   (1) Of all the common building materials, heavy stone is the most difficult to penetrate. The LAW, AT4 or AT8, 90-mm RCLR, and RPG-7 usually will not penetrate a heavy European-style stone wall. Surface cratering is usually the only effect.

(2) Layered brick walls are also difficult to breach with light recoilless weapons. Some brick walls can be penetrated by multiple firings, especially if they are less than three bricks thick. Five LAW rounds fired at the same spot on a 8-inch (double-brick) wall normally produces a loophole. Heavier weapons, such as the AT4 and 90-mm RCLR, may require fewer rounds. The AT8 and SMAW produce a hole in brick walls that is often large enough to be a breach hole.

(3) Wooden structural walls offer little resistance to light recoilless weapons. Even heavy timbered walls are penetrated and splintered. Three LAW rounds fired at the same area of a wood-frame wall usually produce a man-sized hole. The AT8 and SMAW have a devastating effect against a wood-frame wall. A single round produces a breach hole as well as significant spall.

(4) Because of its high velocity, the AT4 may penetrate a soft target, such as a car body or frame building, before exploding.

(5) None of the light recoilless weapons are as effective against structural walls as demolitions or heavier weapons such as tank main gun, field artillery, or combat engineer vehicle demolition guns. Of all the light recoilless weapons, the SMAW and AT8 are the most effective.

8-6. ANTITANK GUIDED MISSILES

Antitank guided missiles (ATGMs) are used mainly to defeat main battle tanks and other armored combat vehicles. They have a moderate capability against bunkers, buildings, and other fortified targets commonly found during combat in built-up areas. This category of weapons includes the TOW and Dragon missiles.

a. **Employment.** TOWs and Dragons provide overwatch antitank fires during the attack of a built-up area and an extended range capability for the engagement of armor during the defense. Within built-up areas, they are best employed along major thoroughfares and from the upper stories of buildings to attain long-range fields of fire. Their minimum firing range of 65 meters could limit firing opportunities in the confines of densely built-up areas.

(1) *Obstacles.* When fired from street level, rubble or other obstacles could interfere with missile flight. At least 3.5 feet (1 meter) of vertical clearance over such obstacles must be maintained. Figure 8-9, page 8-20 shows the most common obstacles to ATGM flights found in built-up areas. Power lines are a special obstacle that present a unique threat to ATGM gunners. If the power in the lines has not been interrupted, the ATGM guidance wires could create a short circuit. This would allow extremely high voltage to pass to the gunner in the brief period before the guidance wires melted. This voltage could either damage the sight and guidance system, or injure the gunner. Before any ATGM is fired over a power line, an attempt must be made to determine whether or not the power has been interrupted.

(2) *Dead space.* Three aspects of dead space that affect ATGM fires are arming distance, maximum depression, and maximum elevation.

(a) Both the Dragon and TOW missiles have a minimum arming distance of 65 meters, which severely limits their use in built-up areas. Few areas in the inner city permit fires much beyond the minimum arming distance—ground-level long-range fires down streets or rail lines and across parks or plazas are possible. ATGMs may be used effectively from upper stories or roofs of buildings to fire into other buildings.

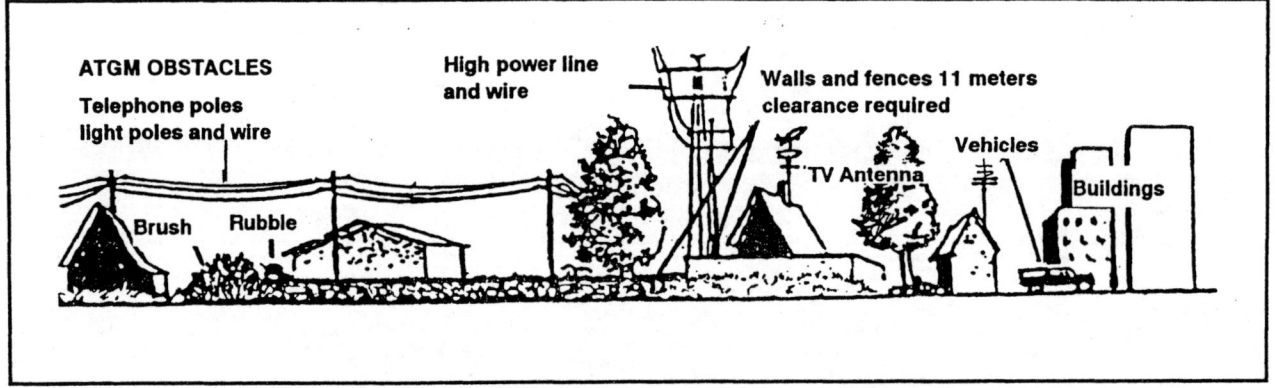

Figure 8-9. Common obstacles to ATGM flights.

(b) The TOW is limited much more than the Dragon by its maximum depression and elevation. The maximum depression and elevation limits of the TOW mount could result in dead space and preclude the engagements of close targets (see Figure 8-10). A target located at the minimum arming range (65 meters) cannot be engaged by a TOW crew located any higher then the sixth floor of a building due to maximum depression limits. At 100 meters the TOW crew can be located as high as the ninth floor and still engage the target.

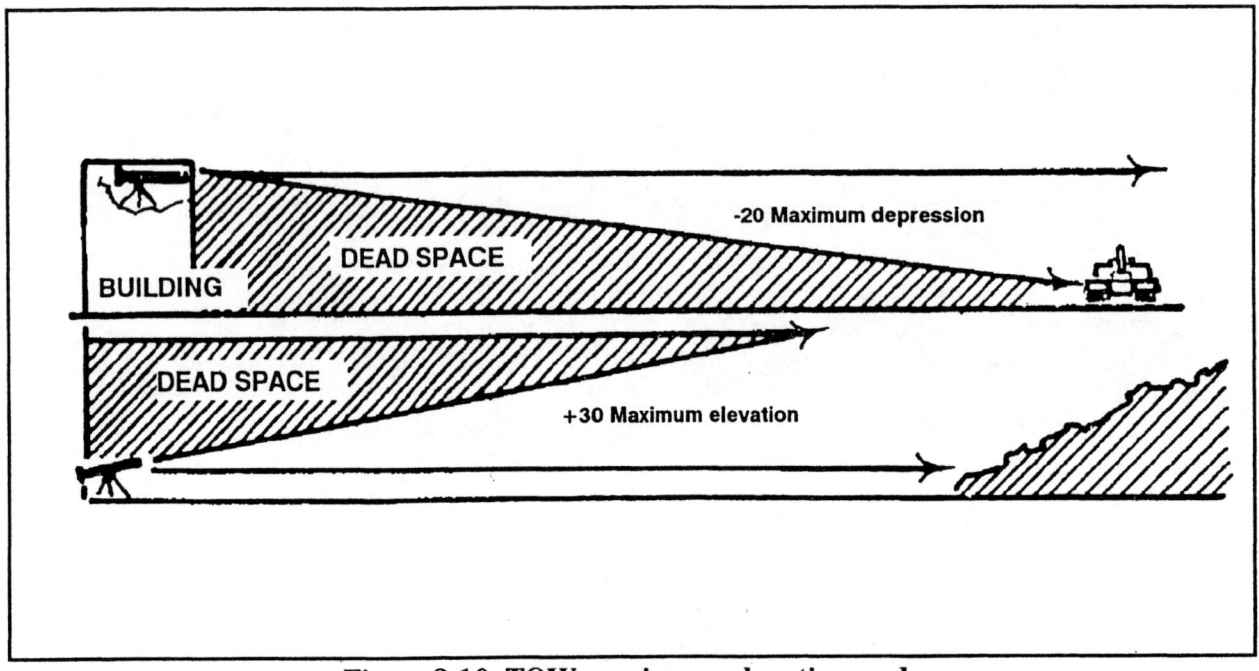

Figure 8-10. TOW maximum elevation and depression limitations.

(3) *Backblast.* As for the light recoilless weapons, backblast for ATGMs is more of a concern during combat in built-up areas then in open country. Any loose rubble in the caution zone could be picked up and thrown by the backblast. The channelling effect of walls and narrow streets is even more pronounced due to the greater backblast. If the ATGM backblast strikes a wall at an angle, it can pick up debris, or be deflected and cause injury to unprotected personnel (Figure 8-11). Both ATGMs can be fired from inside some buildings. In addition to the helmet and protective vest, eye protection and earplugs should be worn by all personnel in the room.

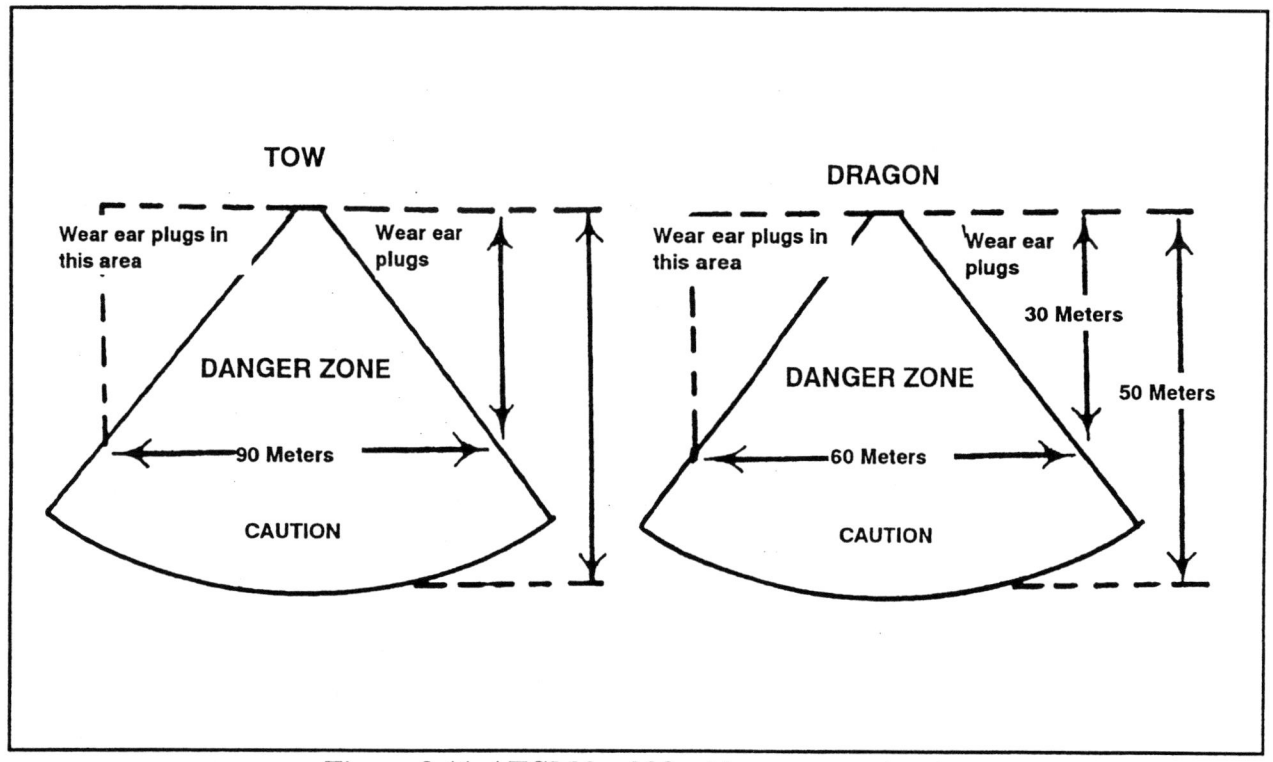

Figure 8-11. ATGM backblast in an open street.

    (a) To fire a TOW from inside a room, the following safety precautions must be taken (Figure 8-12, page 8-22).
- The building must be of sturdy construction.
- The ceiling should be at least 7 feet high.
- The floor size of the room should be at least 15 by 15 feet; larger, if possible.
- At least 20 square feet of room ventilation should exist, preferably to the rear of the weapon. An open 7- by 3-foot door is sufficient. Additional ventilation can be created by removing sections of interior partitions.
- All glass must be removed from the windows, and all small loose objects removed from the room.

- All personnel in the room should be forward of the rear of the TOW.
- All personnel in the room should wear ballistic eye protection and earplugs.
- A clearance of 9 inches (23 centimeters) must be between the launch tube and aperture from which it is fired. (See AR 385-62 and AR 385-63 for more detailed safety information.)

Figure 8-12. TOW fired from inside a room.

(b) To fire a Dragon from inside a room, the following safety precautions must be taken.
- The building must be of sturdy construction.
- The ceiling should be at least 7 feet high.
- The floor size should be at last 15 by 15 feet; larger, if possible.
- At least 20 square feet of ventilation should exist (room openings), preferably to the rear of the weapon. An open 7- by 3-foot door would provide minimum ventilation.
- All glass should be removed from windows, and small loose objects removed from the room.
- The room should be clean or the floors must be wet to prevent dust and dirt (kicked up by the backblast) from obscuring the vision of other soldiers in the room.

- All personnel in the room must be forward of the rear of the weapon.
- All personnel in the room must wear ballistic eye protection and earplugs.
- At least a 6-inch clearance must exist between the launch tube and aperture from which it is fired.

b. Weapon Penetration. ATGMs can penetrate and destroy heavily armored tanks. They have large warheads employing the shape-charge principle. Because of their size, these warheads can achieve significant penetration against typical urban targets. Penetration, however, does not mean a concurrent destruction of the structural integrity of a position. The shaped-charge warhead produces relatively little spall. Enemy personnel not standing directly behind or near the point of impact of an ATGM may escape injury.

(1) *Standard TOW missiles.* The basic TOW missile can penetrate 8 feet of packed earth, 4 feet of reinforced concrete, or 16 inches of steel plate. The improved TOW (ITOW), the TOW 2, and the TOW 2A all have been modified to improve their penetration. They all penetrate better than the basic TOW. All TOW missiles can defeat triple sandbag walls, double layers of earth filled 55-gallon drums, and 18-inch log walls.

(2) *TOW 2B.* The TOW 2B uses a different method of defeating enemy armor. It flies over the target and fires an explosively formed penetrator down onto the top armor, which is thinner. Because of this design feature, the TOW 2B missile cannot be used to attack nonmetallic structural targets. When using the TOW 2B missile against enemy armor, gunners must avoid firing directly over other friendly vehicles, disabled vehicles, or large metal objects such as water or oil tanks.

(3) *Dragon missile.* The Dragon missile can penetrate 8 feet of packed earth, 4 feet of concrete, or 13 inches of steel plate. It can attain effective short-range fire from upper stories, or from the rear or flanks of a vehicle. These engagements are targeted against the most vulnerable parts of tanks, and can entrap tanks in situations where they are unable to counterfire. Elevated firing positions increase the first-round hit probability. Firing down at an angle of 20 degrees increases the chance of a hit by 67 percent at 200 meters. A 45-degree down angle doubles the first-round hit probability, compared to a ground-level shot.

c. Breaching Structural Walls. Firing ATGMs is the least efficient means to defeat structures. Because of their small basic load and high cost, ATGMs are better used against tanks or enemy-fortified firing positions. They can be effective against bunkers or other identified enemy firing positions.

8-7. FLAME WEAPONS

Flame weapons are characterized by both physical and psychological casualty-producing abilities. Flame does not need to be applied with pinpoint accuracy, but it also must not spread to structures needed by friendly forces. Large fires in built-up areas are catastrophic. If they burn out of control, fires can create an impenetrable barrier for hours. The most common United States flame weapons are the M202 FLASH and the M34 white phosphorus

(WP) grenade. The M2A1-7 portable flamethrower is stored in war reserve status as a standard "C" item. Its availability is limited.

   a. **Employment.** Flame weapons used against fortified positions should be aimed directly at the aperture. Even if the round or burst misses, enough flaming material enters the position to cause casualties and to disrupt the enemy occupants. The M34 WP grenade is difficult to throw far or into a small opening such as a bunker aperture. However, its effects are dramatic when thrown into a room or building.

   b. **Effects.** The three standard flame weapons have different effects against typical urban targets.

      (1) *M202 FLASH.* The M202 FLASH can deliver area fire out to 500 meters. In combat in built-up areas, the range to targets is normally much less. Point targets, such as an alleyway or bunker, can usually be hit from 200 meters. Precision fire against a bunker aperture is possible at 50 meters.

      (a) The FLASH warhead contains a thickened flame agent that ignites when exposed to air. The minimum safe combat range is 20 meters, which is the bursting radius of the rocket warhead due to splashback. If the projectile strikes a hard object along its flight path and breaks open, it will burst into flames even if the fuze has not armed. M202 rocket packs must be protected from small-arms fire and shell fragments that could ignite them. The M202 has a backblast that must be considered before firing (see Figure 8-13). Urban conditions affect this backblast exactly the same as the LAW (see paragraph 8-5). The same considerations for firing a LAW from an enclosed area apply to the M202.

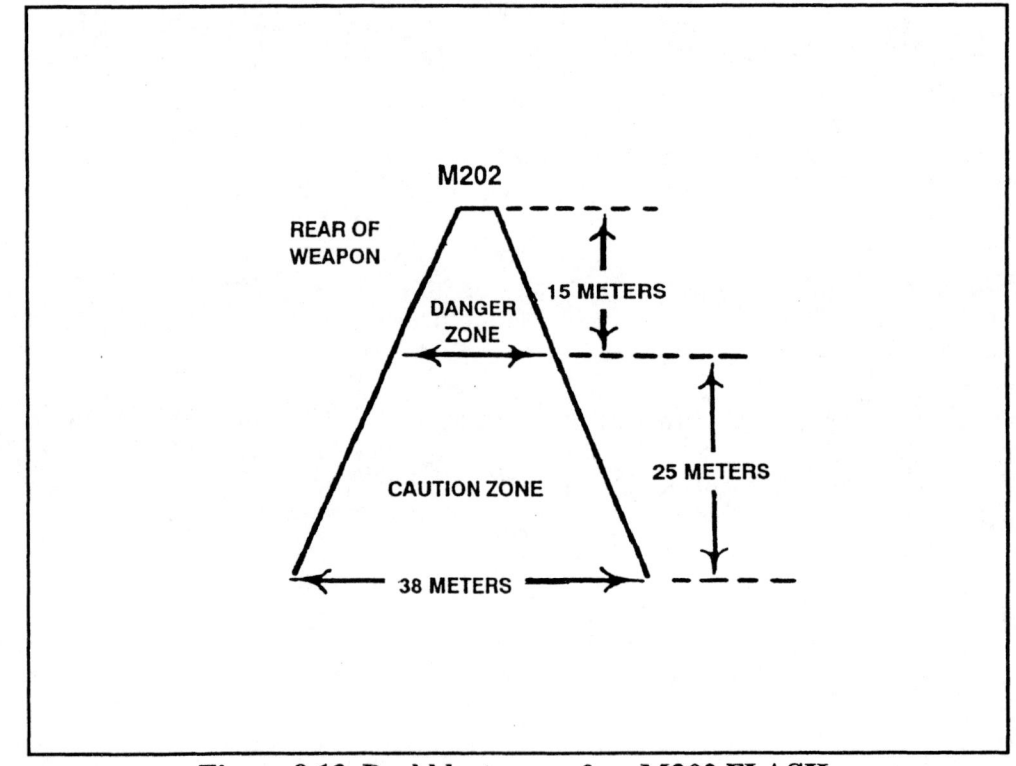

Figure 8-13. Backblast area of an M202 FLASH.

(b) The M202 FLASH is not effective in penetrating typical urban targets. It can penetrate up to 1 inch of plywood at 200 meters, and at close range it can penetrate some wooden doors. The rocket reliably penetrates window glass. The M202 does not damage brick or cinder block construction. The flame agent splattered against the top, flanks, and rear of light armored vehicles can be effective. The psychological effect of hits by flame rockets on closed-in crewmen is significant.

(c) A round detonating near or on a vehicle's rear deck or engine compartment could set the vehicle on fire. A wheeled vehicle, such as the BTR, could have its tires severely damaged by the M202. Modern threat tanks and BMPs have an NBC protective overpressure system that could prevent flame from reaching the vehicle's interior.

(2) *M34 WP Hand grenade.* The M34 is used to ignite and destroy flammable objects, especially wooden structures. It is also used to create an immediate smoke cloud to conceal movement across a narrow open space such as a street. Its smoke is not toxic but can cause choking in heavy concentrations. (a) The grenade's explosion, bright flash, smoke, and burning WP particles all combine to make the M34 one of the most effective psychological weapons available. The M34 hand grenade throws WP fragments up to 35 meters from the point of detonation. These fragments can attach to clothing or skin and continue burning. Because of its weight, most infantrymen can throw this grenade only 30 to 40 meters.

(a) The soldier must avoid injury from friendly use of the M34. As with the M202, the M34 can ignite if the WP inside is exposed to the air. Bullets and shell fragments have been known to strike and rupture M34 grenades, therefore, grenades must be protected from enemy fire.

(b) The M34 WP grenade is an effective weapon against enemy armored vehicles when used in the close confines of combat in built-up areas. It can be thrown or dropped from upper stories onto enemy vehicles. The M34 can be combined with flammable liquids, detonating cords, blasting caps, and fuze igniters to create the eagle fireball, a field-expedient antiarmor device. (See FM 21-75, Appendix H.)

(c) The M34 is also excellent as a screening device. A grenade can be thrown from behind cover into an open street or plaza. When it explodes, the enemy's observation is temporarily obscured. Thus, friendly forces can quickly cross the open area—if the enemy fires, it is unaimed and presents less of a danger. If screening smoke is used to cover a squad's movement across short open areas, it will reduce expected casualties from small-arms fire by about 90 percent.

(3) *M2A1-7 Portable flamethrower.* Portable flamethrowers have a much shorter effective range than the M202 (20 to 50 meters) but require no special backblast consideration. The psychological and physical effects of the portable flamethrower are impressive. When used against troops behind a street barricade, the flamethrower can be fired in a traversing burst to cover a wide frontage. A blind-angle burst can be fired to exploit the splattering effect of the thickened fuel without exposing the gunner (see Figure 8-14, page 8-26).

(a) A burst of unlit fuel (wet shots) can be fired with the flamethrower and ignited with a subsequent shot, creating an intense fireball. This technique is effective in destroying captured equipment or for killing enemy soldiers in sewers. If the enemy has established a position in a wooden building, the building can be burned down. Flame is also effective when fired onto the back deck of tanks or at vision blocks.

(b) Thickened fuel is difficult to extinguish, and, therefore, a commander must decide what will burn before he employs flame. Limits imposed on collateral damage, either political or tactical, are the most serious constraints to the use of flames. If the portable flamethrower is issued in combat in built-up areas, it will probably be used by specially trained personnel. The infantry leader must ensure the flame operator is provided adequate security as he approaches the target. The enemy will concentrate his fire on any flamethrowers he detects.

(c) Although pinpointing targets at night is difficult, commanders should consider using flamethrowers at night for the psychological as well as destructive effect on the enemy.

Figure 8-14. Blind-angle burst.

8-8. HAND GRENADES

Hand grenades are used extensively during combat in built-up areas. Smoke grenades are used for screening and signalling. Riot control grenades are used to drive the enemy out of deep fortifications. Fragmentation and concussion grenades are used to clear the enemy out of rooms and basements. They are the most used explosive munition during intense combat in built-up areas. In World War II, it was common for a battalion fighting in a city to use over 500 hand grenades each day.

a. **Employment.** Smoke and riot control grenades have similar employment techniques. Fragmentation grenades are used to produce enemy casualties.

(1) The AN-M8 HC grenade produces a dense white or grey smoke. It burns intensely and cannot be easily extinguished once it ignites. The smoke can be dangerous in heavy concentrations because it makes breathing difficult and causes choking. The M8 grenade is normally used for screening. It produces a slowly building screen of longer duration then the M34 WP grenade, without the problem of collateral damage caused by scattered burning particles.

(2) The M18-series smoke grenades produce several different colors of smoke, which are used for signalling. Yellow smoke is sometimes difficult to see in built-up areas. Newer versions of yellow smoke grenades are more visible than before.

(3) The M7A3 CS riot control grenade can be used to drive enemy troops out of fortifications when civilian casualties or collateral damage constraints are considerations. Built-up areas often create variable and shifting wind patterns. When using CS grenades, soldiers must prevent the irritating smoke from affecting friendly troops. The CS grenade burns intensely and can ignite flammable structures. Enemy troops wearing even rudimentary chemical protective masks can withstand intense concentrations of CS gas.

(4) The MK3A2 offensive hand grenade, commonly referred to as the concussion grenade, produces casualties during close combat while minimizing the danger to friendly personnel. For this reason, it is the preferred hand grenade during offensive operations in a MOUT environment. The grenade produces severe concussion effects in enclosed areas. It can be used for light blasting and demolitions, and for creating breach holes in interior walls. The concussion produced by the MK3A2 is much greater than that of the fragmentation grenade. It is very effective against enemy soldiers within bunkers, buildings, and underground passages.

(5) The fragmentation grenade is the most commonly available grenade during combat in built-up areas. It provides suppression during room-to-room or house-to-house fighting, and it is used while clearing rooms of enemy personnel. When used at close ranges, it can be cooked off for two seconds to deny the enemy the time to throw it back. The fragmentation grenade can be rolled, bounced, or ricocheted into areas that cannot be reached by 40-mm grenade launchers. Soldiers must be cautious when throwing grenades up stairs. This is <u>not</u> the most desired method of employment.

b. **Effects.** Each type of hand grenade has its own specific effect during combat in built-up areas.

(1) The urban area effects of smoke grenades are nominal. Smoke grenades produce dense clouds of colored or white smoke that remain stationary in the surrounding area. They can cause fires if used indiscriminately. If trapped and concentrated within a small space, their smoke can suffocate soldiers.

(2) The fragmentation grenade has more varied effects in combat in built-up areas. It produces a large amount of small high-velocity fragments, which can penetrate sheetrock partitions and are lethal at short ranges (15

to 20 meters). Fragments lose their velocity quickly and are less effective beyond 25 meters. The fragments from a fragmentation grenade cannot penetrate a single layer of sandbags, a cinder block, or a brick building, but they can perforate wood frame and tin buildings if exploded close to their walls.

(3) Fragmentation barriers inside rooms, consisting of common office furniture, mattress, doors, or books, can be effective against the fragmentation grenade. For this reason, a room should never be considered safe just because one or two grenades have been detonated inside. Fragmentation grenades detonated on the floor not only throw fragments laterally but also send fragments and spall downward to lower floors. Predicting how much spall will occur is difficult since flooring material varies, but wooden floors are usually affected the most.

(4) Some foreign grenades throw fragments much larger than those of the US-made M26. Light barriers and interior walls would probably be less effective against these grenades than against the M26. A major problem with the US-made fragmentation grenade is its tendency to bounce back off hard targets. Grenades are often directed at window openings on the ground floor or second floor. At ranges as close as 20 meters, a thrower's chances of missing a standard 1-meter by 1-meter window are high. The fragmentation grenade normally breaks through standard window glass and enters a room. If the grenade strikes at a sharp angle or the glass is thick plate, the grenade could be deflected without penetrating.

(5) Hand grenades are difficult weapons to use. They involve a high risk of fratracide. Commanders should conduct precombat training with hand grenades as part of normal preparations. Soldiers must be very careful when throwing hand grenades up stairs.

(6) The pins of both fragmentation and concussion grenades can be replaced if the thrower decides not use the weapon. This pin replacement must be done carefully (see FM 23-30).

(7) METT-T and ROE will dictate what type of grenade will be used to clear each room. Because of the high expenditure of grenades, units should use butt packs or assault packs to carry additional grenades of all types. Additional grenades can also be carried in empty ammunition or canteen pouches.

8-9. MORTARS

The urban environment greatly restricts low-angle indirect fires because of overhead masking. While all indirect fire weapons are subject to overhead masking, mortars are less affected than field artillery weapons due to the mortar's higher trajectory. For low-angle artillery fire, dead space is about five times the height of the building behind which the target sits. For mortar fire, dead space is only about one-half the height of the building. Because of these advantages, mortars are even more important to the infantry during combat in built-up areas.

a. **Employment.** Not only can mortars fire into the deep defilade created by tall buildings, but they can also fire out of it. Mortars emplaced behind buildings are difficult for the enemy to accurately locate, and even harder for him to hit with counterfire. Because of their light weight, even heavy mortars can be hand carried to firing positions that may not be accessible to vehicles.

(1) Mortars can be fired through the roof of a ruined building if the ground-level flooring is solid enough to withstand the recoil. If there is only concrete in the mortar platoon's area, mortars can be fired using sandbags as a buffer under the baseplates and curbs as anchors and braces. Aiming posts can be placed in dirt-filled cans.

(2) The 60-mm, 81-mm, and 107-mm mortars of the US Army have limited affect on structural targets. Even with delay fuzes they seldom penetrate more than the upper stories of light buildings. However, their wide area coverage and multioption fuzes make them useful against an enemy force advancing through streets, through other open areas, or over rubble. The 120-mm mortar is moderately effective against structural targets. With a delay fuze setting, it can penetrate deep into a building and create great destruction.

(3) Mortar platoons often operate as separate firing sections during combat in built-up areas. The lack of large open areas can preclude establishing a platoon firing position. Figure 8-15 shows how two mortar sections, which are separated by only one street, can be effective in massing fires and be protected from counter-mortar fire by employing defilade and dispersion.

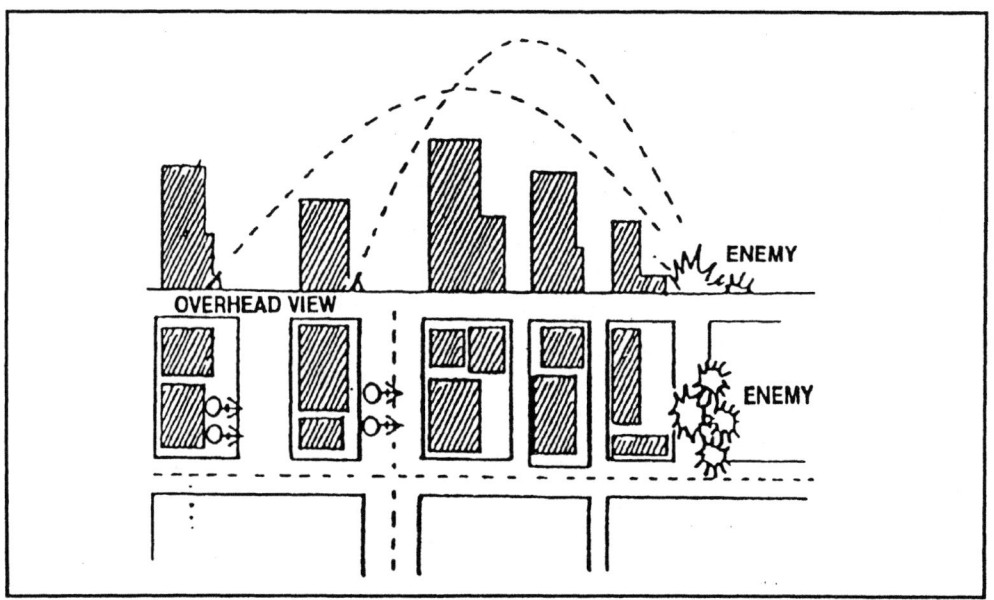

Figure 8-15. Split-section mortar operations on adjacent streets.

(4) All three of the standard mortar projectiles are useful during combat in built-up areas. High-explosive fragmentation is the most commonly used round. WP is effective in starting fires in buildings and forcing the enemy out of cellars and light-frame buildings, and it is also the most effective mortar round against dug-in enemy tanks. Even near-misses blind and suppress the tank crew, forcing them to button up. Hits are difficult to achieve, but are effective.

(5) Because the artificial roughness of urban terrain reduces wind speed and increases atmosphere mixing, mortar smoke tends to persist longer and give greater coverage in built-up areas than in open terrain.

(6) Urban masking impacts on the use of illumination. In built-up areas, it is often necessary to plan illumination behind friendly positions, which places friendly troops in shadows and enemy troops in the light. Illumination rounds are difficult to adjust and are often of limited use because of the deep canyon nature of the urban area. Rapidly shifting wind currents in built-up areas also affect mortar illumination, making it less effective.

b. **Effects of Mortar Fire.** The multioption fuze on newer United States mortar rounds makes them effective weapons on urban terrain. Delay settings can increase penetration slightly, and proximity bursts can increase the lethal area covered by fragments. Tall buildings can cause proximity fuzed mortar rounds to detonate prematurely if they pass too closely.

(1) *60-mm Mortar.* The 60-mm mortar round cannot penetrate most rooftops, even with a delay setting. Small explosive rounds are effective, however, in suppressing snipers on rooftops and preventing roofs from being used by enemy observers. The 60-mm WP round is not normally a good screening round due to its small area of coverage. In combat in built-up areas, however, the tendency of smoke to linger and the small areas to be screened make it more effective. During the battle for Hue in South Vietnam, 60-mm WP rounds were used to create small, short-term, smoke screens to conceal movement across open areas such as parks, plazas, and bridges. Fragments from 60-mm HE rounds, landing as close as 10 feet, cannot penetrate a single sandbag layer or a single-layer brick wall. The effect of a 60-mm mortar HE round that achieves a direct hit on a bunker or fighting position is equivalent to 1 or 2 pounds of TNT. Normally, the blast will not collapse a properly constructed bunker but can cause structural damage. The 60-mm mortar will not normally crater a hard-surfaced road.

(2) *81-mm Mortar.* The 81-mm mortar has much the same effect against urban targets as the 60-mm mortar. It has a slightly greater lethal area and its smoke rounds (WP and RP) are more effective. A direct hit is equivalent to about 2 pounds of TNT. The 81-mm round cannot significantly crater a hard-surfaced road. With a delay setting, the 81-mm round can penetrate the roofs of light buildings.

(3) *107-mm Mortar.* The 107-mm mortar can affect moderately hard urban targets. It is more effective than the 81-mm mortar. Even when fired with a delay fuze setting, the round cannot penetrate deep into typical urban targets. The mortar's lethal fragment area is somewhat increased in built-up areas, because its blast picks up significant amounts of debris and throws it outward. The minimum range of the 107-mm mortar is the main constraint in its employment during battle in a built-up area. Out of all the United States mortars, the 107-mm is the least capable in reaching targets in deep defilade. The 107-mm mortar slightly craters a hard-surfaced road, but not enough to prevent vehicle traffic.

(4) *120-mm Mortar.* The 120-mm mortar is large enough to have a major effect on common urban targets. It can penetrate deep into a building, causing extensive damage because of its explosive power. A minimum of 18 inches of packed earth or sand is needed to stop the fragments from a 120-mm HE round that impacts 10 feet away. The effect from a direct hit from a 120-mm round is equivalent to almost 10 pounds of TNT, which can crush fortifications built with commonly available materials. The 120-mm mortar round can create a large but shallow crater in a road surface, which

is not deep or steep-sided enough to block vehicular movement. However, craters could be deep enough to damage or destroy storm drain systems, water and gas pipes, and electrical or phone cables.

(5) *160-mm Mortar.* The Soviet 160-mm mortar can inflict massive damage to almost any urban structure. Only large buildings and deep cellars offer protection against this weapon. Even well-built bunkers can be crushed by near-misses. The effect from a direct hit by this weapon is equivalent to over 15 pounds of TNT. The 160-mm mortar creates significant craters in urban road surfaces. These craters are several meters wide and are deep enough to interfere with vehicular movement. The 160-mm mortar can destroy storm drainage systems, water mains, and underground power lines.

(6) *240-mm Mortar.* The Soviet 2S4 240-mm mortar is designed to destroy heavy fortifications. Average buildings do not provide certain protection from this mortar. Its HE rounds weigh over 280 pounds. It has a concrete-piercing round for use in urban areas. The 2S4 can fire one round per minute. A round will do massive damage to urban road surfaces, breaking and heaving large slabs of road surface many yards from the point of impact.

8-10. 25-MM AUTOMATIC GUN

The 25-mm automatic gun mounted on the M2/M3 fighting vehicle and on the USMC LAV-25 offers infantrymen a new and effective weapon to aid them during combat in built-up areas. The primary roles of BFVs and LAV-25s during combat in built-up areas are to provide suppressive fire and to breach exterior walls and fortifications. (See paragraph 8-3 for the suppression effects and penetration of the 7.62-mm coaxial machine gun.) The wall and fortification breaching effects of the 25-mm automatic gun are major assets to infantrymen fighting in built-up areas.

a. **Obliquity.** The 25-mm gun produces its best urban target results when fired perpendicular to the hard surface (zero obliquity). In combat in built-up areas, however, finding a covered firing position that permits low obliquity firing is unlikely, unless the streets and gaps between buildings are wide. Most shots impact the target at an angle, which normally reduces penetration. With the APDS-T round, an angle of obliquity of up to 20 degrees can actually improve breaching. The rounds tend to dislodge more wall material for each shot but do not penetrate as deeply into the structure.

b. **Target Types.** The 25-mm gun has different effects when fired against different urban targets.

(1) *Reinforced concrete.* Reinforced concrete walls, which are 12 to 20 inches thick, present problems for the 25-mm gun when trying to create breach holes. It is relatively easy to penetrate, fracture, and clear away the concrete, but the reinforcing rods remain in place. These create a "jail window" effect by preventing entry but allowing grenades or rifle fire to be placed behind the wall. Steel reinforcing rods are normally 3/4 inch thick and 6 to 8 inches apart—there is no quick way of cutting these rods. They can be cut with demolition charges, cutting torches, or special power saws. Firing with either APDS-T or HEI-T rounds from the 25-mm gun will not always cut these rods.

(2) *Brick walls.* Brick walls are more easily defeated by the 25-mm gun regardless of their thickness, and they produce the most spall.

(3) *Bunker walls.* The 25-mm gun is devastating when fired against sandbag bunker walls. Obliquity has the least affect on the penetration of bunker walls. Bunkers with earth walls up to 36 inches thick are easily penetrated. At short ranges typical of combat in built-up areas, defeating a bunker should be easy, especially if the 25-mm gun can fire at an aperture.

c. **Burst Fire.** The 25-mm gun's impact on typical urban targets seem magnified if the firing is in short bursts. At close ranges, the gunner might need to shift his point of aim in a spiral pattern to ensure that the second and third bursts enlarge the hole. Even without burst fire, sustained 25-mm gun fire can defeat almost all urban targets.

d. **Weapon Penetration.** The penetration achieved by the two combat rounds (HEI-T and APDS-T) differ slightly—both are eventually effective. However, the best target results are not achieved with either of the combat rounds. At close range against structural targets, the training round (TP-T) is significantly more effective. The TP-T round, however, has little utility when used against enemy armored vehicles. It will rarely, if ever, be carried into combat.

(1) *APDS-T.* The armor-piercing, discarding, sabot with tracer round penetrates urban targets by retaining its kinetic energy and blasting a small hole deep into the target. The APDS-T round gives the best effects behind the wall, and the armor-piercing core often breaks into two or three fragments, which can create multiple enemy casualties. The APDS-T needs as few as four rounds to achieve lethal results behind walls. Table 8-10 explains the number of APDS-T rounds needed to create different-size holes in common urban walls.

TARGET	LOOPHOLE	BREACHHOLE
3-inch brick wall at 0-degree obliquity	22 rounds	75 rounds
3-inch brick wall at 45-degree obliquity	22 rounds	35* rounds
5-inch brick wall at 0-degree obliquity	32 rounds	50* rounds
8-inch reinforced concrete at 0-degree obliquity	22 rounds	75 rounds (Note: Reinforcing rods still in place)
8-inch reinforced concrete at 45-degree obliquity	22 rounds	40* rounds (Note: Reinforcing rods still in place)

*Obliquity and depth tend to increase the amount of wall material removed.

Table 8-10. Breaching effects of APDS-T rounds.

(a) When firing single rounds, the APDS-T round provides the greatest capability for behind-the-wall incapacitation. The APDS-T round can penetrate over 16 inches of reinforced concrete with enough energy left to cause enemy casualties. It penetrates through both sides of a wood frame or brick veneer building. Field fortifications are easily penetrated by APDS-T rounds. Table 8-11 explains the number of APDS-T rounds needed to create different-size holes in commonly found bunkers.

TYPE BUNKER	OBLIQUITY	PENETRATION	LOOPHOLE	SMALL BREACHHOLE
36-inch sand/timber	0 degree	1 round	25 rounds	40 rounds
36-inch sand/ 6-inch concrete	0 degree	6 rounds	6 rounds	20 rounds

Table 8-11. Number of APDS-T rounds needed to create different size holes in bunkers.

(b) The APDS-T round creates a hazardous situation for exposed personnel because of the pieces of sabot that are thrown off the round. Personnel not under cover forward of the 25-mm gun's muzzle and within the danger zone could be injured or killed by these sabots, even if the penetrator passes overhead to hit the target. The danger zone extends at an angle of about 10 degrees below the muzzle level, out to at least 100 meters, and about 17 degrees left and right of the muzzle. Figure 8-16 shows the hazard area of the APDS-T round.

Figure 8-16. APDS danger zone.

(2) *HEI-T.* The high-explosive, incendiary with tracer round penetrates urban targets by blasting away chunks of material.

(a) The HEI-T round does not penetrate an urban target as well as the APDS-T, but it creates the effect of stripping away a greater amount of material for each round. The HEI-T does more damage to an urban target when fired in multiple short bursts because the accumulative impact of multiple rounds is greater than the sum of individual rounds. Table 8-12 explains the number of HEI-T rounds needed to create different-size holes.

TARGET	LOOPHOLE	BREACHHOLE
3-inch brick wall at 0-degree obliquity	10 rounds	20 rounds
3-inch brick wall at 45-degree obliquity	20 rounds	25 rounds
5-inch brick wall at 0-degree obliquity	30 rounds	60 rounds
8-inch reinforced concrete at 0-degree obliquity	15 rounds	25 rounds
8-inch reinforced concrete at 45-degree obliquity	15 rounds	30 rounds

Table 8-12. Number of HEI-T rounds needed to create different size holes.

(b) The HEI-T round does not provide single-round perforation or incapacitating fragments on any external masonry structural wall. It can create first-round fragments behind wood frame and brick veneer walls. HEI-T rounds cannot penetrate a bunker as quickly as APDS-T, but they can create more damage inside the bunker once the external earth has been stripped away. Against a heavy bunker, about 40 rounds of HEI-T are needed to strip away the external earth shielding and breach the inner lining of concrete or timber. The HEI-T round is also used for suppression against known or suspected firing ports, such as doors, windows, and loopholes.

8-11. TANK CANNON
The powerful, high-velocity cannon mounted on the M1, M1A1, M60, and M48 series tanks provides the infantryman with a key requirement for victory in built-up areas—heavy direct-fire support. Although the infantry assumes the lead role during combat in built-up areas, tanks and infantry work as a close team. Tanks move down streets after the infantry has cleared them of any suspected ATGM positions and, in turn, support the infantry with fire. The tank is one of the most effective weapons for heavy fire against structures. The primary role of the tank cannon during combat in built-up areas is to provide heavy direct-fire against buildings and strongpoints that are identified as targets by the infantry. The wall and fortification breaching effects of the 105-mm and 120-mm tank cannon are major assets to infantrymen fighting in built-up areas.

a. **Obliquity.** Tank cannons produce their best urban target effects when fired perpendicular to the hard surface (zero obliquity). During combat in built-up areas, however, finding a covered firing position that permits low-obliquity firing is unlikely. Most shots strike the target at an angle that would normally reduce penetration. With tank cannon APDS rounds, obliquity angles up to 25 degrees have little affect, but angles greater than 45 degrees greatly reduce penetration. For example, a 105-mm APDS round cannot penetrate a 2-inch reinforced concrete wall at an angle of obliquity greater than 45 degrees due to possible ricochet.

b. **Ammunition.** Armor-piercing, fin-stabilized, discarding sabot (APFSDS) rounds are the most commonly carried tank ammunition. These rounds best against armored vehicles. Other types of ammunition can be carried that are more effective against masonry targets. The 105-mm cannon has HEAT and WP rounds in addition to APDS. The 120-mm cannon has an effective high-explosive, antitank, multipurpose (HEAT-MP) round.

c. **Characteristics.** Both 105-mm and 120-mm tank cannons have two specific characteristics that affect their employment in built-up areas: limited elevation and depression, and short arming ranges. In addition, the M1 and M1A1 tanks have another characteristic not involved with its cannon but affecting infantrymen working with it—extremely hot turbine exhaust.

(1) The cannon of the M1 and M1A1 tank can be elevated +20 degrees or depressed -10 degrees. The M60 and M48-series tanks have upper limits of +19 degrees and lower limits of -10 degrees. The lower depression limit creates a 35-foot (10.8-meter) dead space around a tank. On a 16-meter-wide street (common in Europe) this dead space extends to the buildings on each side (see Figure 8-17). Similarly, there is a zone overhead in which the tank cannot fire (see Figure 8-18, page 8-36). This dead space offers ideal locations for short-range antiarmor weapons and allows hidden enemy gunners to fire at the tank when the tank cannot fire back. It also exposes the tank's most vulnerable areas: the flanks, rear, and top. Infantrymen must move ahead, alongside, and to the rear of tanks to provide close protection. The extreme heat produced immediately to the rear of the M1-series tanks prevents dismounted infantry from following closely, but protection from small-arms fire and fragments is still provided by the tank's bulk and armor. The M1-series tanks also have a blind spot caused by the 0-degree of depression available over part of the back deck. To engage any target in this area, the tank must pivot to convert the rear target to a flank target.

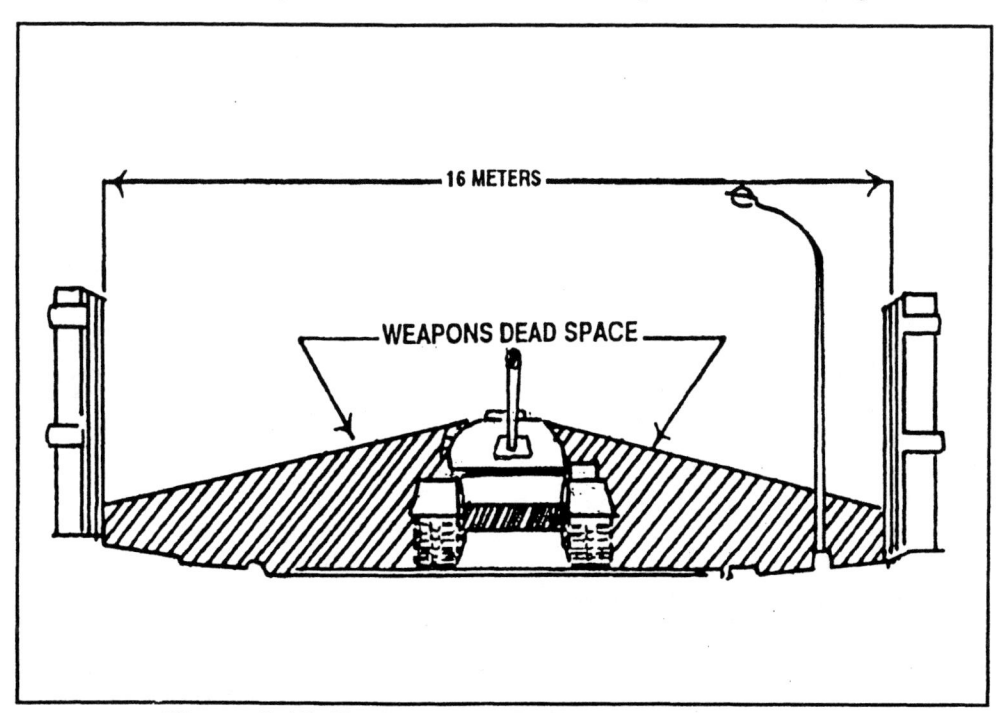

Figure 8-17. Tank cannon dead space at street level.

FM 90-10-1

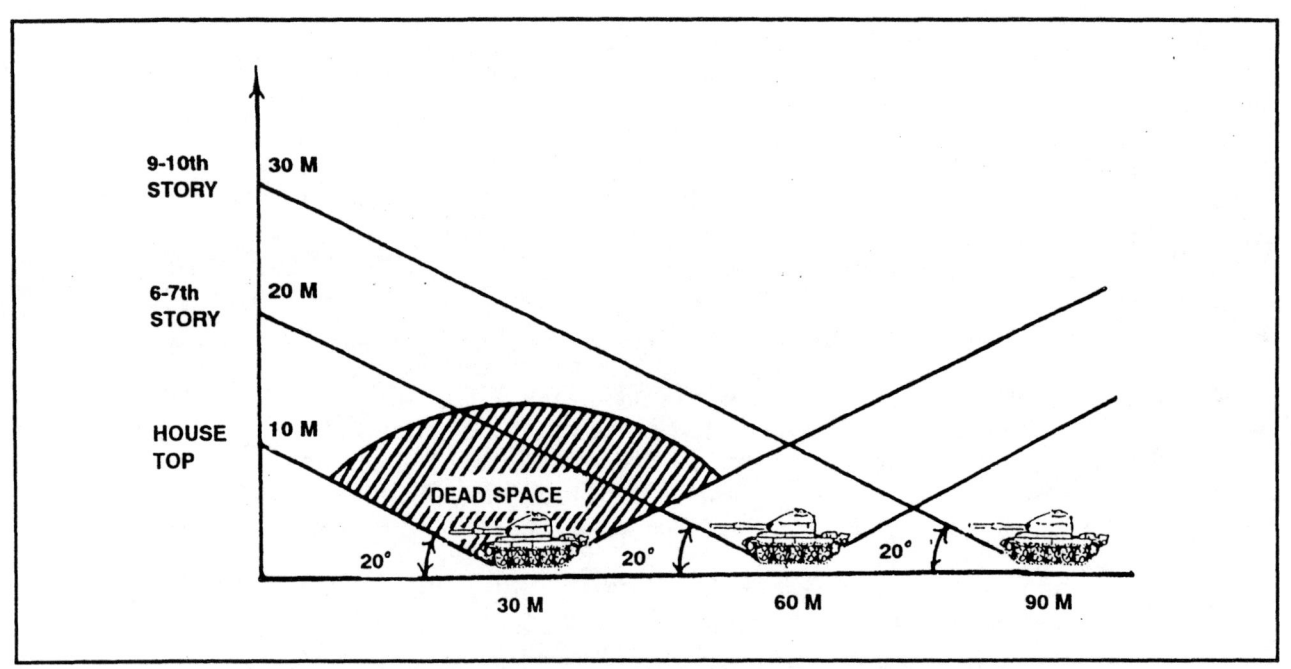

Figure 8-18. Tank cannon dead space above street level.

(2) The 105-mm HEAT round arms within 25 to 30 feet, and the 120-mm HEAT-MP round arms at about 36 feet. These arming distances allow the tank to engage targets from short ranges. The armor of the tank protects the crew from both the blowback effects of the round and enemy return fire. The APFSDS round does not need to arm and can, therefore, be fired at almost any range. The discarding portions of the round can be lethal to exposed infantry forward of the tank.

d. **Target Effects.** High-explosive, antitank rounds are most effective against masonry walls. The APFSDS round can penetrate deeply into a structure but does not create as large a hole or displace as much spall behind the target. In contrast to lighter HEAT rounds, tank HEAT rounds are large enough to displace enough spall to inflict casualties inside a building. One HEAT round normally creates a breach hole in all but the thickest masonry construction—brick veneer and wood frame construction are demolished by a single round. Even the 120-mm HEAT round cannot cut all the reinforcing rods, which are usually left in place, often hindering entry through the breach hole (see Figure 8-19). Both HEAT and APFSDS rounds are effective against all field fortifications. Only large earth berms and heavy mass construction buildings can provide protection against tank fire.

e. **Employment.** Tank-heavy forces could be at a severe disadvantage during combat in built-up areas, but a few tanks working with the infantry can be most effective, especially if they work well together at the small-unit level. Tank, infantry, and engineer task forces are normally formed to attack a fortified area. Individual tanks or pairs of tanks can work together with rifle squads or platoons.

(1) Tanks need infantry on the ground to provide security in built-up areas and to designate targets. Against targets protected by structures, tanks should be escorted forward to the most covered location that provides a clear

shot. On-the-spot instructions by the infantry unit leader ensures that the tank's fire is accurate and its exposure is limited. The tank commander may have to halt in a covered position, dismount, and reconnoiter his route forward into a firing position.

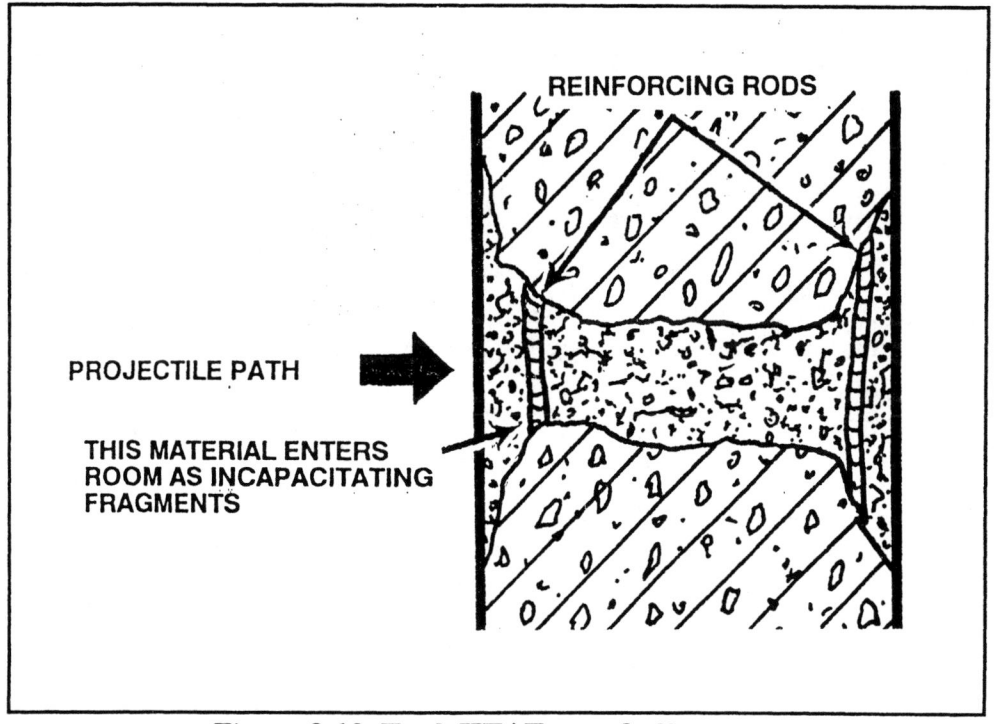

Figure 8-19. Tank HEAT round effects on reinforced concrete walls.

(2) When the tank main gun fires, it creates a large fireball and smoke cloud. In the confines of a built-up area, dirt and masonry dust are also picked up and add to this cloud. The target is further obscured by the smoke and dust of the explosion. Depending on the local conditions, this obscuration could last as long as two or three minutes. Infantry can use this period to reposition or advance unseen by the enemy. Caution must be exercised, however, because the enemy might also move.

(3) Tank cannon creates an overpressure and noise hazard to exposed infantrymen. All dismounted troops working near tanks should wear their Kevlar helmet and protective vest, as well as ballistic eye protection. If possible, they should also wear earplus and avoid the tank's frontal 60-degree arc during firing.

(4) Tanks are equipped with powerful thermal sights that can be used to detect enemy personnel and weapons that are hidden in shadows and behind openings. Dust, fires, and thick smoke significantly degrade these sights.

(5) Tanks have turret-mounted grenade launchers that project screening smoke grenades. The grenades use a bursting charge and burning red phosphorous particles to create this screen. Burning particles can easily start uncontrolled fires and are hazadous to dismounted infantry near the tank. The tank commander and the infantry small-unit leader must coordinate

when and under what conditions these launchers can be used. Grenade launchers are a useful feature to protect the tank but can cause significant problems if unwisely employed.

(6) The tank's size and armor can provide dismounted infantry with cover from direct-fire weapons and fragments. With coordination, tanks can provide moving cover for infantrymen as they advance across small open areas. However, enemy fire striking a tank but not penetrating is a major threat to nearby infantry. Fragmentation generated by antitank rounds and ricochets off tank armor have historically been a prime cause of infantry casualties while working with tanks in built-up areas.

(7) Some tanks are equipped with dozer blades that can be used to remove rubble barriers under fire, breach obstacles, or seal exits.

8-12. COMBAT ENGINEER VEHICLE DEMOLITION GUN

The CEV is a special-purpose engineer equipment vehicle. It provides a heavy demolition capability. The CEV has a 7.62-mm machine gun that is coaxially mounted. It also has a .50-caliber machine gun in the commander's cupola, and a 165-mm main gun. The main gun fires a high-explosive plastic (HEP) round of great power. The weapon's maximum range is 925 meters.

a. Target Effects. The HEP round is very effective against masonry and concrete targets. The pushing and heaving effects caused by the HEP round's base detonating fuze and large amount of explosive can demolish barriers and knock down walls. One round produces a 1-foot diameter hole in a 7-inch thick reinforced concrete wall. The round's effects against bunkers and field fortifications are dramatic, often crushing or smashing entire walls.

b. Employment. The CEV is normally used for special engineer tasks in direct support of infantry battalions. It must be given the same close infantry protection and target designation as tanks. Although the CEV consists of a tank hull and a short-barrelled turret, it is not a tank and should not be routinely used against enemy tanks. It is an excellent heavy assault support vehicle when used as part of a combined engineer-infantry team.

8-13. ARTILLERY AND NAVAL GUNFIRE

A major source of fire support for infantry forces fighting in built-up areas is the fire of field artillery weapons. If the built-up area is near the coast, naval gunfire can be used. Field artillery employment can be in either the indirect- or direct-fire mode.

a. Indirect Fire. Indirect artillery fire is not effective in attacking targets within walls and masonry structures. It tends to impact on roofs or upper stories rather than structurally critical wall areas or pillars.

(1) Weapons of at least 155-mm are necessary against thick reinforced concrete, stone, or brick walls. Even with heavy artillery, large expenditures of ammunition are required to knock down buildings of any size. Tall buildings also create areas of indirect-fire dead space, which are areas that cannot be engaged by indirect fire due to a combination of building height and angle of fall of the projectile (see Figure 8-20). Usually the dead space for low-angle indirect fire is about five times the height of the highest building over which the rounds must pass.

(2) Even when it is theoretically possible to hit a target in a street over a tall building, another problem arises because of range probable error (PE).

Only 50 percent of the rounds fired on the same data can be expected to fall within one range PE of the target. This means that when firing indirect fire into built-up areas with tall buildings, it is necessary to double the normal ammunition expenditure to overcome the problem of a reduced target area and range PE. Also, up to 25 percent of all HE rounds are duds due to glancing off hard surfaces.

(3) Naval gunfire, because of its flat trajectory, is even more affected by terrain masking. It is usually difficult to adjust onto the target, because the gun-target line is constantly changing.

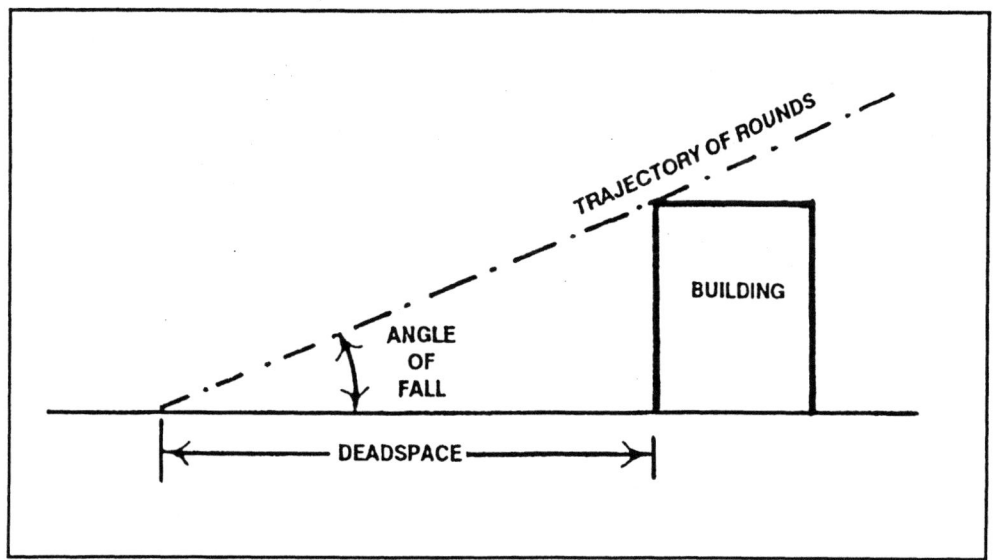

Figure 8-20. Indirect-fire dead space (low angle).

b. **Direct Fire.** Self-propelled artillery pieces are not as heavily armored as tanks, but they can still be used during combat in built-up areas if adequately secured by infantry. The most likely use of US artillery in an urban direct-fire role is to reinforce tank fires against tough or important urban targets. Because of their availability and habitual relationship with infantry, tanks remain a more common direct-fire support means than self-propelled artillery. Self-propelled artillery should be used in this role only after an analysis of the need for heavy direct fire and the tradeoff involved in the extreme decentralization of artillery firepower. It has the same need for close security and target designation as tanks.

c. **Target Effects.** Medium caliber (155-mm) and heavy caliber (203-mm) direct fire has a devastating affect against masonry construction and field fortifications. Smaller artillery pieces (105-mm) are normally towed and, therefore, are difficult to employ in the direct-fire mode. Their target effects are much less destructive than the larger caliber weapons.

(1) *155-mm Howitzers.* The 155-mm self-propelled howitzer offers its crew mobility and limited protection in built-up areas. It is effective due to its rate of fire and penetration. High-explosive rounds can penetrate up to 38 inches of brick and unreinforced concrete. Projectiles can penetrate up to 28 inches of reinforced concrete with considerable damage beyond the wall. HE rounds fuzed with concrete-piercing fuzes provide an excellent

means of penetrating strong reinforced concrete structures. One round can penetrate up to 46 inches. Five rounds are needed to reliably create a 1.5-meter breach in a 1-meter thick wall. About 10 rounds are needed to create such a large breach in a wall 1.5 meters thick. Superquick fuzing causes the rubble to be blown into the building, whereas delay fuzing tends to blow the rubble outward into the street.

(2) *203-mm Howitzers.* The 203-mm howitzer is the most powerful direct-fire weapon available to the Army. It has a slow rate of fire, but its projectile has excellent penetration abilities. One round normally creates a breach hole in walls up to 56 inches thick. The howitzer crew is exposed to enemy fire. The vehicle only carries three rounds on board, which limits its use.

(3) *Naval cannon.* The most common naval cannon used to support ground troops is the 5-inch 54 caliber gun. In either single or double mounts, this weapon has a high rate of fire and is roughly equivalent to the 155-mm howitzer in target effect. The heaviest guns used to engage land targets are the 16-inch guns of the recently renovated Iowa-class battleships. When used singly or in salvo, these massive guns can penetrate any structure common to a built-up area. Their blast effect is destructive to buildings up to a block away from the point of impact. Battleship gunfire rarely, if ever, is used for close support of ground troops. Its long range and destructive power can be controlled and adjusted ahead of advancing forces by aerial observers to clear or destroy enemy strongpoints and supporting artillery.

8-14. AERIAL WEAPONS

Both rotary- and fixed-wing aircraft can quickly deliver large volumes of firepower over large built-up areas. Specific targets are hard to distinguish from the air. Good ground-to-air communications are vital to successfully employing aerial firepower. Aviators have historically tended to overestimate the effects on defenders of high-explosive ordnance. Modern, large buildings are remarkably resistant to damage from bombs and rocket fire.

a. **Rotary-Winged Aircraft.** Armed attack helicopters can be used to engage targets in built-up areas. Enemy armored vehicles in small parks, boulevards, or other open areas are good targets for attack helicopters.

NOTE: The target effects of TOW missiles and 40-mm grenades carried by attack helicopters have already been discussed.

(1) The HELLFIRE missile has a larger warhead and greater range than the TOW, but it too is a shaped-charge warhead and is not specifically designed for use against masonry targets. Laser target designation for the HELLFIRE may not be possible due to laser reflections off glass and shiny metal surfaces. The use of attack helicopters to deliver ATGMs against targets in the upper stories of high buildings is sometimes desirable.

(2) The 2.75-inch folding fin aerial rocket and the 20-mm cannon common to some attack helicopters are good area weapons to use against enemy forces in the open or under light cover. They are usually ineffective against a large masonry target. The 20-mm cannon produces many ricochets, especially if AP ammunition is fired into built-up areas.

(3) The 30-mm cannon carried by the Apache helicopter is an accurate weapon. It penetrates masonry better than the 20-mm cannon.

b. **Fixed-Wing Aircraft.** Close air support to ground forces fighting in built-up areas is a difficult mission for fixed-wing aircraft. Targets are hard to locate and identify, enemy and friendly forces could be intermingled, and enemy short-range air defense weapons are hard to suppress.

(1) Because enemy and friendly forces can be separated by only one building, accurate delivery of ordnance is required. Marking panels, lights, electronic beacons, smoke, or some other positive identification of friendly forces is needed.

(2) General-purpose bombs from 500 to 2,000 pounds are moderately effective in creating casualties among enemy troops located in large buildings. High-dive angle bomb runs increase accuracy and penetration but also increase the aircraft's exposure to antiaircraft weapons. Low-dive angle bomb runs using high drag (retarded) bombs can be used to get bombs into upper stories. Penetration is not good with high-drag bombs. Sometimes aerial bombs pass completely through light-clad buildings and explode on the outside.

(3) Aerial rockets and 20-mm cannons are only moderately effective against enemy soldiers in built-up areas since rockets lack the accuracy to concentrate their effects. The 20-mm cannon rounds penetrate only slightly better than the .50-caliber round; 20-mm AP rounds can ricochet badly; and tracers can start fires.

(4) The 30-mm cannon fired from the A-10 aircraft is an accurate weapon. It is moderately effective against targets in built-up areas, penetrating masonry better than the 20-mm cannon.

(5) The AC-130 aircraft has weapons that can be most effective during combat in built-up areas. This aircraft can deliver accurate fire from a 20-mm Vulcan cannon, 40-mm rapid-fire cannon, and 105-mm howitzer. The 105-mm howitzer round is effective against the roof and upper floors of buildings. The AC-130 is accurate enough to concentrate its 40-mm cannon and 105-mm howitzer fire onto a single spot to create a rooftop breach, which allows fire to be directed deep into the building.

(6) Laser and optically guided munitions can be effective against high-value targets. The USAF has developed special, heavy, laser-guided bombs to penetrate hardened weapons emplacements. Problems associated with dense smoke and dust clouds hanging over the built-up area and laser scatter can restrict their use. If the launching aircraft can achieve a successful laser designation and lock-on, these weapons have devastating effects, penetrating deep into reinforced concrete before exploding with great force. If launched without a lock-on, or if the laser spot is lost, these weapons are unpredictable and can travel long distances before they impact.

8-15. DEMOLITIONS

Combat in built-up areas requires the extensive use of demolitions. All soldiers, not just engineer troops, should be trained to employ demolitions. (See FM 5-25 for specific information on the safe use of demolitions.)

a. **Bulk Demolitions.** Bulk demolitions come in two types, TNT and C4. Exposed soldiers must take cover or move at least 300 meters away from bulk explosives that are being used to breach walls.

(1) TNT comes in 1/4-, 1/2-, and 1-pound blocks. About 5 pounds of TNT are needed to breach a nonreinforced concrete wall 12 inches thick if

the explosives are laid next to the wall and are not tamped. If the explosives are tamped, about 2 pounds are sufficient.

(2) C4 comes in many different sized blocks. About 10 pounds of C4 placed between waist and chest high will blow a hole in the average masonry wall large enough for a man to walk through.

b. **Shaped Charges.** There are two sizes of US Army shaped charges, a 15-pound M2A3 and a 40-pound M3A3. The M3A3 is the most likely shaped charge to be used in built-up areas. It can penetrate 5 feet of reinforced concrete. The hole tapers from 5 inches down to 2 inches. The amount of spall thrown behind the target wall is considerable. There is also a large safety hazard area for friendly soldiers.

c. **Satchel Charges.** There are two standard US Army satchel charges: the M183 and the M37. Both come in their own carrying satchel with detonators and blasting cords. Each weighs 20 pounds. The M183 has 16 individual 1 1/4-pound blocks that can be used separately. When used untamped, a satchel breaches a 3-foot thick concrete wall. Satchel charges are very powerful. Debris is thrown great distances. Friendly troops must move away and take cover before detonation.

d. **Cratering Charges.** The standard US Army cratering charge is a 43-pound cylinder of ammonium nitrate. This explosive does not have the shattering effect of bulk TNT or C4. It is more useful in deliberate demolitions than in hasty ones.

APPENDIX A

NUCLEAR, BIOLOGICAL, AND CHEMICAL CONSIDERATIONS

Current US policy regarding lethal or incapacitating agents is that their use against an armed enemy requires approval at the national command authority level. Potential enemies may not operate under the same restrictions. Field commanders must be prepared to assume an adequate NBC defensive posture when engaged in urban fighting. Leaders must be aware of how the built-up environment affects the protection, detection, and decontamination process. Buildings are usually not strong enough to provide shelter from a nuclear explosion but do provide some protection against fallout. They also have unique characteristics concerning the use of biological and chemical agents. Personnel who must move through a contaminated built-up area should employ the procedures outlined in FM 3-3, FM 3-4, FM 3-5, and FM 3-100.

A-1. PROTECTION FROM NBC

The lowest floor or basement of a reinforced concrete or steel-formed building offers good protection from nuclear hazards and liquid chemical contamination. Tunnels, storm drains, subway tubes, and sewers provide better protection than buildings. Tanks, BFVs, and APCs also provide protection.

    a. Biological attacks are difficult to detect or recognize. Biological agents can be disseminated by using aerosols, vectors, and covert methods (see FM 3-3 for more detailed information.) Since biological agents can be sprayed or dropped in bomblets, personnel who observe such indicators should promptly report them. Prompt reporting and treatment of the sick speeds the employment of medical countermeasures. Although buildings and shelters provide some protection against spraying, they provide little protection against biological agents.

    b. Chemical agents cause casualties by being inhaled or by being absorbed through the skin. They may afford soldiers a few seconds to mask. Buildings have a channeling effect and tend to contain the effects of an agent, causing great variation in chemical concentration from room to room or from building to building. Chemical agents usually settle in low places, making sewers and subways hazardous hiding places. A prepared defender should include some collective protective measures in the defensive network. Personnel using fans may be able to put enough overpressure into tunnels to keep some chemical agents from entering. The individual protective mask and battle dress overgarment provide the best protection against chemical agents.

    c. Personal hygiene is a critical defensive measure against infection and disease. Unfortunately, built-up areas are characterized by sophisticated sanitation systems. When those systems are destroyed, the resulting sanitary conditions become much worse than those in areas where sanitary facilities do not exist. Commanders must ensure that personnel employ sanitation measures and that their immunizations are current.

d. Commanders should plan their MOPP realizing that built-up area logistics also apply to NBC equipment. Protective clothing, detection and decontamination equipment, and sealed containers of food and water must be stockpiled the same as other supplies. When operating in protective clothing, commanders must make allowances for the strenuous activities normally associated with combat in built-up areas.

(1) *Detection.* After an NBC attack, battalions should dispatch their detection and survey teams. Detection in built-up areas is complicated by the containing nature of buildings. Teams should conduct tests and surveys of major streets, intersections, and buildings in their area for inclusion in initial NBC reports. A systematic survey of all buildings, rooms, and underground facilities must be accomplished before occupation by unmasked personnel. All data should be forwarded using the appropriate NBC report.

(2) *Decontamination.* Personnel must begin decontamination operations as soon after an NBC attack as the mission allows. Personnel should conduct individual decontamination of themselves and their personal equipment. Unit commanders must determine the need for MOPP gear exchange and the requirements for a hasty or deliberate decontamination operation.

(a) *Radiological.* Personnel should wear wet-weather gear for certain decontamination operations (hosing down buildings) to prevent radioactive material from touching the skin.

(b) *Chemical and biological.* Roads, sidewalks, and other hard surfaces are best decontaminated by weathering, if time permits. Agents can also be covered with several inches of dirt or sand to provide protection. Fragment testing should be conducted periodically to ensure that the agent has not seeped through the covering. For critical sections of roads a truck-mounted M12A1 (power-driven decontaminating apparatus [PDDA]) can be used to spray STB slurry; this aids rapid decontamination. Buildings are difficult to decontaminate, especially wooden ones. Some techniques for their decontamination are scrubbing with STB slurry; washing with hot, soapy water; washing or spraying with a soda solution; and airing.

A-2. SMOKE OPERATIONS

To stay combat effective when faced with many complications caused by NBC operations, commanders must plan before combat. The use of smoke as an integral part of either offensive or defensive operations can complement missions in built-up areas. Chemical support could be needed from smoke generator units for both offensive and defensive operations. In the offense, smoke can support the maneuver of combat elements and deception operations. Smoke employed in the defense obscures enemy air and ground observation, limiting the accuracy of enemy fires and target intelligence.

a. Smoke should not be used when it degrades the effectiveness of friendly forces. Likewise, an extremely dense concentration of smoke in a closed room displaces the oxygen in the room, smothering soldiers even when they are wearing protective masks.

b. Smoke pots, generators, or artillery smoke munitions should be used to cover the withdrawal of defending forces or the movement of attacking forces. Artillery-delivered white phosphorus can also be effective on enemy forces by causing casualties and fires. The incendiary effects of both white phosphorus and base ejection munitions on the litter and debris of built-up areas must be considered.

c. Smoke grenades can be massed to provide a hasty screen for concealing personnel movement across streets and alleys. Smoke grenades can also be used for signalling; those launched by an M203 can be used to mark targets for attack helicopters or tactical air.

d. The use of smoke in built-up areas is affected by complex wind patterns caused from buildings. When covering a built-up area with a smoke haze or blanket, personnel must include all buildings. Failure to obscure tall buildings, towers, and steeples provides enemy observers with reference points for fire placement within the built-up area.

A-3. RIOT CONTROL AGENTS

Riot control agents, such as CS and CN, can be used to drive enemy troops from proposed positions or to deny them areas for occupation. Riot control agents are incapacitating but have no lasting effects. They are appropriate when preventing civilian casualties is a planning consideration. However, riot control agents are not effective against an enemy well trained in chemical defense.

APPENDIX B

BRADLEY FIGHTING VEHICLE

Bradley fighting vehicle platoons and squads seldom fight alone in built-up areas. They normally fight as part of their company or in a company team.

Section I. EMPLOYMENT

Fighting in built-up areas is centered around prepared positions in houses and buildings. Such positions cover street approaches and are protected by mines, obstacles, and booby traps. Therefore, bridges, overpasses, and buildings must be inspected and cleared of mines before they are used. Reconnaissance parties must ascertain the weight-supporting capacity of roads, bridges, and floors to determine if they can support the weight of BFVs and tanks.

B-1. TARGET ENGAGEMENT

Streets and alleys are ready-made firing lanes and killing zones. Because of this, all vehicle traffic is greatly restricted and canalized, and subject to ambush and short-range attack. Tanks are at a disadvantage because their main guns cannot be elevated enough to engage targets on the upper floors of tall buildings. The BFV, with +60 to -10 degrees elevation of the 25-mm gun and 7.62-mm coax machine gun, has a much greater ability in this role. With firing port weapons, the BFV can also place suppressive fire at ground level at the same time to the flanks and rear of enemy vehicles. A tank is restricted in its ability to provide this support.

B-2. BRADLEY FIGHTING VEHICLES AND TANKS

BFVs and tanks are not employed alone. Working as a team, dismounted infantrymen (the rifle team) provide security. In turn, the BFVs and tanks provide critical fire support for the rifle teams.

    a. When moving, BFVs should stay close to a building on either side of the street. This allows each BFV to cover the opposite side of the street. BFVs can button up for protection, but the BFV crew must remain alert for signals from dismounted infantry. Cooperation between the rifle team and the BFV team in a built-up environment is critical. Visual signals should be developed, telephones used where possible, and rehearsals and training conducted.

    b. Commanders should consider using the long-range fires of the tank's main armament from overwatch positions. The BFV, with its greater capability to depress and elevate the 25-mm gun, can provide some of the support previously derived from tanks within the built-up area.

    c. Because the BFV, while having better armor protection than the M113, lacks adequate armor protection to withstand medium to heavy ATGM fire, it is normally employed after the area has been cleared of ATGM positions or on terrain dominating the city to provide long-range antiarmor support. A great portion of the platoon's short-range antiarmor fires in built-up areas is provided by LAWs and Dragons. The BFVs 25-mm gun and machine gun are employed while providing direct fire support.

Section II. OFFENSE

Because of the nature of the terrain, fighting in built-up areas is usually conducted by dismounted troops. BFVs are employed as much as possible in close support of dismounted teams. Tanks follow and are brought to locations secured by leading infantry to provide heavy direct fire support. (Figure B-1 shows the organization of the BFV.)

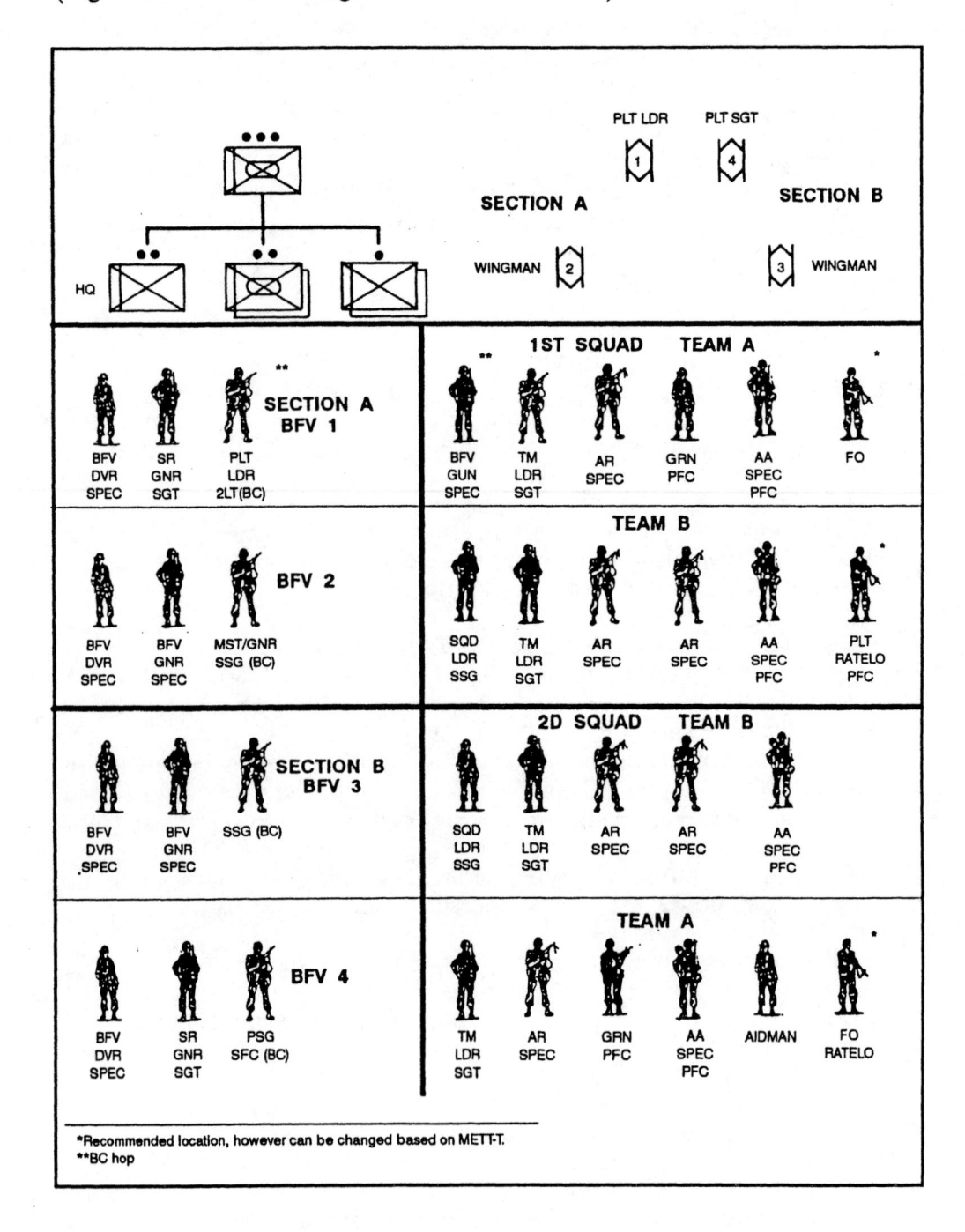

Figure B-1. Bradley fighting vehicle organization.

B-3. MISSIONS

Missions common to the rifle team and the BFV team are explained herein.

   a. The missions of the rifle team during attacks in built-up areas are:

- Assaulting and reducing enemy positions and clearing buildings under the covering fires of tanks and BFVs.
- Neutralizing and destroying enemy antiarmor weapons.
- Locating targets for engagement by tank or BFV weapons.
- Protecting tanks and BFVs against enemy individual antiarmor measures and surprise.
- Securing and defending an area after it is cleared.

   b. The missions of the BFV team during attacks in built-up areas are:

- Destroying enemy positions within a building with the direct fire of the 25-mm gun (using APDS ammunition) and the 7.62-mm coax machine gun (when the wall is constructed of light material).
- Suppressing enemy gunners within the objective building and adjacent structures. This is accomplished with the 25-mm gun (Figure B-2) and 7.62-mm coax machine gun.

Figure B-2. Suppression by 25-mm gun.

- Isolating the objective building with direct fire to prevent enemy withdrawal, reinforcement, or counterattack.
- Breaching walls en route to and in the objective structure. This is best accomplished with the 25-mm gun, using a spiral firing pattern (see Figure B-3).
- Establishing a roadblock or barricade.
- Securing cleared portions of the objective.
- Obscuring the enemy's observation using the BFV's smoke system.
- Evacuating casualties from areas of direct fire.

Figure B-3. Spiral firing pattern.

B-4. RIFLE TEAM

The BFV is best used to provide direct fire support to the rifle team. The BFV team should move behind the rifle team, when required, to engage targets located by the rifle team (Figure B-4). The dash speed (acceleration) of the BFV enables the team to rapidly cross streets, open areas, or alleys.

  a. The BFV teams provide fire with their 25-mm gun and 7.62-mm coax machine gun for the rifle teams on the opposite side of the street. The 25-mm gun is the most effective weapon on BFVs while fighting in built-up terrain (Figure B-5).

  b. The use of the 25-mm gun in support of rifle teams requires safety considerations.

Figure B-4. Cover for rifle teams.

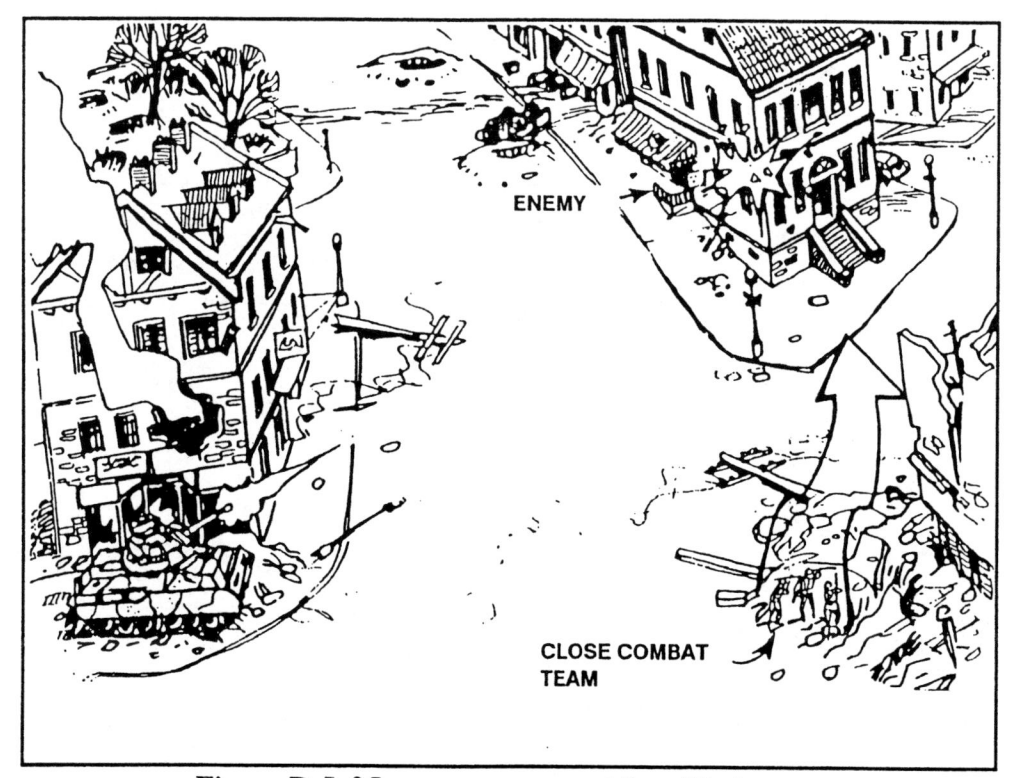

Figure B-5. 25-mm gun support for rifle teams.

- High-explosive 25-mm rounds arm 10 meters from the gun and explode on contact.
- APDS rounds discard their plastic sabots to the front of the gun when fired. This requires a 100-meter safety fan to the front of 25-mm gun (Figure B-6).

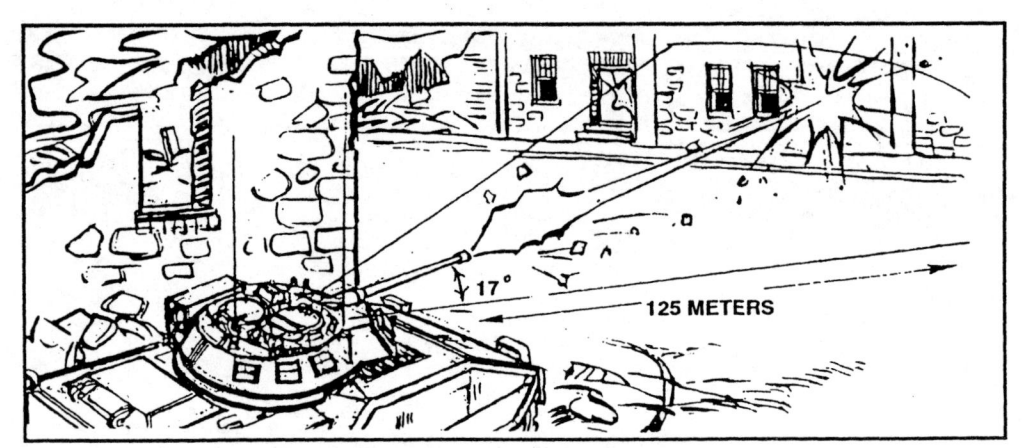

Figure B-6. Safety fan for 25-mm gun.

c. The BFVs' engine exhaust smoke system can be used in built-up areas to cover the movement of the rifle team. The BFV can also provide a smoke screen by using its smoke grenade launchers. This requires careful analysis of wind conditions to ensure the enemy, and not friendly units, is affected by the smoke. This is a difficult task since wind currents tend to be erratic between buildings. The smoke can also screen the movements of the BFVs once the rifle teams cross the danger area (Figure B-7).

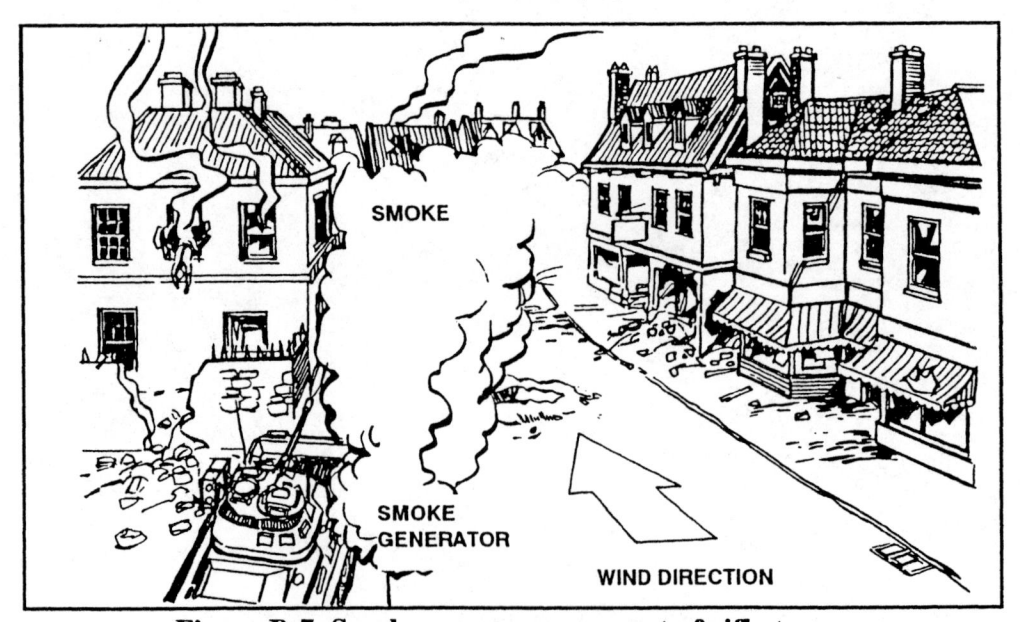

Figure B-7. Smoke screens movement of rifle team.

B-5. COMMUNICATIONS

Communication between the rifle team and BFV crew is critical. These communications can be visual or voice signals, radio, or telephone.

B-6. CLEARING OF A BUILDING

The most common mission of the platoon in offensive operations is to attack and clear a building or a group of buildings. The platoon leader first designates the composition of the rifle teams and BFV teams. The composition of those teams changes with the situation. Normally, the BFV teams and the men needed for security compose the BFV element. Each squad rifle team is organized into two- or three-man assault parties. The platoon leader can designate a part of the platoon as a demolition team. The entire platoon rifle team is normally used to attack one building at a time. In smaller buildings, the platoon leader could have a single rifle team conduct a three-step attack:

STEP 1: The fighting vehicle element supported by indirect and direct fire isolates the building.

STEP 2: The rifle team enters the building to seize a foothold.

STEP 3: The rifle team clears the building room by room.

a. To isolate the building, the BFV element takes an overwatch position (Figure B-8). It fires the 25-mm gun and 7.62-mm coax machine gun, and adjusts indirect fire to suppress enemy troops in the building and in nearby buildings who can fire at the dismounted element.

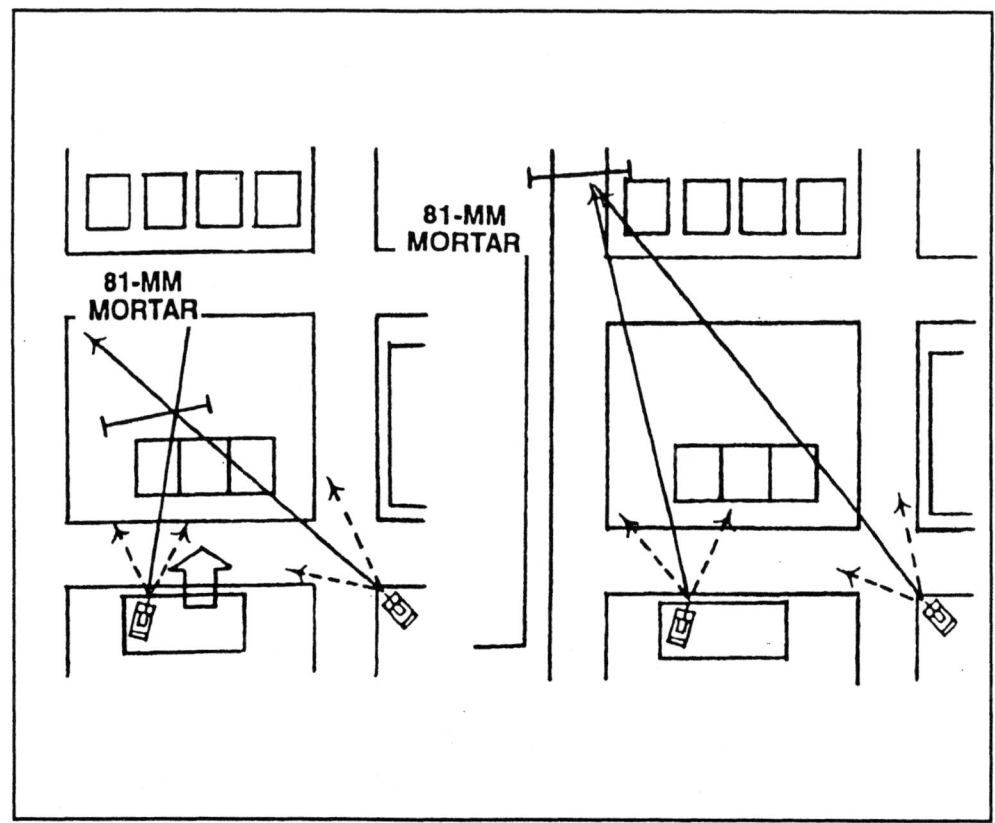

Figure B-8. Isolation of a building and shifting of fires.

b. The rifle teams move to the building on covered and concealed routes. Smoke grenades, smoke pots, and the smoke system of the BFV can provide added concealment. The rifle teams enter the building at the highest point they can because—
- The ground floor and basement are usually the enemy's strongest defenses.
- The roof of a building is normally weaker than the walls.
- It is easier to fight down stairs than up stairs.

c. If there is no covered route to the roof, the rifle teams could encounter the enemy at a lower story or at ground level. They should seize a foothold quickly, fight to the highest story, and then clear room by room, floor by floor, from the top down.

Section III. DEFENSE

Most defensive fighting is performed by the rifle teams. It is harder to build the platoon's defense around the BFV in cities than in other types of terrain, but the BFV element's role is still important. A platoon normally defends from positions in one to three buildings, depending on the size and strength of the buildings, the size of the platoon, and the disposition of the buildings.

B-7. MISSIONS

Defensive missions of the rifle teams and BFV teams are discussed herein.

a. The following are typical missions of rifle teams in the defense:
- Preparing defensive positions.
- Providing observation and security to prevent enemy infiltration.
- Engaging and defeating assaulting enemy forces.
- Acquiring targets for engagement by tanks and BFV weapons.
- Protecting tanks and BFVs from close antiarm or weapons.
- Emplacing demolitions and obstacles (supported by combat engineers).

b. The following are typical missions of BFV teams in the defense:
- Providing fire support for the rifle teams and mutual support to other BFV teams.
- Destroying enemy armored vehicles and direct fire artillery pieces.
- Neutralizing or suppressing enemy positions with 25-mm gun and 7.62-mm coaxial machine gun fire in support of local counterattacks.
- Destroying or making enemy footholds untenable by fire using the 25-mm gun.
- Providing rapid, protected transport for the rifle teams.

- Reinforcing threatened areas by movement through covered and concealed routes to new firing positions.
- Providing mutual support to other antiarmor fires.
- Providing resupply of ammunition and other supplies to the dismount teams.
- Evacuating casualties from the area of direct fire.

NOTE: In the last two missions, the overall value of the BFV to the defense must be weighed against the need to resupply or to evacuate casualties.

B-8. DEVELOPMENT OF THE DEFENSE

The platoon leader must consider the following when developing his defense:

   a. **Dispersion.** Defensive positions in **two** mutually supporting buildings is better than having positions in **one** building that can be bypassed.

   b. **Fields of Fire.** Positions should have good fields of fire in all directions. Broad streets and open areas, such as parks, offer excellent fields of fire.

   c. **Observation.** The buildings selected should permit observation into the adjacent sector. The higher stories offer the best observation but also attract enemy fire.

   d. **Concealment.** City buildings provide excellent concealment. Obvious positions, especially at the edge of a built-up area, should be avoided since they are the most likely to receive the heaviest enemy fire.

   e. **Covered Routes.** These are used for movement and resupply, and are best when they go through or behind buildings.

   f. **Fire Hazard.** Buildings that burn easily should be avoided.

   g. **Time.** Buildings that need extensive preparations are undesirable when time is a factor.

   h. **Strength.** Buildings in which BFVs or tanks are to be placed must withstand the weight of the vehicles and the effects of their weapon systems.

B-9. FIRE PLAN POSITIONS

The BFV should be integrated into the platoon fire plan. The 25-mm gun and 7.62-mm coax machine gun fields of fire should cover streets and open areas. Once placed in position, BFVs should not be moved for logistical or administrative functions. Other vehicles should accomplish these functions, when possible.

   a. Once the platoon leader chooses the building(s) he will defend, he positions his BFV teams and rifle teams. BFVs and dismounted machine guns should be positioned to have grazing fire. Dragons should be positioned on upper stories for longer range and to permit firing at the tops of tanks. Squads should be assigned primary and, if feasible, supplementary and alternate positions for their rifle teams and BFV teams. These positions should permit continuous coverage of the primary sectors and all-round defense.

   b. None of the platoon's antiarmor weapons can be fired from unvented or enclosed rooms. However, the TOW can be fired from any room that a BFV can be placed in, as long as all hatches are closed and there are no dismounted troops in the room.

c. Due to the close engagement ranges on urban terrain, the 25-mm gun and 7.62-mm coaxial machine gun are used more than the TOW. The antiarmor capability of the BFV is degraded by short ranges and must be supplemented by Dragons and LAWs (Figure B-9). Dragon and LAW positions should be placed where they can support the BFV but must not attract enemy attention to the BFV location. Dragons and LAWs are much more effective against the flanks, rear, and tops of threat armored vehicles and should be positioned to attack those areas. The TOWs should be conserved and employed against threat tanks as their primary target.

Figure B-9. Dragon position supporting BFV.

B-10. BRADLEY-EQUIPPED ECHO COMPANY

The Chief of Staff of the Army approved the replacement of the M901 ITV with the BFV in mechanized infantry battalions. While this change provides a much greater improvement in mobility, survivability, and firepower over the M901, the basic mission of the Echo Company remains the same. (For more information on the employment of the Bradley-equipped Echo Company [BFV E CO], see FM 7-91 and Change 1 to FM 71-2.)

a. **Considerations.**

(1) Due to the lack of a dismounted element, the Bradley-equipped Echo Company must rely on attached and or supporting dismounted infantry to provide local security.

(2) It should be employed at the very least in sections or pairs (the wingman concept). This provides for some degree of mutual support.

(3) The Bradley-equipped Echo Company consumes slightly more fuel than a M901 ITV company. This does not present much of a problem since

the additional fuel requirements easily fall within the battalion's organic fuel hauling capacity.

b. Offensive Employment.

(1) The organization of the Bradley-equipped Echo Company makes it a likely choice to provide a base of fire for the battalion task force in the attack. The battalion commander can assign support by fire or attack by fire missions.

(2) It can conduct a guard or screen mission much more effectively than a M901-equipped company since the Bradley-equipped Echo Company has better mobility and firepower.

c. Defensive Employment.

(1) The vehicles should have multiple firing positions. The Bradley-equipped Echo Company commander can control the massing of long-range antiarmor fires into the battalion's engagement area(s) by firing from several directions at once. This has two major effects: the enemy's ability to maneuver his mechanized infantry and armored forces will be destroyed and or degraded; and the survivability of the Bradley-equipped Echo Company's vehicles and crews will be greatly enhanced.

(2) The Bradley-equipped Echo Company, teamed with an armor counterattack unit, can suppress or destroy enemy ATGMs so the armor unit may decisively maneuver.

(3) As with offensive operations, the Bradley-equipped Echo Company can conduct guard and screening operations.

(4) The Bradley-equipped Echo Company can also conduct counter-reconnaissance operations to deprive the enemy of information about the friendly forces' disposition and composition.

APPENDIX C

OBSTACLES, MINES, AND DEMOLITIONS

Obstacles and mines are used extensively in combat in built-up areas to allow the defender to canalize the enemy, impede his movement, and disrupt his attack.

Section I. OBSTACLES

Obstacles are designed to prevent movement by personnel, to separate infantry from tanks, and to slow or stop vehicles.

C-1. TYPES

Antipersonnel mines, barbed wire, booby traps, and exploding flame devices are used to construct antipersonnel obstacles (Figure C-1, page C-2). (See FM 5-25 for more detailed information.) These obstacles are used to block the following infantry approaches:

- Streets.
- Buildings.
- Roofs.
- Open spaces.
- Dead space.
- Underground systems.

   a. The approval authority to booby trap buildings is the Corps commander; however, this authority may be delegated to brigade level. (See FM 20-32 for more information.)

   b. The three types of obstacles used in defensive operations are protective, tactical, and supplementary.

      (1) Protective obstacles are usually located beyond hand-grenade range (40 to 100 meters) from the defensive position.

      (2) Tactical obstacles are positioned to increase the effectiveness of the friendly weapons fire. The tactical wire is usually positioned on the friendly side of the machine gun FPLs.

      (3) Supporting obstacles are used to break up the pattern of tactical wire to prevent the enemy from locating friendly weapons.

   c. Dead space obstacles are designed and built to restrict infantry movement in areas that cannot be observed and are protected from direct fires.

   d. Antiarmor obstacles are restricted to streets (Figure C-2, page C-4).

FM 90-10-1

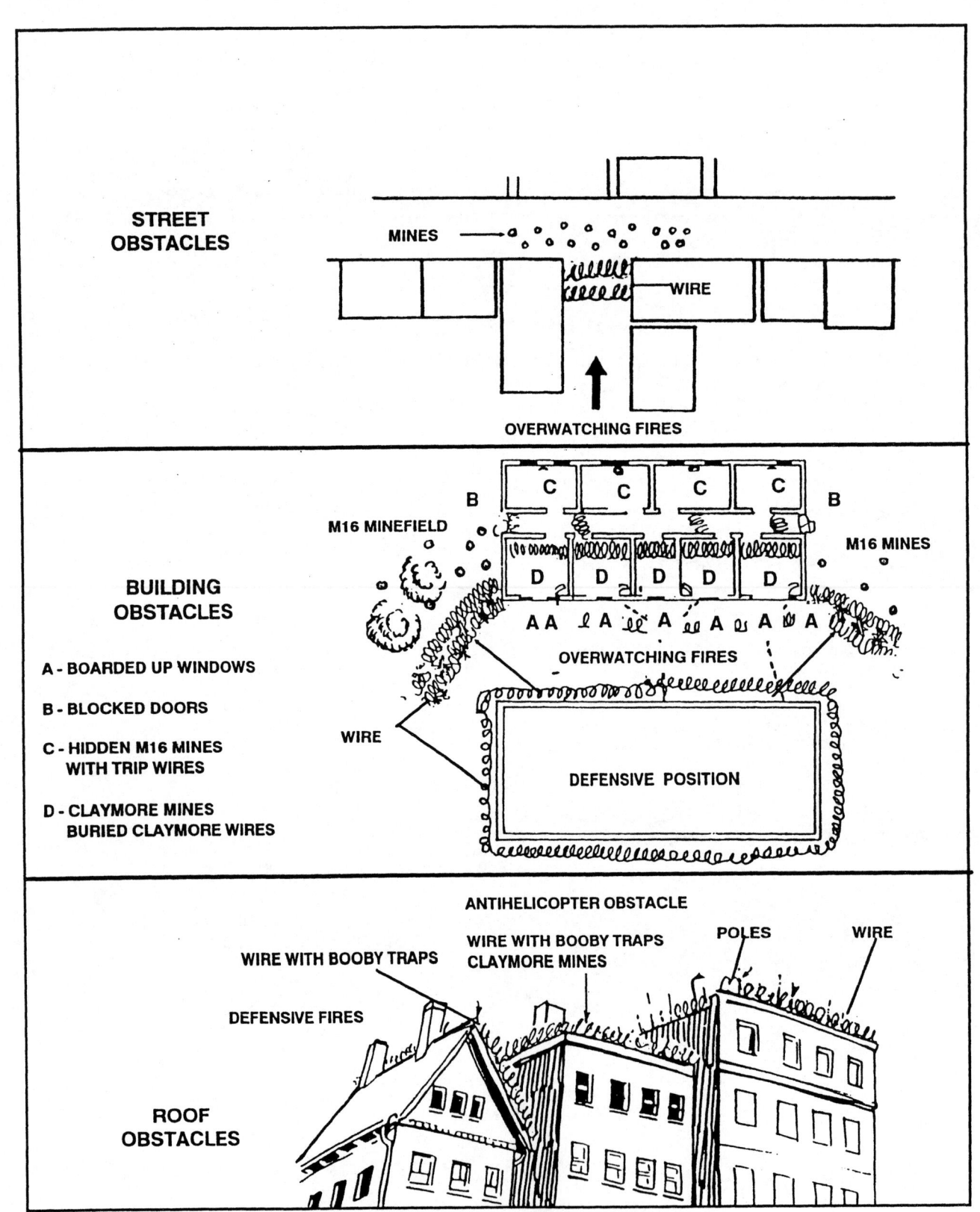

STREET OBSTACLES

BUILDING OBSTACLES

A - BOARDED UP WINDOWS
B - BLOCKED DOORS
C - HIDDEN M16 MINES WITH TRIP WIRES
D - CLAYMORE MINES BURIED CLAYMORE WIRES

ROOF OBSTACLES

Figure C-1. Antipersonnel obstacles.

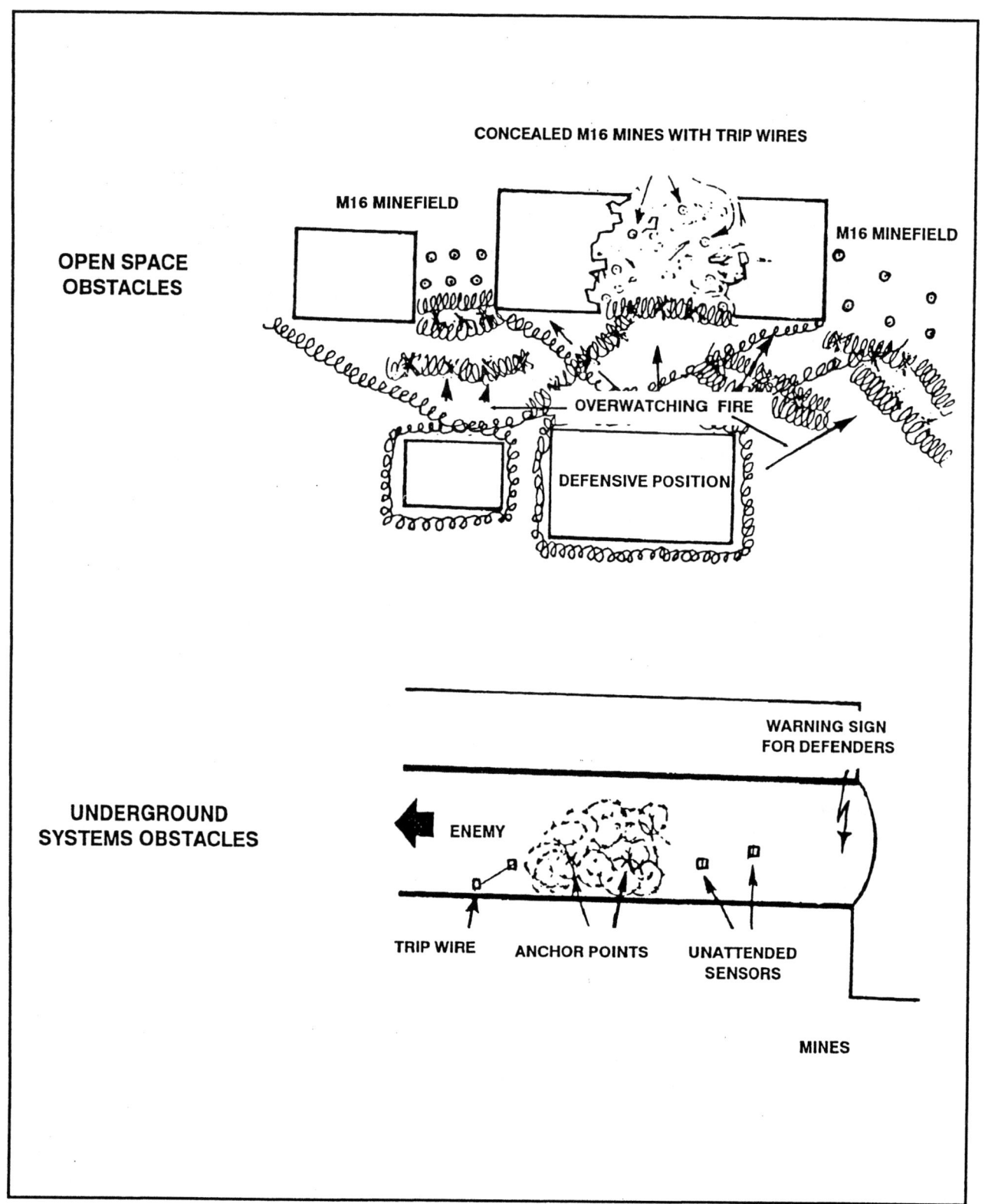

Figure C-1. Antipersonnel obstacles (continued).

FM 90-10-1

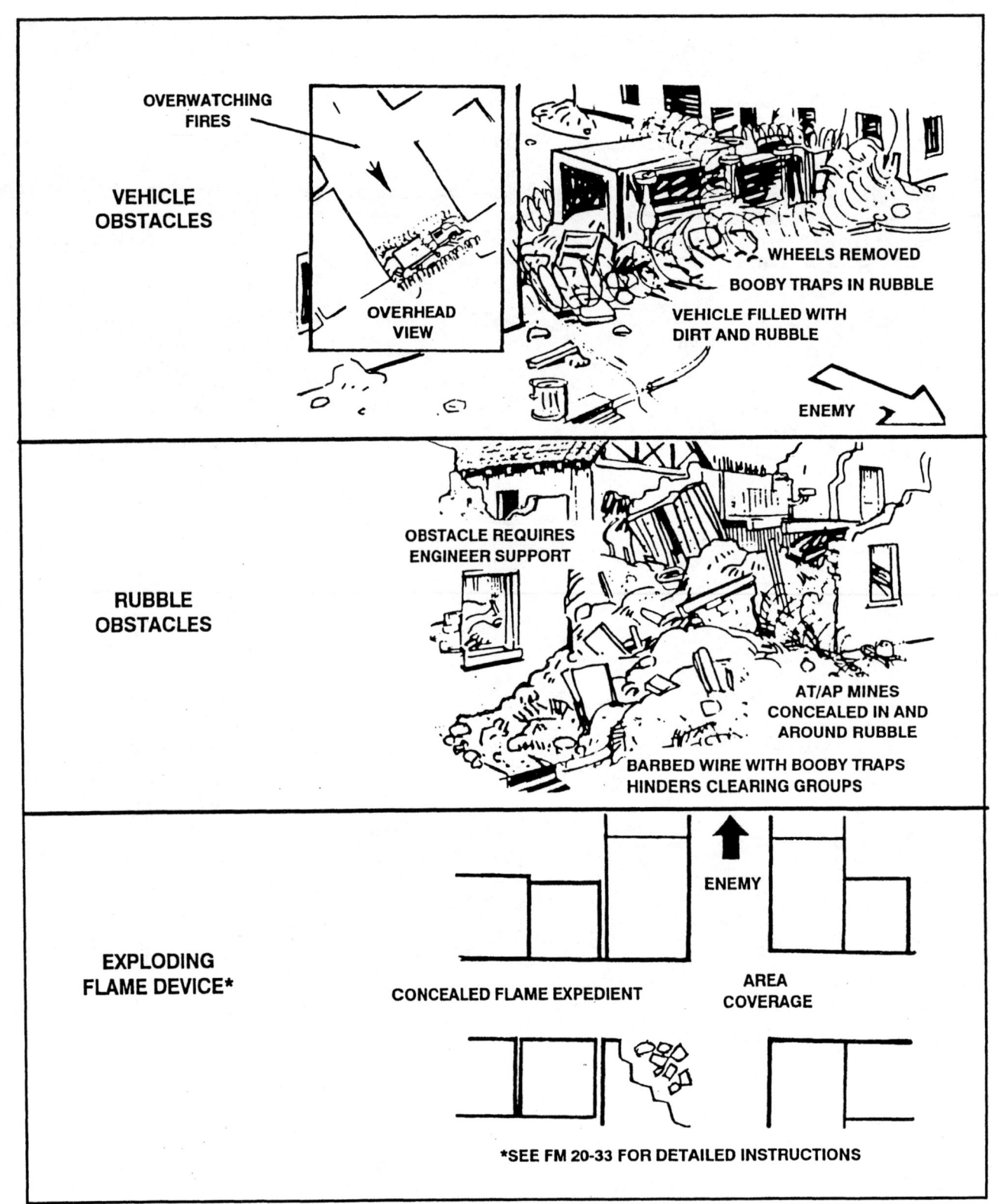

Figure C-2. Antiarmor obstacles.

STEEL HEDGEHOG

TIE IN WITH BUILDING OR OTHER OBSTACLES

CRATERED ROAD

SIZE OF CRATER MUST BE LARGE ENOUGH TO PREVENT BYPASS

AT/AP MINES ARE CONCEALED IN RUBBLE AROUND CRATER

THIS OBSTACLE WILL REQUIRE ENGINEER ASSISTANCE

CONCEALED EXPLOSIVES

REQUIRES ENGINEER SUPPORT

EXPLOSIVE CONCEALED IN BUILDING

EXPLOSIVE CONCEALED IN SEWER

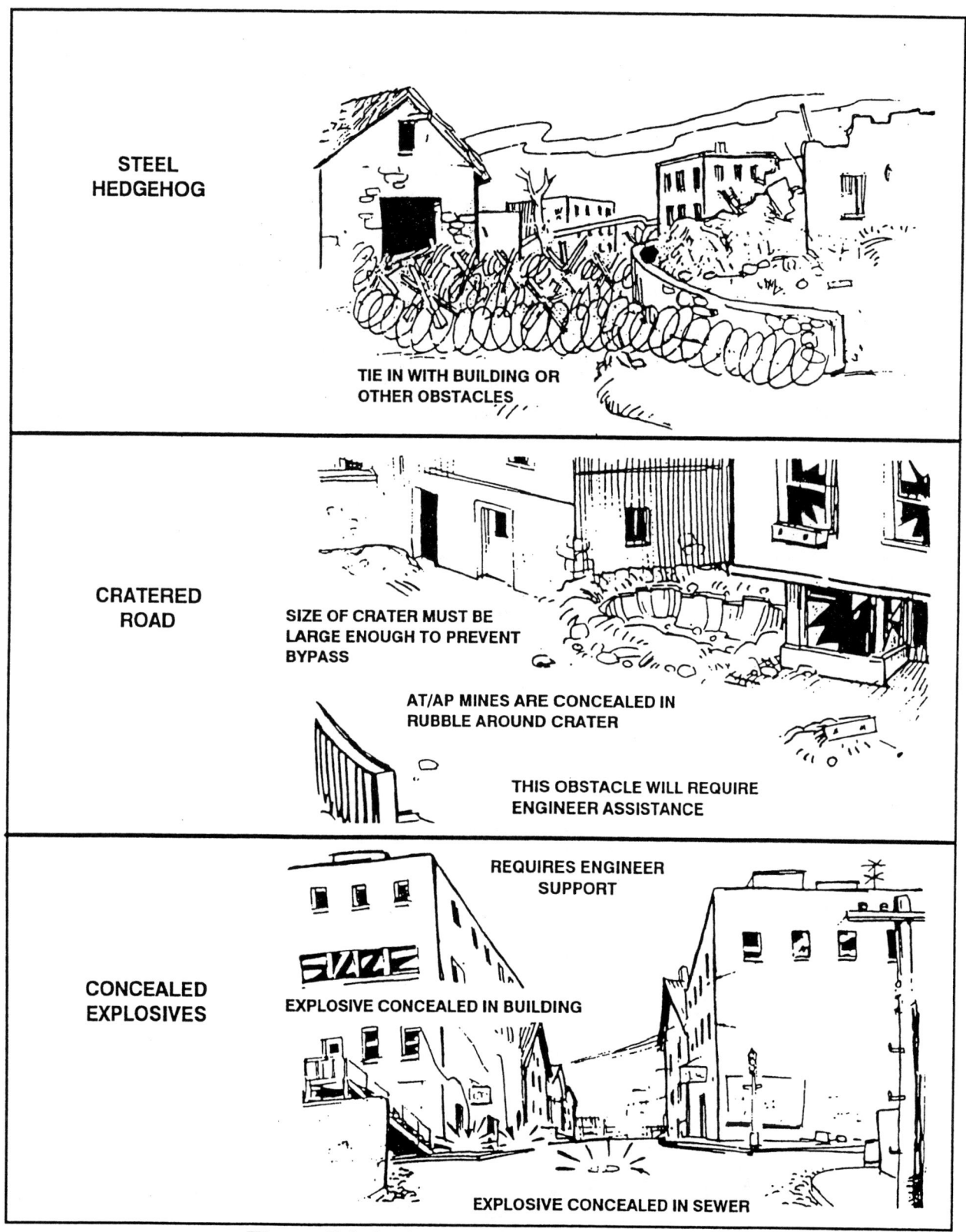

Figure C-2. Antiarmor obstacles (continued).

C-2. CONSTRUCTION OF OBSTACLES

Obstacles are constructed in buildings to deny enemy infantry covered routes and weapons positions close to friendly defensive positions. They can be constructed by rubbling with explosives or flame, by using wire, or by using booby traps within buildings. The building can be prepared as an explosive or flame trap for execution after enemy occupation.

Section II. MINES

Mines in built-up areas should be recorded on a building sketch (Figure C-3) and on a DA Form 1355 or DA Form 1355-1-R. (See FM 20-32 for instructions on how to complete these forms.) The sketch should include the number of the building (taken from a city map) and all floor plans. It should also include the type of mine and firing device. When possible, mined buildings should be marked on the friendly side (Figure C-4). Clearing areas or buildings that have been mined is extremely difficult. Therefore, they should be considered "NO GO" areas. This factor must be carefully considered when planning and authorizing the placement of mines. (See Table C-1 for the approving authority for minefields.)

TYPE MINEFIELD	APPROVING AUTHORITY
Protective hasty	Brigade commander (may be delegated down to battalion or company level on a mission basis).
Deliberate	Division or installation commander.
Tactical	Division commander (may be delegated to brigade level).
Point	Brigade commander (may be delegated to battalion level).
Interdiction	Corps commander (may be delegated to division level).
Phony	Corps commander (may be delegated to division level).
Scatterable long duration (24 hours or more)	Corps commander (may be delegated to division or brigade level).
Short duration (less than 24 hours)	Corps commander (may be delegated to division, brigade, or battalion level).

Table C-1. Minefield employment authority.

FM 90-10-1

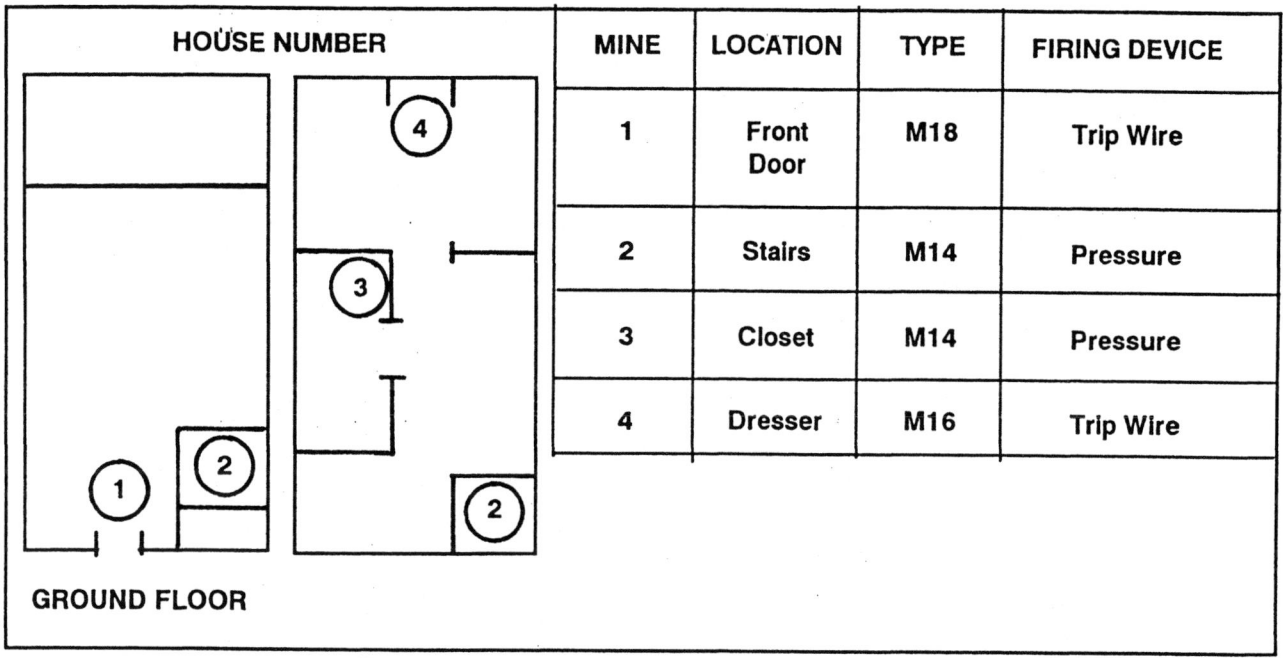

HOUSE NUMBER	MINE	LOCATION	TYPE	FIRING DEVICE
	1	Front Door	M18	Trip Wire
	2	Stairs	M14	Pressure
	3	Closet	M14	Pressure
	4	Dresser	M16	Trip Wire

Figure C-3. Building sketch showing mines.

Figure C-4. Marking of mined buildings.

C-3. TYPES
Several types of mines can be employed in built-up areas.

a. The M14 mine should be used with metallic antipersonnel, antitank, or chemical mines to confuse and hinder enemy breaching attempts. It must be carefully employed because its light weight makes it easy to displace (Figure C-5, page C-8). However, its size makes it ideal for obscure places such as stairs and cellars.

b. The M16 mine is ideal for covering large areas such as rooftops, backyards, parks, and cellars. It should be expediently rigged for command detonation by attaching a rope or piece of communications wire to the release pin ring (Figure C-6, page C-8).

C-7

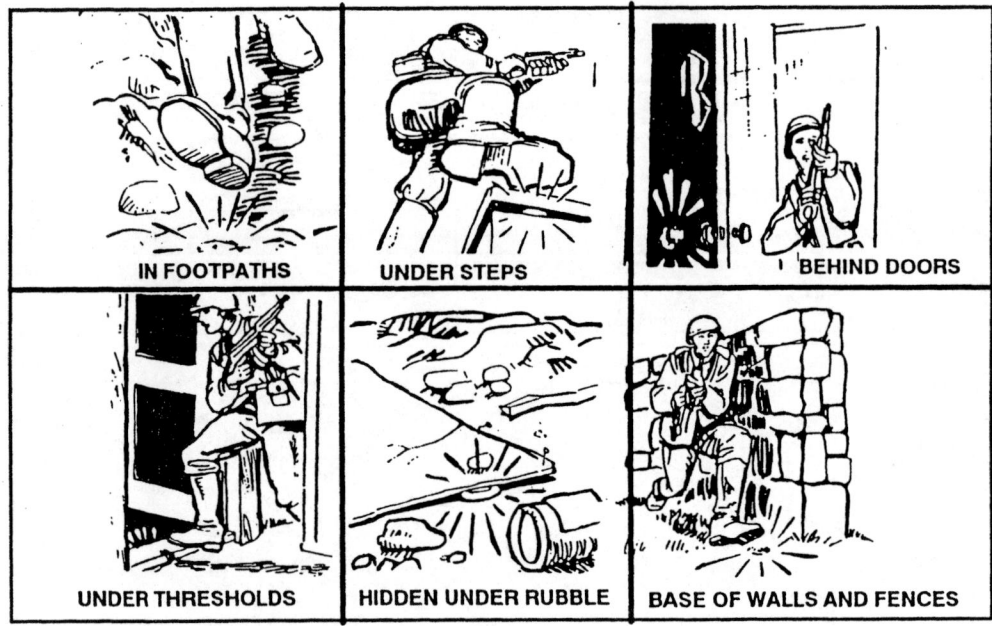

Figure C-5. Emplacing the M14 antipersonnel mine.

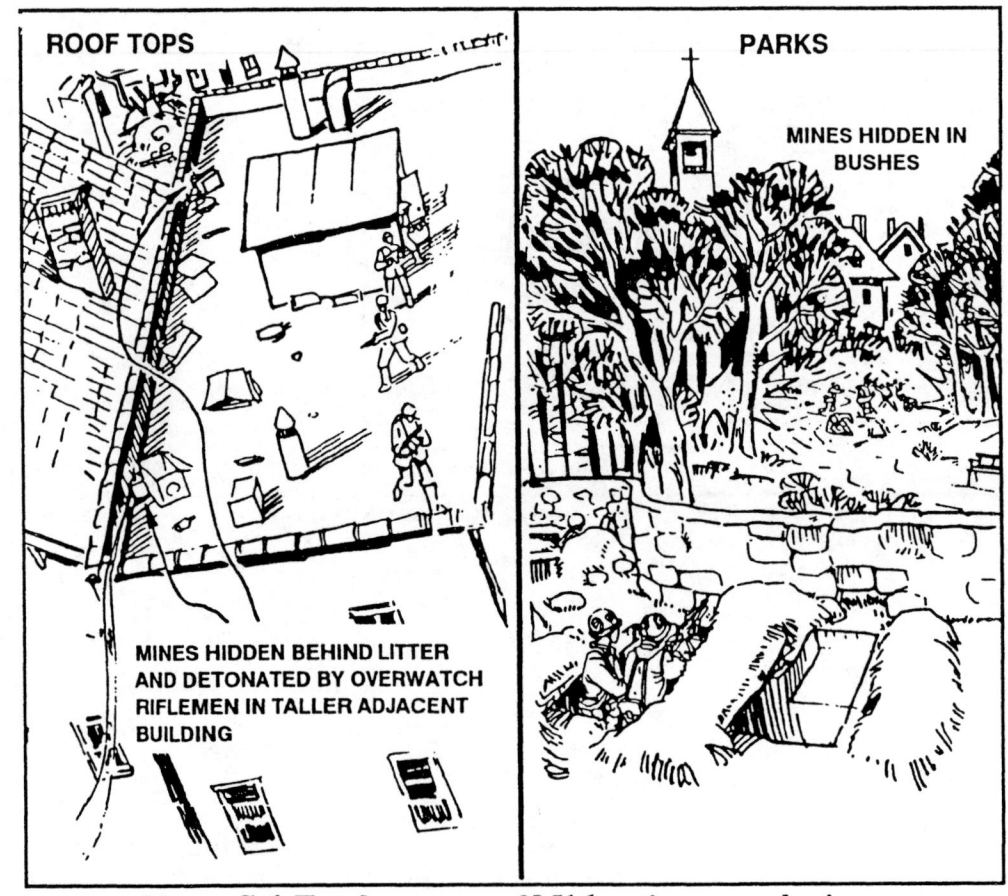

Figure C-6. Emplacements of M16 antipersonnel mines.

c. The M18A1 Claymore mine can be employed during the reorganization and consolidation phase on likely enemy avenues of approach. It does not have to be installed in the street but can be employed on the sides of buildings or any other sturdy structure.

(1) Claymore mines can be used for demolition against thin-skinned buildings and walls, or the 1 1/2 pounds of composition C4 can be removed from the mine and used as an explosive, if authorized.

(2) Claymore mines arranged for detonation by trip wire can be mixed with antipersonnel and antitank mines in nuisance minefields. They can fill the dead space in the final protective fires of automatic weapons (Figure C-7).

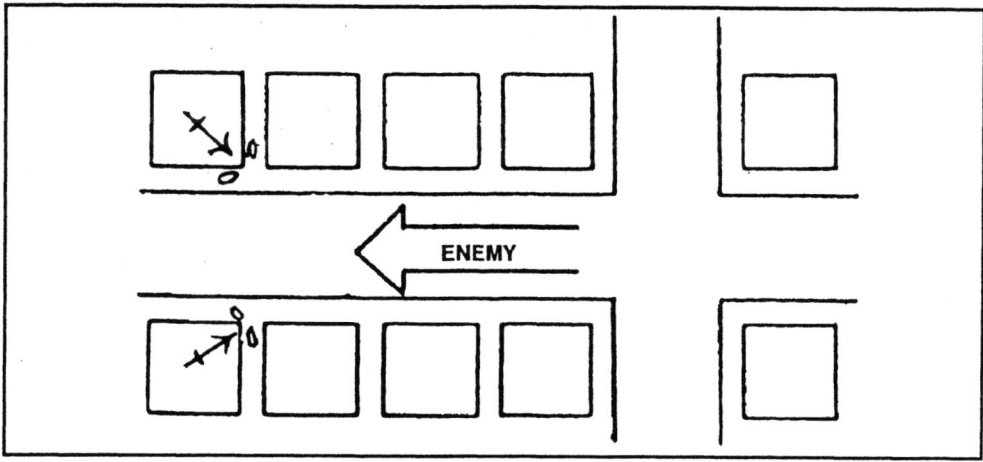

Figure C-7. Claymore mine used to cover dead space of automatic weapons.

(3) Claymore mines can be used several ways in the offense. For example, if friendly troops are advancing on a city, Claymore mines can be used in conjunction with blocking positions to cut off enemy avenues of escape (Figure C-8).

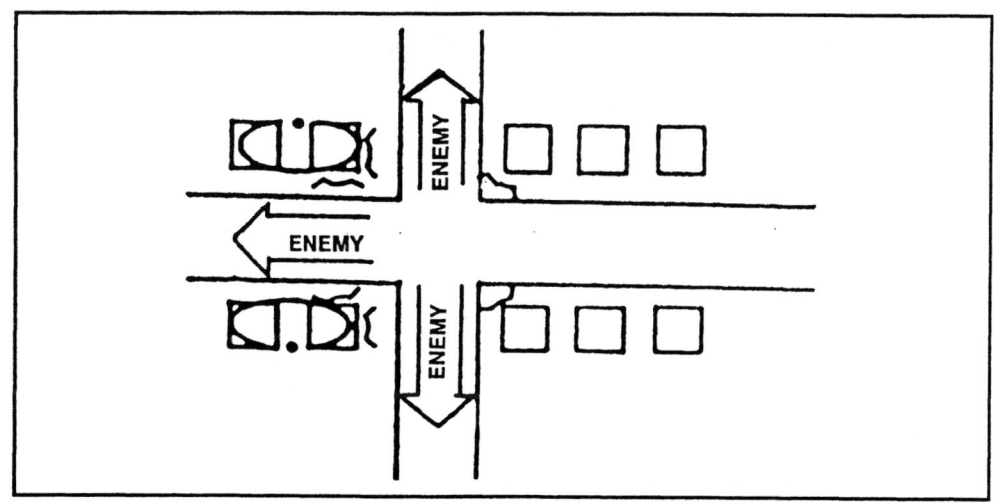

Figure C-8. Claymore mines used to block enemy avenues of escape.

d. The M15, M19, and M21 antitank mines are employed (Figure C-9)—
- In conjunction with other man-made obstacles and covered with fire.
- As standard minefields in large open areas with the aid of the M57 dispenser.
- In streets or alleys to block routes of advance in narrow defiles.
- As command detonated mines with other demolitions.

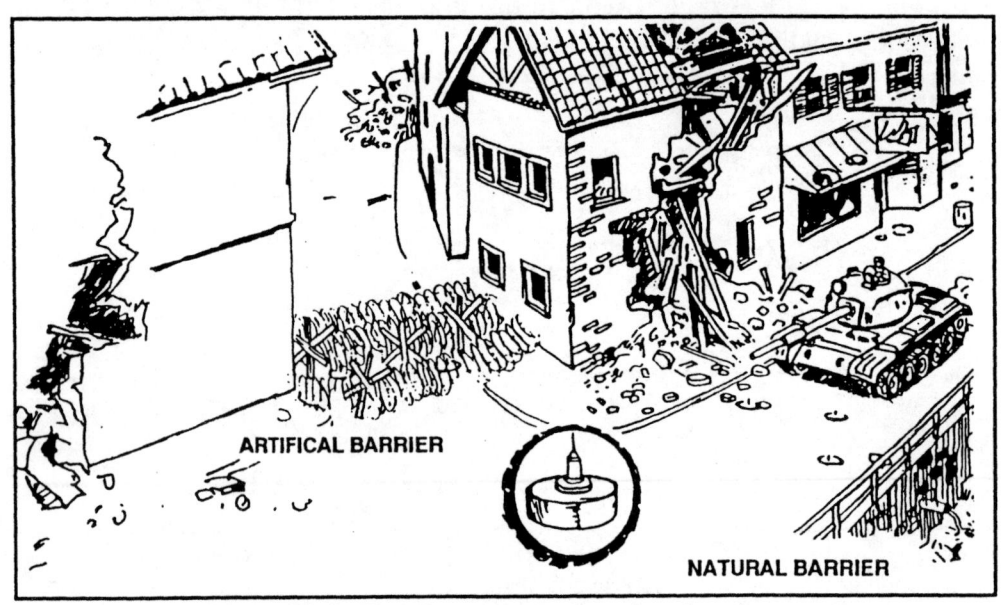

Figure C-9. Emplacement of antitank mines.

C-4. ENEMY MINES AND BOOBY TRAPS

Buildings contain many areas and items that are potential hiding places for booby traps—for example, doors, windows, telephones, stairs, books, canteens, and so on.

When moving through a building, soldiers must not pick up anything—food, valuables, enemy weapons, and so on. Such items could be rigged with pressure devices that explode when they are moved. Soldiers must be well dispersed so that if a booby trap explodes, the number of casualties will be few. Many different types of mines and booby traps could be encountered during combat in built-up areas (Figure C-10).

a. The equipment used in clearing operations is—
- Mine detectors.
- Probes.
- Grappling hooks.
- Ropes.
- C4 explosives.
- Flak vests.
- Eye protection.
- Engineer tape.

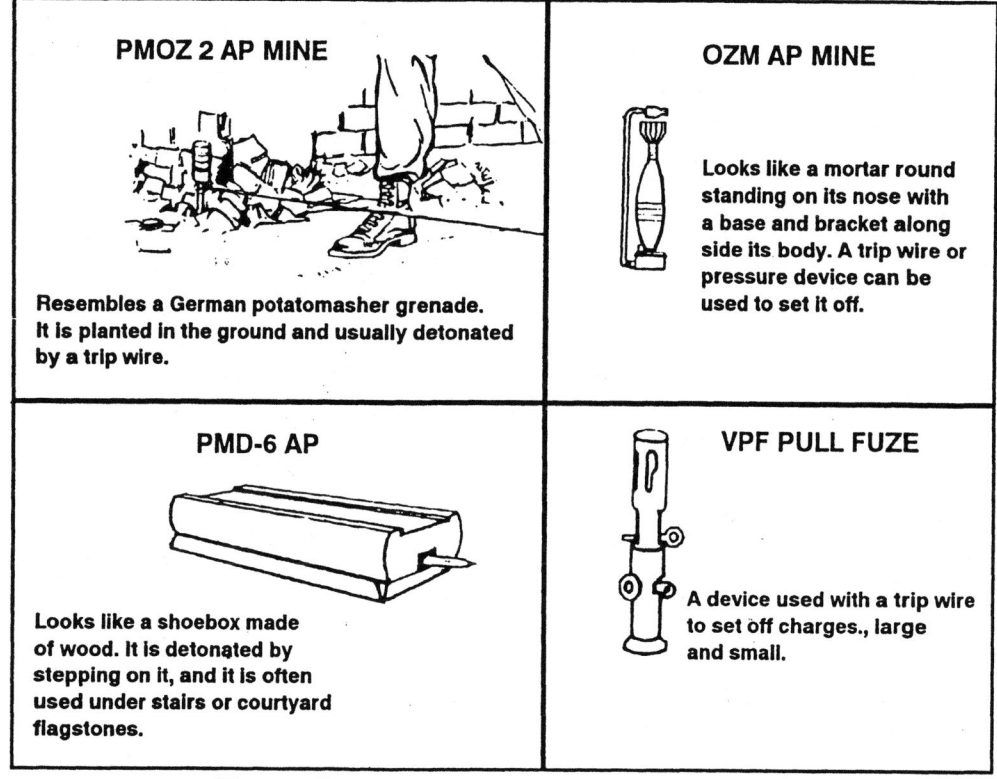

PMOZ 2 AP MINE	OZM AP MINE
Resembles a German potatomasher grenade. It is planted in the ground and usually detonated by a trip wire.	Looks like a mortar round standing on its nose with a base and bracket along side its body. A trip wire or pressure device can be used to set it off.
PMD-6 AP	**VPF PULL FUZE**
Looks like a shoebox made of wood. It is detonated by stepping on it, and it is often used under stairs or courtyard flagstones.	A device used with a trip wire to set off charges., large and small.

Figure C-10. Threat mines and booby traps.

    b. If available, scout dogs should be used to "alert" soldiers to trip wires or mines.

    c. To detect trip wires, soldiers can use a 10-foot pole with 5 feet of string tied on one end. He attaches a weight to the loose end of the string, which snaps on the trip wire. This allows the lead man to easily detect a trip wire (Figure C-11).

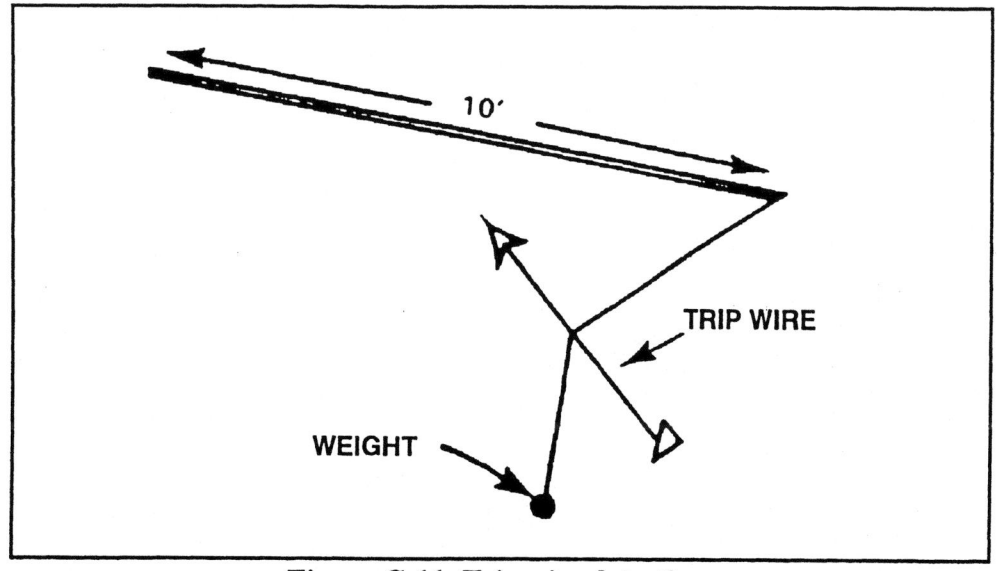

Figure C-11. Trip wire detection.

d. Many standard antipersonnel mines are packed in boxes and crates. If a soldier discovers explosive storage boxes, he should sketch them and turn the sketch over to the platoon leader or S2.

e. Most booby traps should be neutralized by explosive ordnance disposal (EOD) personnel. If EOD teams are not available, booby traps can be blown in place. Personnel should be protected by adequate cover. If the booby trap is in a building, all personnel should go outside before the booby trap is destroyed. Engineer tape placed around the danger area is one method of marking booby traps. If tape is not available, strips ripped from bedsheets can be used. If possible, a guide should lead personnel through booby-trapped areas. Prisoners and civilians can be a good source of information on where and how booby traps are employed. Figure C-10 shows some of the types of Threat mines and booby traps that could be encountered.

Section III. DEMOLITIONS

Demolitions are used more often during combat in built-up areas than during operations in open natural terrain. Demolition operations should be enforced by the engineers that support the brigade, battalion task force, and company team. However, if engineers are involved in the preparation and execution of the barrier plan, infantrymen can prepare mouseholes, breach walls, and rubble buildings themselves, assisted and advised by the brigade, task force, or team engineer.

C-5. OFFENSIVE USE

When assaulting or clearing a built-up area, demolitions enable the maneuver commander to create an avenue of approach through buildings. As discussed earlier in the text, the infantry commander forms his personnel into assault teams and overwatch teams for seizing and clearing buildings.

a. Every other man in an assault team should carry demolitions, and other selected personnel should carry blasting caps. In a fire team, one man should carry the demolitions. The same man should not carry both the explosives and the blasting caps. As the demolitions are expended by the assault teams, they should be replaced by explosives carried by the overwatch force.

b. One of the most difficult breaching operations faced by the assault team is the breaching of masonry and reinforced concrete walls. When demolitions must be used, composition C4 is the ideal charge to use. Normally, building walls are 15 inches thick or less. Assuming that all outer walls are constructed of reinforced concrete, a rule of thumb for breaching is to place 10 pounds of C4 against the target between waist and chest height. When detonated, this normally blows a hole large enough for a man to go through. The amounts of TNT required to breach concrete are shown in Figure C-12.

c. However, metal reinforcing rods cannot be cut by this charge. Once exposed, they can be removed by using saddle or diamond charges on the rods. Hand grenades should be thrown into the opening to clear the area of enemy (see FM 5-25, Chapter 3).

REINFORCED CONCRETE		
THICKNESS OF MATERIAL	TNT	SIZE OF OPENING
Up to 10 CM (4 inches)	5 KG (11 lbs)	10 to 15 CM (4 to 6 inches)
10 to 15 CM (4 to 6 inches)	10 KG (22 lbs)	15 to 25 CM (6 to 10 inches)
15 to 20 CM (6 to 8 inches)	20 KG (44 lbs)	20 to 30 CM (8 to 12 inches)
NONREINFORCED CONCRETE MASONRY		
THICKNESS OF MATERIAL	TNT	SIZE OF OPENING
Up to 35 CM (14 inches)	1 KG (2.2 lbs)	35 CM (14 inches)
35 to 45 CM (14 to 18 inches)	2 KG (4.4 lbs)	45 CM (18 inches)
45 to 50 CM (18 to 20 inches)	3 KG (6.6 lbs)	50 CM (20 inches)

Figure C-12. TNT required to breach concrete.

d. Mouseholes provide the safest method of moving between rooms and floors. They can be created with C4. Since C4 comes packaged with an adhesive backing or can be emplaced using pressure-sensitive tape, it is ideal for this purpose. When using C4 to blow a mousehole in a lath and plaster wall, one block or a strip of blocks should be placed on the wall from neck-to-knee height. Charges should be primed with detonating cord or electrical blasting caps to obtain simultaneous detonation, which will blow a hole large enough for a man to fit through.

C-6. DEFENSIVE USE

The use of demolitions in defensive operations is the same as in offensive operations. When defending a built-up area, demolitions are used to create covered and concealed routes through walls and buildings that can be used for withdrawals, reinforcements, or counterattacks. Demolitions are also used to create obstacles and clear fields of fire.

a. Infantrymen use demolitions for creating mouseholes and constructing command-detonated mines. Expedient C4 satchel charges can be concealed in likely enemy weapons, in firing positions, or on movement routes. Expedient-shaped charges (effective against lightly armored vehicles) can also be emplaced on routes of mounted movement when integrated into antiarmor ambushes.

b. The engineers must furnish technical assistance for selective rubbling. Normally, buildings can be rubbled by using shaped charges or C4 on the supports and major beams of buildings.

c. Charges should be placed directly against the surface that is to be breached (Figure C-12), unless a shaped charge is used. Whenever possible, demolitions should be tamped to increase their effectiveness. Tamping materials could be sandbags, rubble, or desks and chairs (Figure C-13).

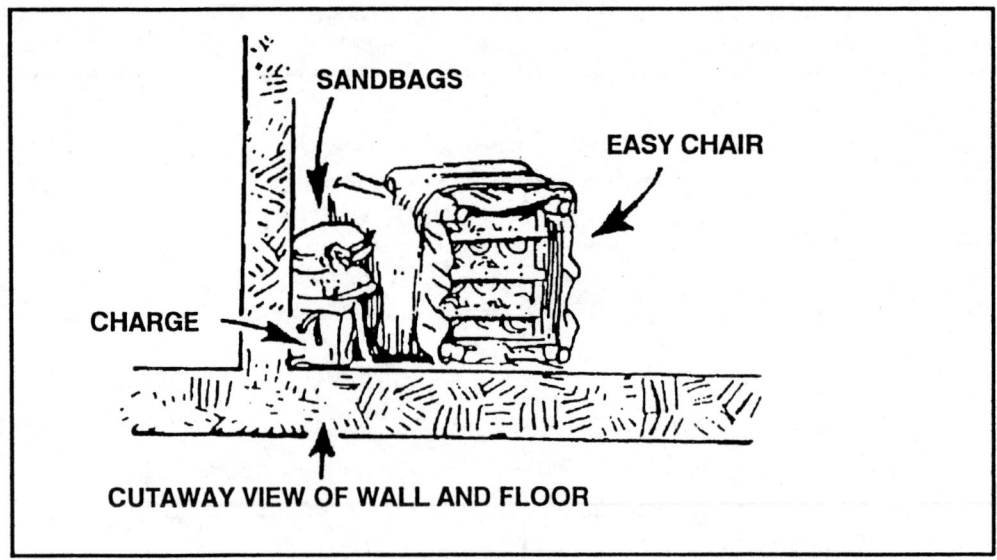

Figure C-13. Chair used to tamp breaching charge.

d. For most exterior walls, tamping of breaching charges could be impossible due to enemy fire. Thus, the untamped charge requires twice the explosive charge to produce the same effect as an elevated charge (Figure C-14).

THICKNESS OF CONCRETE	METHODS OF PLACEMENT			
FEET	POUNDS OF TNT	POUNDS OF C4	POUNDS OF TNT	POUNDS OF C4
2	14	11	28	21
2 1/2	27	21	54	41
3	39	30	78	59
3 1/2	62	47	124	93
4	93	70	185	138
4 1/2	132	99	263	196
5	147	106	284	211
5 1/2	189	141	376	282
6	245	186	490	366

Figure C-14. Breaching reinforced concrete.

e. When enemy fire prevents an approach to the wall, the breaching charge must be attached to a pole and slid into position for detonation at the base of the wall untamped (Figure C-15). Small-arms fire will not detonate C4 or TNT; the charge must be primed with detonating cord. Soldiers must take cover before detonating the charge.

Figure C-15. Charge placement when small-arms fire cannot be suppressed.

f. The internal walls of most buildings function as partitions rather than load-bearing members. Therefore, smaller explosive charges can be used to breach them. In the absence of C4 or other military explosives, internal walls can be breached by using one or more hand grenades or a Claymore mine (Figure C-16). These devices should be tamped to increase their effectiveness and to reduce the amount of explosive force directed to the rear.

Figure C-16. Tamping of a Claymore mine and hand grenades to breach internal walls.

FM 90-10-1

g. The Molotov cocktail (Figure C-17) is an expedient device for disabling both wheeled and tracked vehicles. It is easy to make since the materials are readily available. The results are most effective because of the close engagement in built-up areas. The objective is to ignite a flammable portion of the vehicle such as the fuel or ammunition that it is transporting. The following materials are needed to make a Molotov cocktail:
- Container-bottle or glass container.
- Gas (60 percent).
- Oil (40 percent).
- Rag for use as a wick.

The gas and oil are mixed thoroughly (60 percent gas to 40 percent oil). The rag is soaked with the mixture, and then the mixture is placed into the bottle. The rag is then inserted in the opening of the bottle as a wick. When a target is sighted, the wick is lit and the bottle is thrown hard enough to break.

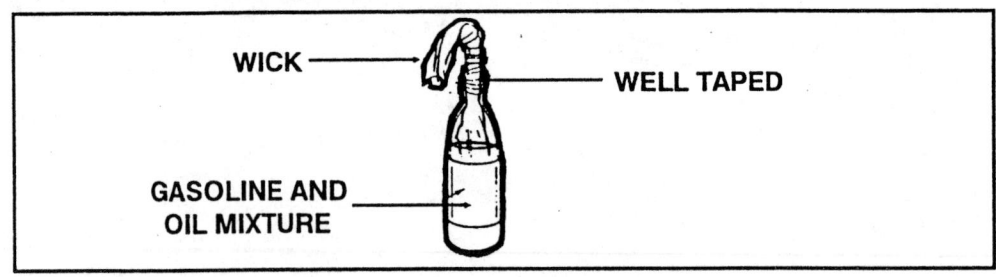

Figure C-17. Molotov cocktail.

WARNING

Ensure that a safe distance is maintained when throwing the Molotov cocktail. Caution troops against dropping the device. Throw it in the opposite direction of personnel and flammable materials. Do not smoke while making this device.

h. The bunker bomb (Figure C-18) is an expedient explosive flame weapon best used against fortified positions or rooms. This expedient munition should be used with a mechanical rather than an electrical firing system. The following materials are required to make a bunker bomb:
- 1 small-arms ammunition container.
- 1 gallon of gasoline.
- 50 feet of detonating cord.
- 1 nonelectric blasting cap.
- 1 M60 fuse igniter.
- 7 1/2 feet of M700 time fuse.
- 3 ounces of M4 thickening compound.
- 1 M49 trip flare or M34 WP grenade.

(1) **Step 1.** Fill the ammunition can 3/4 full with thickened flame fuel and secure the lid.
(2) **Step 2.** "Hasty whip" the device with 15 turns around the center of the container using 44 feet of detonating cord. Leave 2-foot "pigtails" for attaching the igniter and fuse igniter.
(3) **Step 3.** Tape the igniter (M49 trip flare or M34 WP grenade) to the container handle.
(4) **Step 4.** Place one detonating cord pigtail end under the igniter spoon handle. Tape it in place.
(5) **Step 5.** Attach the M60 fuse igniter and the nonelectrical blasting cap to the M700 time fuse.
(6) **Step 6.** Attach the nonelectrical firing system to the other pigtail by making a loop in the detonating cord and attaching the blasting cap to it.
(7) **Step 7.** Remove the safety pin from the igniter (M49 trip flare or M34 WP grenade). The device is ready to be fired.

WARNING

Never carry the device by the handle or igniter. Remove the igniter safety pin only when it is time to use the device. Use extreme care when handling or carrying nonelectrical firing systems. Protect blasting caps from shock and extreme heat. Do not allow the time fuse to kink or become knotted. Doing so may disrupt the powder train and may cause a misfire. Prime detonating cord and remove the time fuse igniter safety pin only when it is time to use the device.

---

1. Ammunition can.
2. 50 feet of detonating cord.
3. Igniter.
4. Detonating cord pigtail.
5. M60 fuse igniter and M700 time fuse.
6. Detonating cord pigtail.
7. Igniter safety pin.

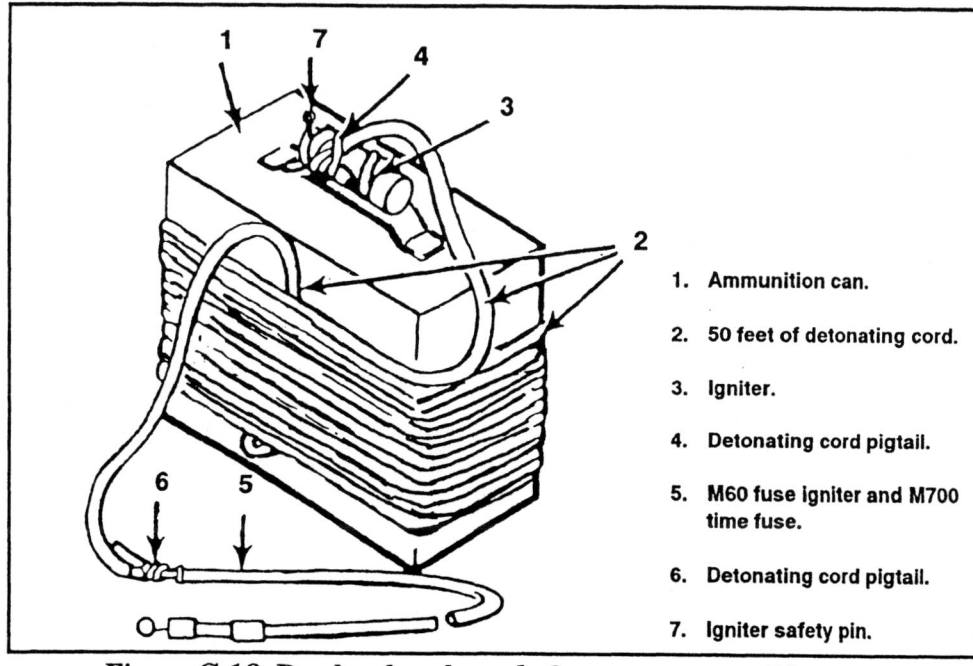

Figure C-18. Bunker bomb made from an ammunition can.

C-7. SAFETY

The greatest danger to friendly personnel from demolitions is the debris thrown by the explosion. Leaders must ensure that protective measures are enforced. The safe distance listed in Table C-2 indicates the danger of demolition effects.

   a. The following are the rules for using demolitions:
- Keep the blasting machine under the control of an NCO.
- Wear helmets at all times while firing explosives.
- Handle misfires with extreme care.
- Clear the room and protect personnel when blowing interior walls.

   b. Some charges should be prepared, minus detonators, beforehand to save time—for example, 10- or 20-pound breaching charges of C4, expedient-shaped charges in No. 10 cans.
- Use C4 to breach hard targets (masonry construction).
- Do not take chances.
- Do not divide responsibility for explosive work.
- Do not mix explosives and detonators.
- Do not carry explosives and caps together.

POUNDS OF EXPLOSIVE	SAFE DISTANCE IN METERS	POUNDS OF EXPLOSIVE	SAFE DISTANCE IN METERS
1 to 29	300	150	514
30	311	175	560
35	327	200	585
40	342	225	609
45	356	250	630
50	369	275	651
60	392	300	670
70	413	325	688
80	431	350	705
90	449	375	722
100	465	400	737
125	500	425 and over	750

NOTE: These distances will be modified in combat when troops are in other buildings, around corners, or behind intervening walls.

Table C-2. Minimum safe distances for personnel in the open.

APPENDIX D

SUBTERRANEAN OPERATIONS

Knowledge of the nature and location of underground facilities is of great value to both the urban attacker and defender. To exploit the advantages of underground facilities, a thorough reconnaissance is required. This appendix describes the techniques used to deny the enemy use of these features, the tactical value of subterranean passage techniques, and the psychological aspects of extended operations in subterranean passages.

D-1. TACTICAL VALUE

In larger cities, subterranean features include sunken garages, underground passages, subway lines, utility tunnels (Figure D-1), sewers, and storm drains. Most of these features allow the movement of many troops. Even in smaller European towns, sewers and storm drains permit soldiers to move beneath the fighting and surface behind the enemy.

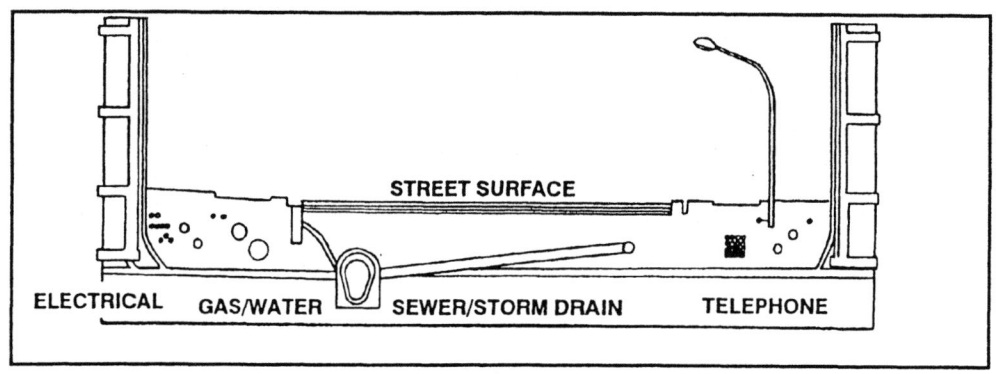

Figure D-1. Tunnels.

   a. Subterranean passages provide the attacker with covered and concealed routes into and through built-up areas. This enables the enemy to launch his attack along roads that lead into the city while infiltrating a smaller force in the defender's rear. The objective of this attack is to quickly insert a unit into the defenders rear, thereby, disrupting his defense and obstructing the avenues of withdrawal for his forward defense.

   b. Depending upon the strength and depth of the defense, the attack along the subterranean avenue of approach could easily become the main attack. Even if the subterranean effort is not immediately successful, it forces the defender to fight on two levels and to extend his resources to more than just street-level fighting.

   c. The existence of subterranean passages forces the defender to defend the built-up area above and below ground. Passages are more of a disadvantage to the defender than the attacker. However, subterranean passages also offer some advantages. When thoroughly reconnoitered and controlled by the defender, subterranean passages provide excellent covered and concealed routes to move reinforcements or to launch counterattacks. They also provide ready-made lines of communication for the movement of supplies and evacuation of casualties, and provide places to cache supplies for forward companies. Subterranean passages also offer the defender a ready-made conduit for communications wire, which protects it from tracked vehicles and indirect fires.

D-2. DENIAL TO THE ENEMY
Subterranean passages are useful to the defender only to the extent that the attacker can be denied their use. The defender has an advantage in that, given the confining, dark environment of these passages, a small group of determined soldiers in a prepared position can defeat a numerically superior force.

　　a. Tunnels afford the attacker little cover and concealment except for the darkness and any man-made barriers. The passageways provide tight fields of fire and amplify the effect of grenades. Obstacles at intersections in the tunnels set up excellent ambush sites and turn the subterranean passages into a deadly maze. These obstacles can be quickly created using chunks of rubble, furniture, and parts of abandoned vehicles interspersed with M18A1 Claymore mines.

　　b. A thorough reconnaissance of the subterranean or sewer system must be made first. To be effective, obstacles must be located at critical intersections in the passage network so that they trap attackers in a kill zone but allow defenders freedom of movement (Figure D-2).

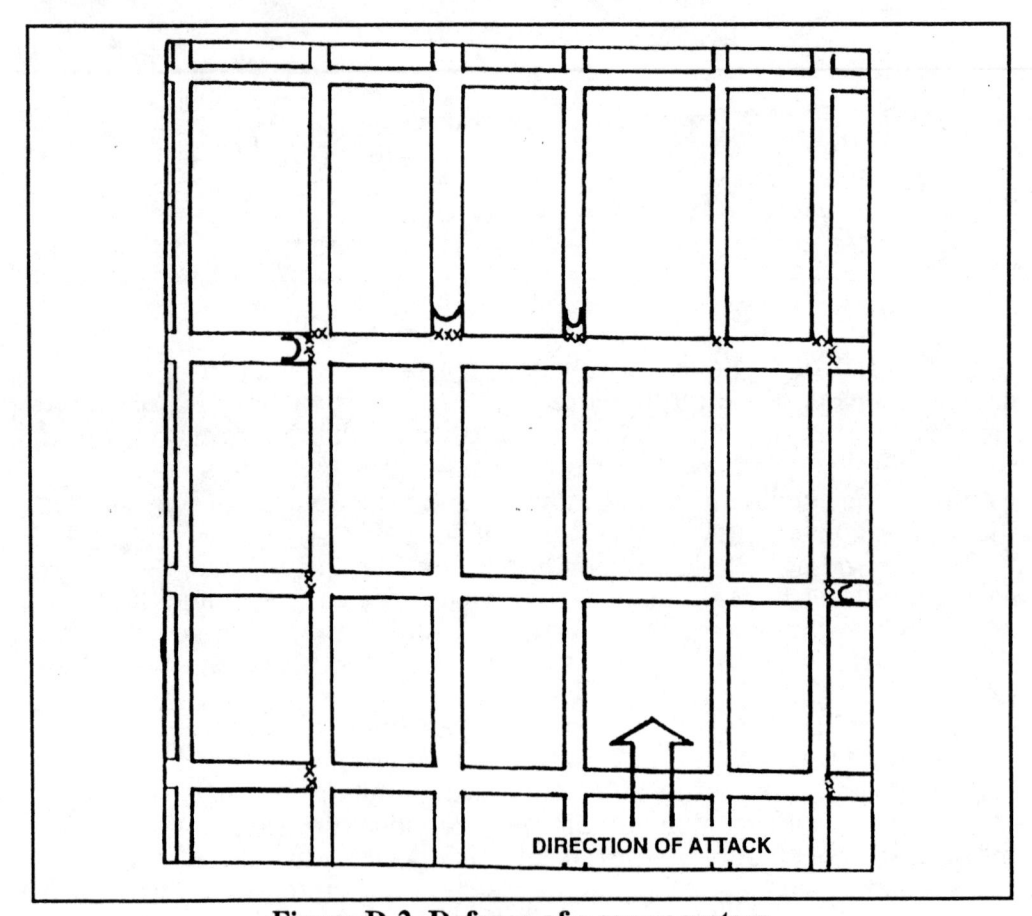

Figure D-2. Defense of a sewer system.

D-3. SUBTERRANEAN RECONNAISSANCE TECHNIQUES

The local reconnaissance mission (platoon or company area of responsibility) should be given to a squad-size element (six or seven personnel). Enough soldiers are in a squad to gather the required data without getting in each other's way in the confines of the tunnel. Only in extremely large subterranean features should the size of a patrol be increased.

   a. The patrol unit leader should organize his patrol with two riflemen—one tasked with security to the front (the point man) and one tasked with security to the rear (Figure D-3). The patrol leader moves directly behind the point man, and navigates and records data collected by the patrol. The grenadier should follow the patrol unit leader, and the demolitions man should follow the grenadier. Two riflemen should be left as a security post at the point of entry. They are responsible for detecting enemy who come upon the patrol unit's rear and for serving as the communications link between the patrol unit leader and his higher headquarters.

Figure D-3. Organization of the patrol unit.

   b. The patrol unit leader should carry a map, compass, street plan, and notebook in which he has written the information he must gather for the platoon leader. The grenadier should carry the tools needed to open manhole covers. If the patrol is to move more than 200 to 300 meters or if the platoon leader directs, the grenadier should also carry a sound-powered telephone (TA-1) and wire dispenser (XM-306A) for communications. (Radios are unreliable in this environment). The point man should be equipped with night vision goggles to maintain surveillance within the sewer.

   c. All soldiers entering the sewer should carry a sketch of the sewer system to include magnetic north, azimuths, distances, and manholes. They should also carry protective masks, flashlights, gloves, and chalk for marking features along the route. The patrol should also be equipped with a 120-foot safety rope to which each man is tied. To improve their footing in slippery sewers and storm drains, the members of the patrol unit should wrap chicken wire or screen wire around their boots.

d. A constant concern to troops conducting a subterranean patrol is chemical defense. Enemy chemical agents used in tunnels are encountered in dense concentrations, with no chance of dispersement. The M8 automatic chemical agent alarm system, carried by the point man, provides instant warning of the presence of chemical agents. M8 detector paper can also be used to detect chemical agents. At the first indication that harmful gases are present, the patrol should mask.

e. In addition to enemy chemical agents, noxious gases from decomposing sewage can also pose a threat. These gases are not detected by the M8 chemical agent alarm system, nor are they completely filtered by the protective mask. Physical signs that indicate their presence in harmful quantities are nausea and dizziness. The patrol leader should be constantly alert to these signs and know the shortest route to the surface for fresh air.

f. Once the patrol is organized and equipped, it moves to the entrance of the tunnel, which is usually a manhole. With the manhole cover removed, the patrol waits 15 minutes before entry to allow any gases to dissipate. Then the point man descends into the tunnel to determine whether the air is safe to breathe and if movement is restricted. The point man should remain in the tunnel for 10 minutes before the rest of the patrol follows. If he becomes ill or is exposed to danger, he can be pulled out by the safety rope.

g. When the patrol is moving through the tunnel, the point man moves about 10 meters in front of the patrol leader. Other patrol members maintain 5-meter intervals. If the water in the tunnel is flowing faster than 2.5 meters per second or if the sewer contains slippery obstacles, those intervals should be increased to prevent all patrol members from falling if one man slips. All patrol members should stay tied into the safety rope so that they can easily be retrieved from danger. The rear security man marks the route with chalk so other troops can find the patrol.

h. The patrol leader should note the azimuth and pace count of each turn he takes in the tunnel. When he encounters a manhole to the surface, the point man should open it and determine the location, which the patrol leader then records. The use of recognition signals (Figure D-4) prevents friendly troops from accidentally shooting the point man as he appears at a manhole.

i. Once the patrol has returned and submitted its report, the platoon leader must decide how to use the tunnel. In the offense, the tunnel could provide a covered route to move behind the the enemy's defenses. In the defense, the tunnel could provide a covered passage between positions. In either case, the patrol unit members should act as guides along the route.

j. If the tunnel is to be blocked, the platoon should emplace concertina wire, early warning devices, and antipersonnel mines. A two-man position established at the entrance of the sewer (Figure D-5) provides security against enemy trying to approach the platoon's defense and should be abandoned when the water rises. It should be equipped with command-detonated illumination. While listening for the enemy, soldiers manning this position should not wear earplugs (they should be put in ears just before firing). The confined space amplifies the sounds of weapons firing to a dangerous level. The overpressure from grenades, mines, and booby traps exploding in a sewer or tunnel can have adverse effects on friendly troops such as ruptured eardrums and wounds from flying debris. Also, gases found in sewers can be ignited by the blast effects of these munitions. For these reasons, small-arms weapons should be employed in tunnels and sewers. Friendly personnel should be outside of tunnels or out of range of the effects when mines or demolitions are detonated. Soldiers should mask at the first sign of a chemical threat.

Figure D-4. Recognition signals.

Figure D-5. Two-man position established at the entrance to a sewer.

D-4. PSYCHOLOGICAL CONSIDERATIONS

Combat operations in subterranean passages are much like night combat operations. The psychological factors that affect soldiers during night operations reduce confidence, cause fear, and increase a feeling of isolation. This feeling of isolation is further magnified by the tight confines of the tunnels. The layout of tunnels could require greater dispersion between positions than is usual for operations in wooded terrain.

a. Leaders must enforce measures to dispel the feelings of fear and isolation experienced by soldiers in tunnels. These measures include leadership training, physical and mental fitness, sleep discipline, and stress management.

b. Leaders maintain communication with soldiers manning positions in the tunnels either by personal visits or by field telephone. Communications inform leaders of the tactical situation as well as the mental state of their soldiers. Training during combat operations is limited; however, soldiers manning positions below ground should be given as much information as possible on the organization of the tunnels and the importance of the mission. They should be briefed on contingency plans and alternate positions should their primary positions become untenable.

c. Physical and mental fitness can be maintained by periodically rotating soldiers out of tunnels so they can stand and walk in fresh air and sunlight. Stress management is also a factor of operations in tunnels. Historically, combat in built-up areas has been one of the most stressful forms of combat. Continuous darkness and restricted maneuver space cause more stress to soldiers than street fighting.

APPENDIX E

FIGHTING POSITIONS

A critical platoon- and squad-level defensive task in combat in built-up areas is the preparation of fighting positions. Fighting positions in built-up areas are usually constructed inside buildings and are selected based on an analysis of the area in which the building is located and the individual characteristics of the building.

E-1. CONSIDERATIONS

Leaders should consider the following factors when establishing fighting positions.

   a. **Protection.** Leaders should select buildings that provide protection from direct and indirect fires. Reinforced concrete buildings with three or more floors provide suitable protection, while buildings constructed of wood, paneling, or other light material must be reinforced to gain sufficient protection. One- to two-story buildings without a strongly constructed cellar are vulnerable to indirect fires and require construction of overhead protection for each firing position.

   b. **Dispersion.** A position should not be established in a single building when it is possible to occupy two or more buildings that permit mutually supporting fires. A position in one building, without mutual support, is vulnerable to bypass, isolation, and subsequent destruction from any direction.

   c. **Concealment.** Buildings that are obvious defensive positions (easily targeted by the enemy) should not be selected. Requirements for security and fields of fire could require the occupation of exposed buildings. Therefore, reinforcements provide suitable protection within the building.

   d. **Fields of Fire.** To prevent isolation, positions should be mutually supporting and have fields of fire in all directions. Clearing fields of fire could require the destruction of adjacent buildings using explosives, engineer equipment, and field expedients.

   e. **Covered Routes.** Defensive positions should have at least one covered route that permits resupply, medical evacuation, reinforcement, or withdrawal from the building. The route can be established by one of the following:

- Through walls to adjacent buildings.
- Through underground systems.
- Through communications trenches.
- Behind protective buildings.

   f. **Observation.** The building should permit observation of enemy avenues of approach and adjacent defensive sectors.

   g. **Fire Hazard.** Leaders should avoid selecting positions in buildings that are a fire hazard. If flammable structures must be occupied, the danger of fire can be reduced by wetting down the immediate environment, laying an inch of sand on the floors, and providing fire extinguishers and fire fighting equipment. Also, routes of escape must be prepared in case of fire.

h. **Time.** Time available to prepare the defense could be the most critical factor. If enough time is not available, buildings that require extensive preparation should not be used. Conversely, buildings located in less desirable areas that require little improvement could probably become the centers of defense.

E-2. PREPARATION
Preparation of fighting positions depends upon proper selection and construction.

a. **Selecting Positions.** Each weapon should be assigned a primary sector of fire to cover enemy approaches. Alternate positions that overwatch the primary sector should also be selected. These positions are usually located in an adjacent room on the same floor. Each weapon must be assigned a supplementary position to engage attacks from other directions, and an FPL (Figure E-1).

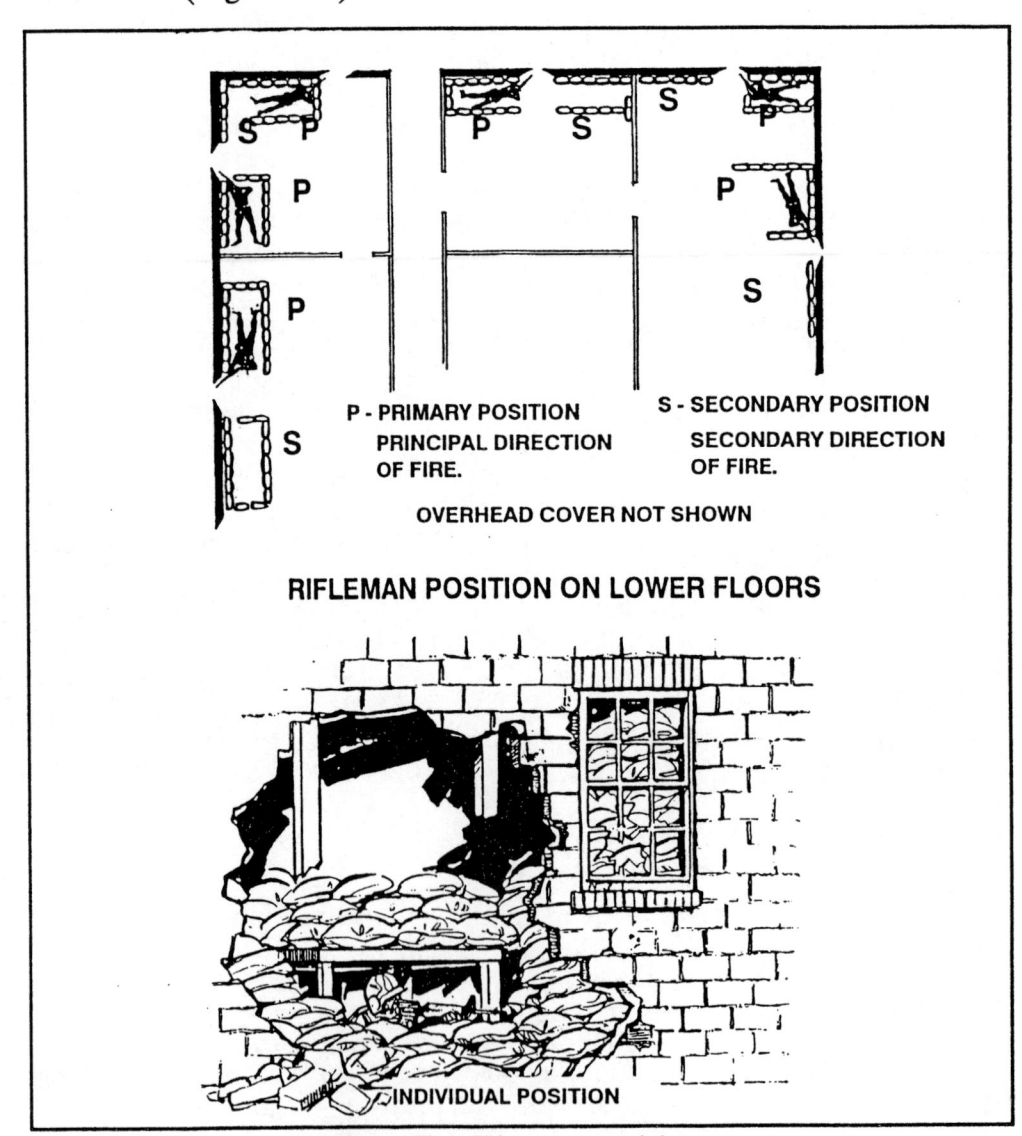

Figure E-1. Weapon positions.

FM 90-10-1

PRIMARY POSITION

SANDBAGS

SANDBAGS — SUPPLEMENTARY POSITION

Wet down muzzle blast area.
Weapon is fired at an angle through firing port.
Muzzle/blast should not protrude beyond the wall.

MACHINE GUN POSITION ON FIRST FLOOR

WIRE MESH

GRENADE SUMP

SHELTER

CELLAR FIRING POSITION

BUILDING SUPPORT

BLOCK WHEN NOT IN USE

PLACE AWAY FROM SUPPORT

CORNER FIRING POSITION

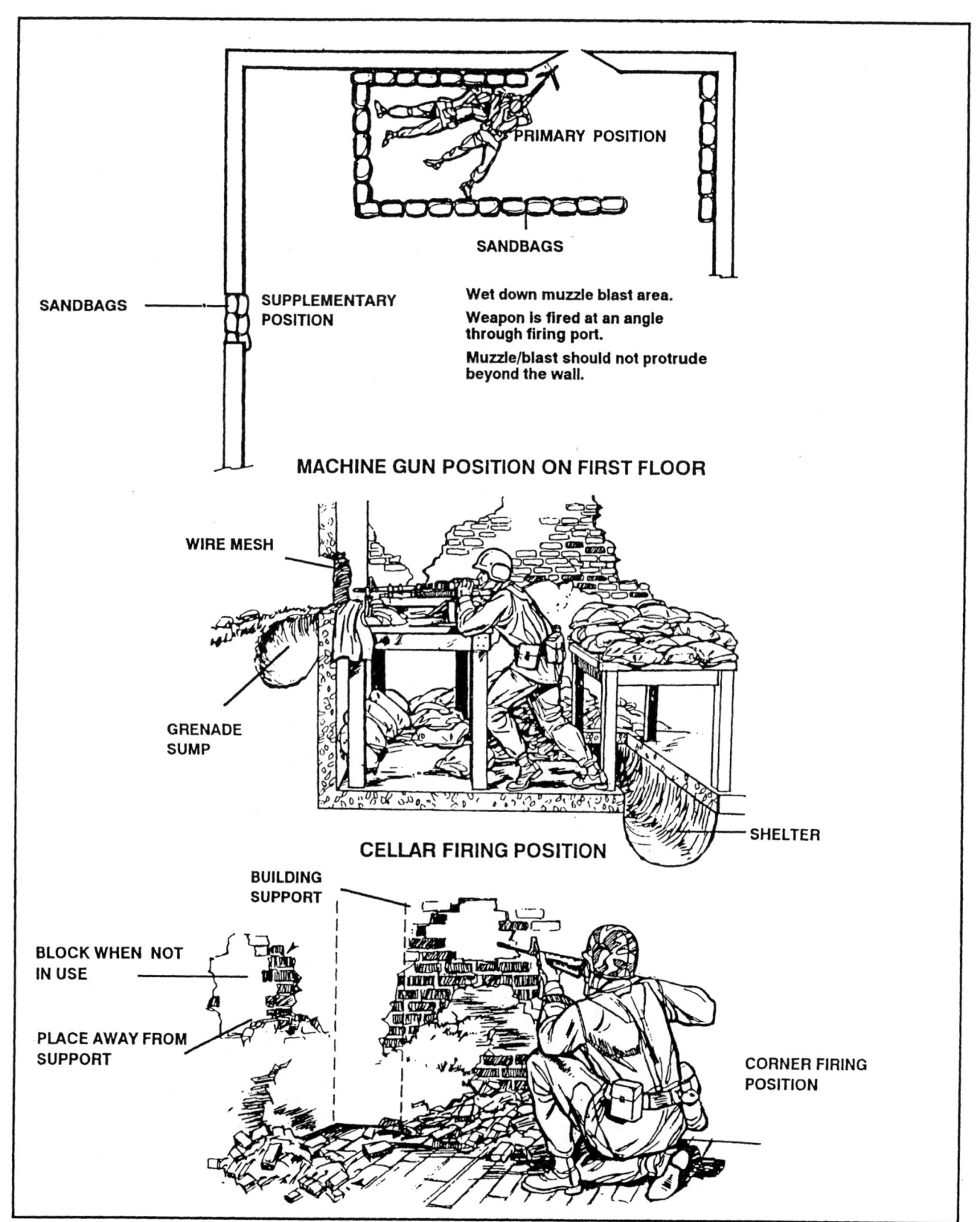

Figure E-1. Weapon positions (continued).

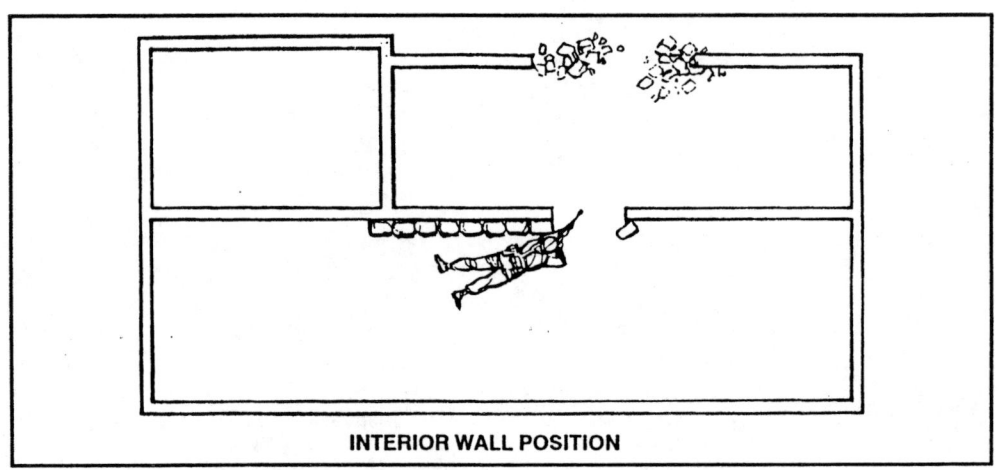

Figure E-1. Weapon positions (continued).

b. **Building Positions.** There are many ways to establish a fighting position in a building.

(1) *Window position.* Soldiers should kneel or stand on either side of a window. To fire downward from upper floors, tables or similar objects can be placed against the wall to provide additional elevation, but they must be positioned to prevent the weapon from protruding through the window. Leaders should inspect positions to determine the width of sector that each position can engage (Figure E-2).

Figure E-2. Window position.

(2) *Loopholes.* To avoid establishing a pattern of always firing from windows, loopholes should be prepared in walls. Soldiers should avoid firing directly through loopholes to enhance individual protection.

(a) Several loopholes are usually required for each weapon (primary, alternate, and supplementary positions). The number of loopholes should be carefully considered because they can weaken walls and reduce protection. Engineers should be consulted before an excessive number of loopholes are made. Loopholes should be made by punching or drilling holes in walls and should be placed where they are concealed. Blasting loopholes can result in a large hole, easily seen by the enemy.

(b) Loopholes should be cone-shaped to obtain a wide arc of fire, to facilitate engagement of high and low targets, and to reduce the size of the exterior aperture (Figure E-3). The edges of a loophole splinter when hit by bullets, therefore, protective linings, such as an empty sandbag held in place by wire mesh, will reduce spalling effects. When not in use, loopholes should be covered with sandbags to prevent the enemy from firing into or observing through them.

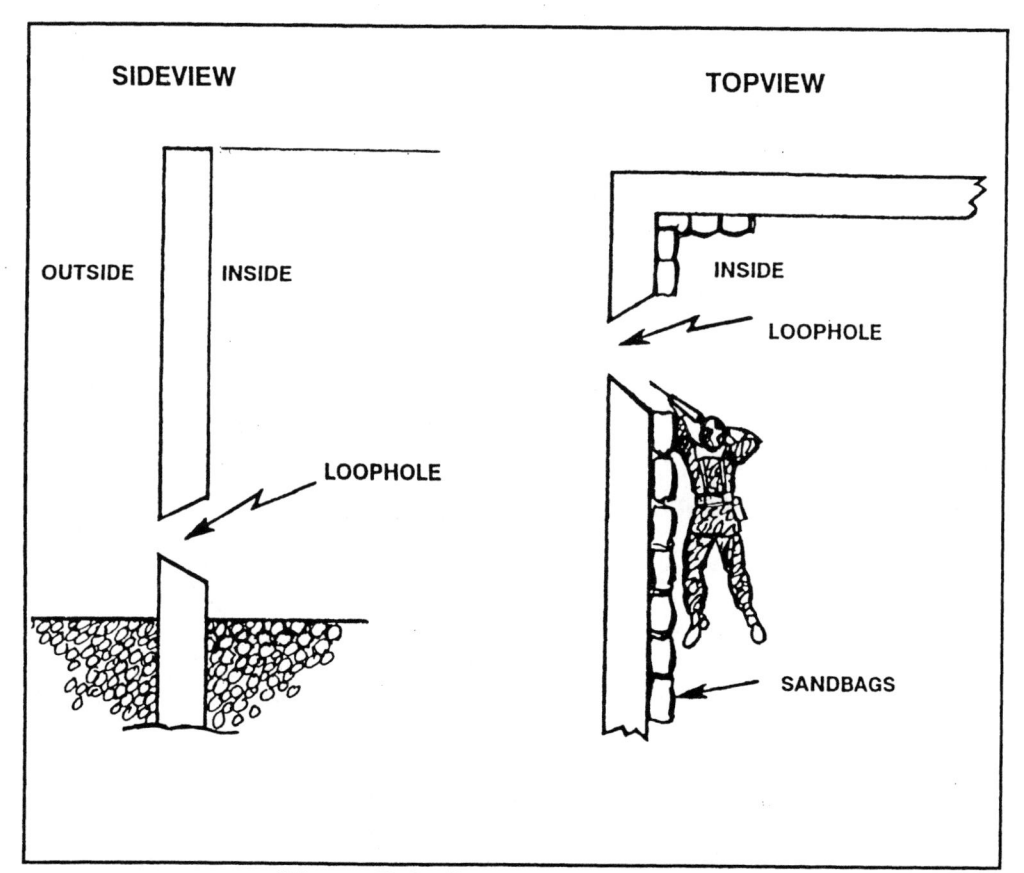

Figure E-3. Cone-shaped loopholes.

(c) Loopholes should also be prepared in interior walls and ceilings of buildings to permit fighting within the position. Interior loopholes should overwatch stairs, halls, and unoccupied rooms, and be concealed by pictures, drapes, or furniture. Loopholes in floors permit the defender to engage enemy personnel on lower floors with small-arms fire and grenades.

(d) Although walls provide some frontal protection, they should be reinforced with sandbags, furniture filled with dirt, or other expedients. Each position should have overhead and all-round protection (Figure E-4).

Figure E-4. Position with overhead and all-round protection.

c. **Other Construction Tasks.** Other construction tasks in basements, on ground floors, and on upper floors will need to be performed.

(1) *Basements and ground floors.* Basements require preparation similar to that of the ground floor. Any underground system not used by the defender that could provide enemy access to the position must be blocked.

(a) *Doors.* Unused doors should be locked, nailed shut, and blocked and reinforced with furniture, sandbags, or other field expedients. Outside doors can be booby trapped by engineers or other training personnel.

(b) *Hallways.* If not required for the defender's movement, hallways should be blocked with furniture and tactical wire (Figure E-5). If authorized, booby traps should be employed.

(c) *Stairs.* Defenders should block stairs not used by the defense with furniture and tactical wire (see Figure E-5) or remove them. If possible, all stairs should be blocked and ladders should be used to move from floor to floor and then removed when not being used. Booby traps should also be employed on stairs.

(d) *Windows.* All glass should be removed. Windows not used should be blocked with boards or sandbags.

(e) *Fighting positions.* Fighting positions should be made in floors. If there is no basement, fighting positions can give additional protection from heavy direct-fire weapons.

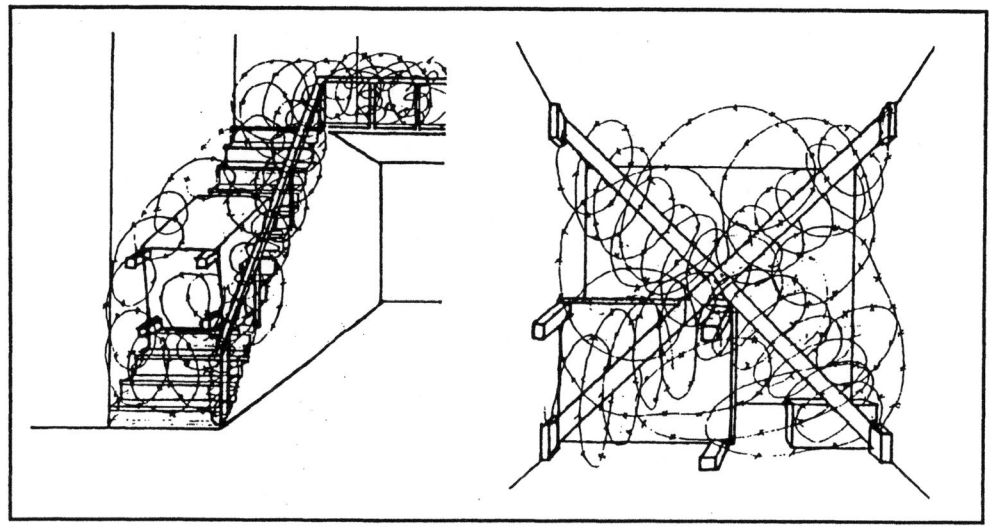

Figure E-5. Blocking hallways and stairs.

(f) *Ceilings.* Support that can withstand the weight of rubble from upper floors should be placed under ceilings (Figure E-6).

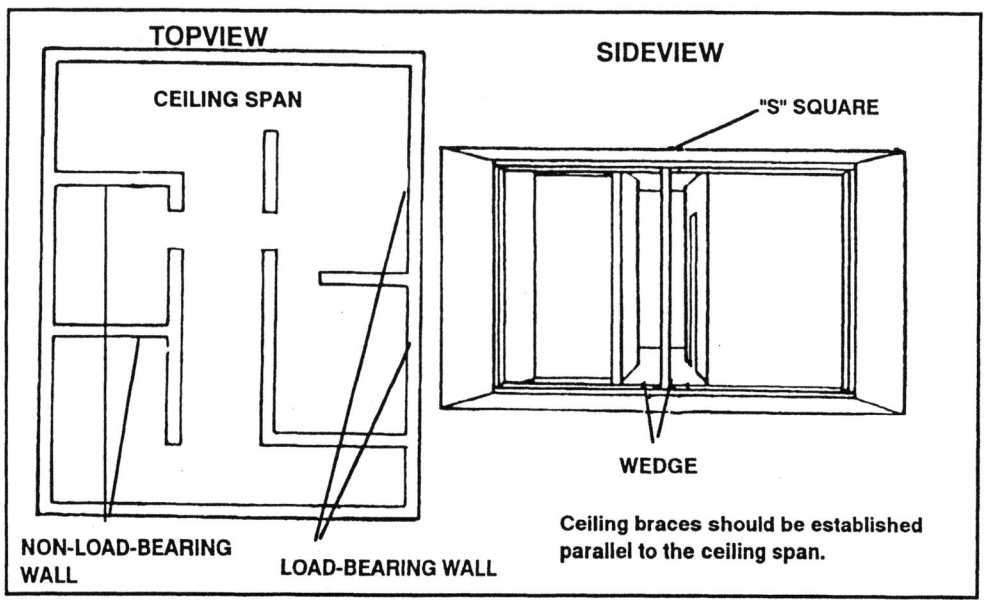

Figure E-6. Ceiling reinforcement.

(g) *Unoccupied rooms.* Rooms not required for defense should be blocked with tactical wire or booby trapped.

(2) ***Upper floors.*** Upper floors require the same preparation as ground floors. Windows need not be blocked, but they should be covered with wire mesh, which blocks grenades thrown from the outside. The wire should be loose at the bottom to permit the defender to drop grenades.

(3) ***Interior routes.*** Routes are required that permit defending forces to move within the building to engage enemy forces from any direction. Escape

routes should also be planned and constructed to permit rapid evacuation of a room or the building. Small holes (called mouse holes) should be made through interior walls to permit movement between rooms. Once the defender has withdrawn to another level, such holes should be clearly marked for both day and night identification. All personnel must be briefed as to where the various routes are located. Rehearsals should be conducted so that everyone becomes familiar with the routes (Figure E-7).

Figure E-7. Movement between floors.

(4) *Fire prevention.* Buildings that have wooden floors and raftered ceilings require extensive fire prevention measures. The attic and other wooden floors should be covered with about 1 inch of sand or dirt, and buckets of water should be positioned for immediate use. Firefighting materials (dirt, sand, fire extinguishers, and blankets) should be placed on each floor for immediate use. Water basins and bathtubs should be filled as a reserve for firefighting. All electricity and gas should be turned off. Fire breaks can be created by destroying buildings adjacent to the defensive position.

(5) *Communications.* Telephone lines should be laid through adjacent buildings or underground systems, or buried in shallow trenches. Radio antennas can be concealed by placing them among civilian television antennas, along the sides of chimneys and steeples, or out windows that direct FM communications away from enemy early-warning sources and ground observation. Telephone lines within the building should be laid through walls and floors.

(6) Rubbling. Rubbling parts of the building provides additional cover and concealment for weapons emplacements, and should be performed only by trained engineers.

(7) Rooftops. Positions in flat-roofed buildings require obstacles that restrict helicopter landings. Rooftops that are accessible from adjacent structures should be covered with tactical wire or other expedients, and must be guarded. Entrances to buildings from rooftops can be blocked if compatible with the overall defensive plan. Any structure on the outside of a building that could assist scaling the buildings to gain access to upper floors, or to the rooftop, should be removed or blocked.

(8) Obstacles. Obstacles should be positioned adjacent to buildings in order to stop tanks and to delay infantry.

(9) Fields of fire. Fields of fire should be improved around the defensive position. Selected buildings can be destroyed to enlarge fields of fire. Obstacles to antitank guided missiles, such as telephone wires, should be cleared. Dead space should be covered with mines and obstacles.

E-3. ARMORED VEHICLE POSITIONS

Fighting positions for tanks and infantry fighting vehicles are essential to a complete and effective defensive plan in built-up areas.

a. **Armored Vehicle Positions.** Armored vehicle positions are selected and developed to obtain the best cover, concealment, observation, and fields of fire, while retaining the vehicle's ability to move.

(1) If fields of fire are restricted to streets, hull-down positions should be used to gain cover and to fire directly down streets (Figure E-8). From those positions, tanks and BFVs are protected and can rapidly move to alternate positions. Buildings collapsing from enemy fires are a minimal hazard to the armored vehicle and crew.

Figure E-8. Hull-down position.

(2) The hide position (Figure E-9) covers and conceals the vehicle until time to move into position for engagement of targets. Since the crew will not be able to see advancing enemy forces, an observer from the vehicle or a nearby infantry unit must be concealed in an adjacent building to alert the crew. The observer acquires the target and signals the armored vehicle to move to the firing position and to fire. After firing, the tank or BFV moves to an alternate position to avoid compromising one location.

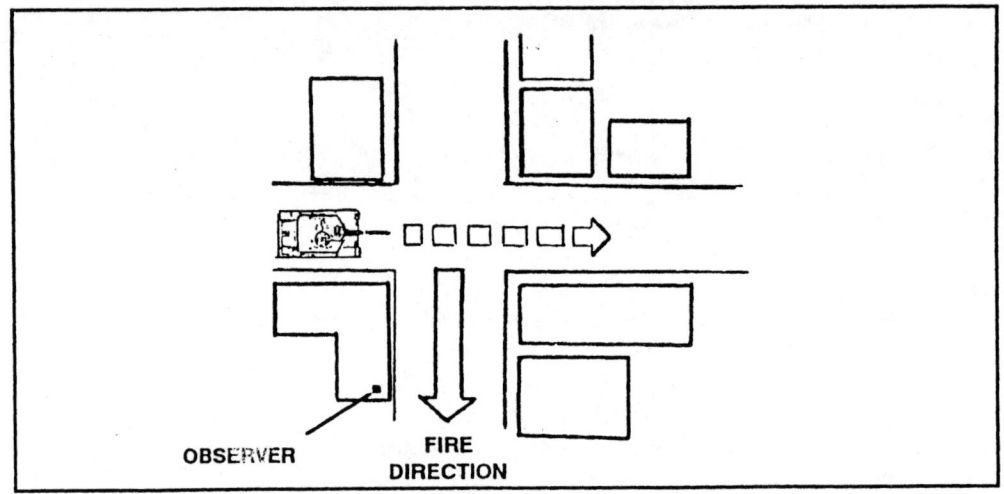

Figure E-9. Hide position.

(3) The building hide position (Figure E-10) conceals the vehicle inside a building. If basement hide positions are inaccessible, engineers must evaluate the building's floor strength and prepare for the vehicle. Once the position is detected, it should be evacuated to avoid enemy fires.

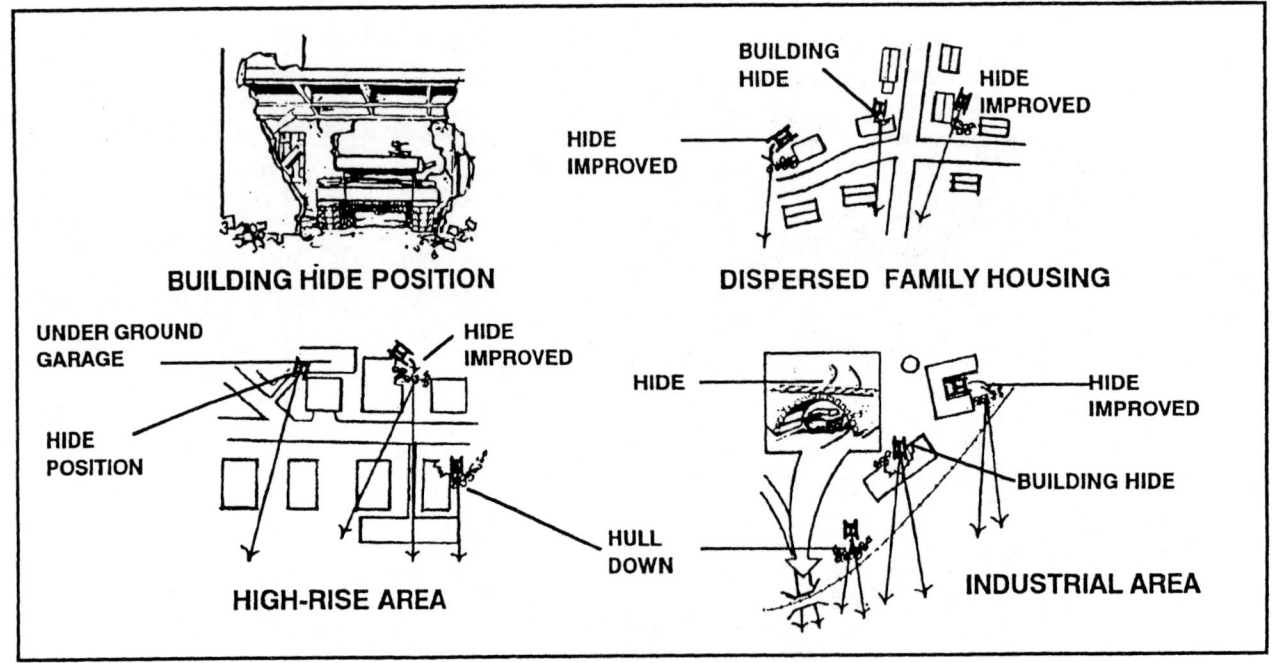

Figure E-10. Building hide position.

E-4. ANTITANK GUIDED MISSILE POSITIONS

Antitank guided missiles must be employed in areas that maximize their capabilities in the built-up area. The lack of a protective transport could require the weapon to be fired from inside or behind a building, or behind the cover of protective terrain (Figure E-11).

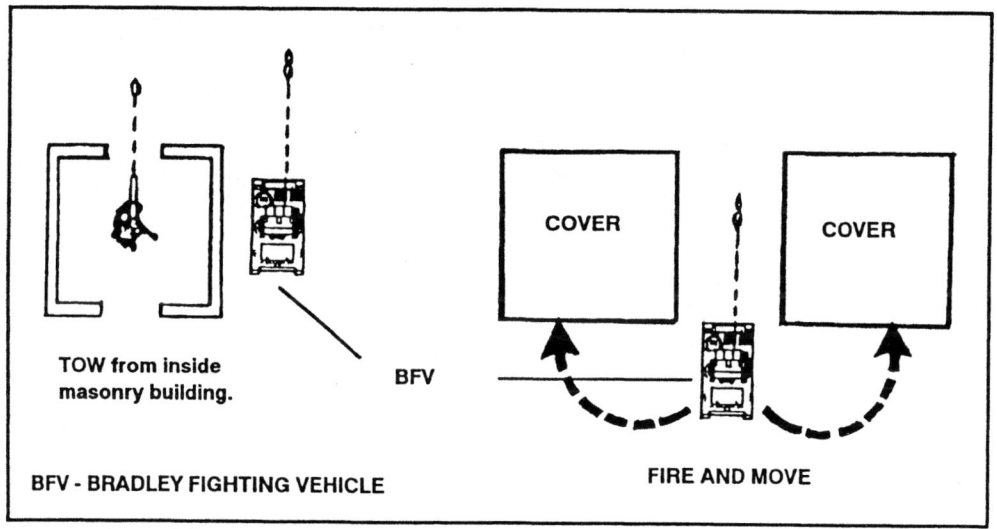

Figure E-11. Antitank guided missiles positions.

a. When ATGMs are fired from a vehicle or from street level or bottom floor fighting positions, rubble can interfere with missile flight. When firing down streets, missiles must have at least 30 inches of clearance over rubble. Other obstacles to missile flight include trees and brush, vehicles, television antennas, buildings, power lines and wires, walls, and fences.

b. A LAW is best suited for built-up areas because its 10-meter minimum arming distance allows employment at close range. LAWs and other light and medium antitank weapons are not effective against the front of modern battle tanks. Because tanks have the least armor protection on the top and rear deck, and the tank presents a larger target when engaged from above, LAWs should fire down onto tanks.

E-6. SNIPER POSITIONS

Snipers contribute to combat in built-up areas by firing on selected enemy soldiers. An effective sniper organization can trouble the enemy far more than its cost in the number of friendly soldiers employed.

a. General areas (a building or group of buildings) are designated as sniper positions (Figure E-12, page E-12), but the sniper selects the best position for engagement. Masonry buildings that offer the best protection, long-range fields of fire, and all-round observation are preferred. The sniper also selects several secondary and supplementary positions to cover his areas of responsibility.

b. Engagement priorities for snipers are determined by the relative importance of the targets to the effective operations of the enemy. Sniper targets usually include tank commanders, direct fire support weapons' crewmen, crew-served weapons' crewmen, officers, forward observers, and radiotelephone operators.

c. Built-up areas often limit snipers to firing down or across streets, but open parts permit engagements at long ranges. Snipers can be employed to cover rooftops, obstacles, dead space, and gaps in FPFs.

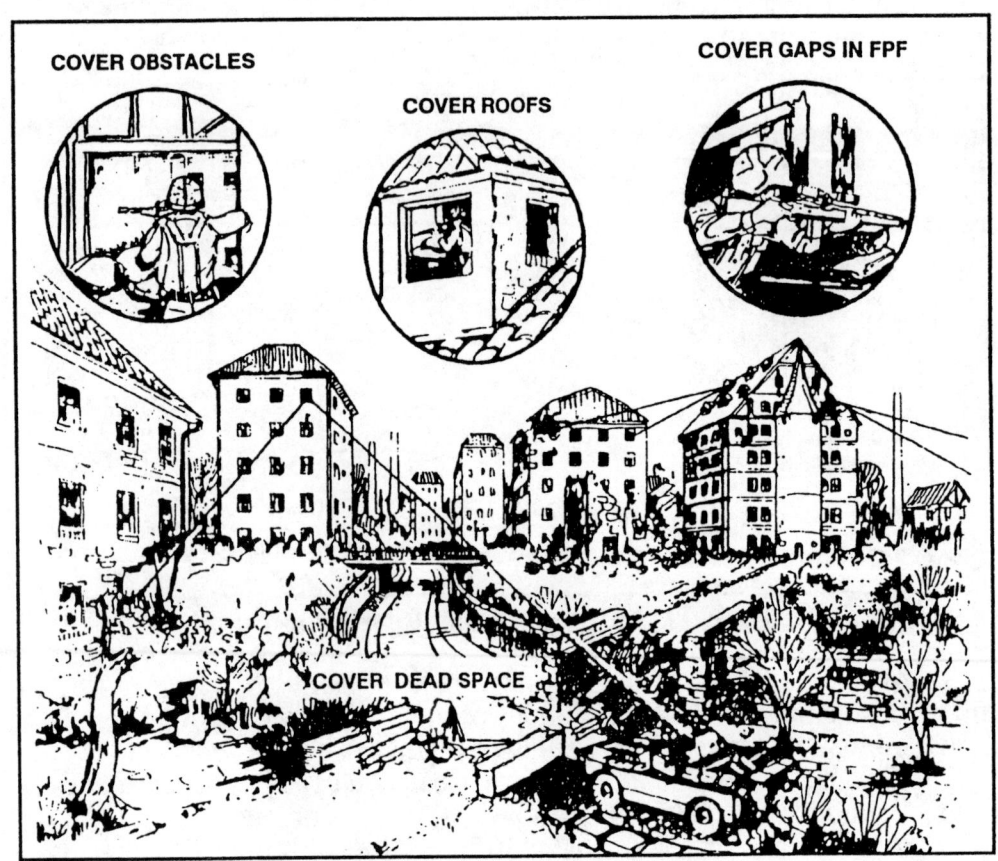

Figure E-12. Sniper positions.

APPENDIX F

ATTACKING AND CLEARING BUILDINGS

At platoon and squad level, the major offensive tasks for combat in built-up areas are attacking and clearing buildings, which involves suppressing fires, advancing infantry assault forces, assaulting a building, and reorganizing the assault force.

F-1. REQUIREMENTS
Regardless of a structure's characteristics or the type of built-up area, there are four interrelated requirements for attacking a defended building: fire support, movement, assault, and reorganization. Proper application and integration of these requirements reduce casualties and hasten accomplishment of the mission. The application is determined by the type of building to be attacked and the nature of the surrounding built-up area. For example, medium-size towns have numerous open spaces, and larger cities have high-rise apartments and industrial and transportation areas, which are separated by parking areas or parks. Increased fire support is required to suppress and obscure enemy gunners covering the open terrain and spaces between buildings. Conversely, the centers of small- and medium-size towns, with twisting alleys and country roads or adjoining buildings, provide numerous covered routes that can decrease fire support requirements.

F-2. FIRE SUPPORT
Fire support and other assistance to advance the assault force are provided by a support force. This assistance includes—

- Suppressing and obscuring enemy gunners within the objective building(s) and adjacent structures.
- Isolating the objective building(s) with direct or indirect fires to prevent enemy withdrawal, reinforcement, or counterattack.
- Breaching walls en route to and in the objective structure.
- Destroying enemy positions with direct-fire weapons.
- Securing cleared portions of the objective.
- Providing replacements for the assault force.
- Providing resupply of ammunition and explosives.
- Evacuating casualties and prisoners.

   a. The size of the support force is determined by the type and size of the objective building(s); whether the adjacent terrain provides open or covered approaches; and the organization and strength of enemy defenses.

   b. The support force could consist of only one infantry fire team with M60 machine guns, M249s, M203 grenade launchers, and M202 multishot flame weapons. In the case of Bradley-equipped units, the BFV may provide support with the 25-mm gun as the rifle team assaults. In situations involving a larger assault force, a platoon or company reinforced with tanks, engineers, and self-propelled artillery may be required to support movement and assault by an adjacent platoon or company.

c. After seizing objective buildings, the assault force reorganizes and may be required to provide supporting fires for a subsequent assault. Each weapon is assigned a target or area to cover. Individual small-arms weapons place fires on likely enemy weapon positions—loopholes, windows, roof areas. Snipers are best employed in placing accurate fire through loopholes or engaging long-range targets. The M202s and M203s direct their fires through windows or loopholes.

d. LAWs and demolitions are employed to breach walls, doors, barricades, and window barriers on the ground level of structures. Tank main guns and BFV 25-mm guns engage first-floor targets and breach walls for attacking infantry. Tank machine guns engage suspected positions on upper floors and in adjacent structures. In addition to destroying or weakening structures, tank main gun projectiles cause casualties by explosive effects and by hurling debris throughout the interior of structures.

e. Artillery and mortars use time fuzes to initially clear exposed personnel, weapons, observation posts, and radio sites from rooftops. They then use delayed fuze action to cause casualties among the defenders inside the structure from shrapnel and falling debris. Artillery can also be used in the direct-fire mode much like the tank and CEV.

F-3. MOVEMENT

The assault force (squad, platoon, or company) minimizes enemy defensive fires during movement by—

- Using covered routes.
- Moving only after defensive fires have been suppressed or obscured.
- Moving at night or during other periods of reduced visibility.
- Selecting routes that will not mask friendly suppressive fires.
- Crossing open areas (streets, space between buildings) quickly under the concealment of smoke and suppression provided by support forces.
- Moving on rooftops that are not covered by enemy direct fires.

a. In lightly defended areas, the requirement for speed may dictate moving through the streets and alleys without clearing all buildings. Thus, the maneuver element should employ infantry to lead the column, closely followed and supported by BFVs or tanks.

b. When dismounted, rifle elements move along each side of the street, with leading squads keeping almost abreast of the lead tanks. When not accompanied by tanks or BFVs, rifle elements move single file along one side of the street under cover of fires from supporting weapons. They are dispersed and move along quickly. Each man is detailed to observe and cover a certain area such as second-floor windows on the opposite side of the street.

F-4. ASSAULT
The assault force, regardless of size, must quickly and violently execute its assault and subsequent clearing operations. Once momentum has been gained, it is maintained to prevent the enemy from organizing a more determined resistance on other floors or in other rooms. The small-unit leaders should keep the assault force moving, yet not allow the operation to become disorganized.

   a. An assault in a built-up area involves the elementary skills of close combat. Leaders must—

   - Be trained in the required techniques to defeat the enemy in a face-to-face encounter.
   - Keep themselves in excellent physical condition.
   - Have confidence in their abilities.

   b. The composition of the assault force varies depending on the situation; however, the considerations for equipping the force remain the same. The criteria for the size of any party are the availability of equipment and personnel, and the tactical situation. The assault force for a squad should consist of 2 three-man teams carrying only a fighting load of equipment and as much ammunition as possible, especially grenades (Figure F-1). A three-man support team provides suppressive fire for the assault force. The assault teams use maneuver techniques to clear a building room by room.

Figure F-1. Rifle squad.

   c. The M249 is normally employed with the support element but can also be used with the assault force to gain the advantages of its more powerful round. The Dragon may not be carried by the assault force due to its weight versus its expected effectiveness against the building being assaulted. The squad leader is located with the element from which he can best control the squad. If the squad is understrength or suffers casualties, priority is given to keeping the assault force up to strength at the expense of the support force (see Tables F-1 and F-2).

SUPPORT FORCE	ASSAULT FORCE
3 - 7.62-mm (Coaxial)	(Each squad organized into two- or three-man assault/support parties).
2 - M249s	
1 - Dragon	2 - 7.62-mm (4 - M249s)
1 - M202	* 2 - Dragons
LAWs	LAWs
4 - 25-mm guns	Hand grenades

* Dependent upon Dragon's effectiveness against building being attacked.

Table F-1. Bradley platoon.

SUPPORT FORCE	ASSAULT FORCE
2 - 7.62-mm	LAWs
2 - Dragons	Hand grenades
1 - M202	* 1 - Dragon
4 - M249s	2 - M203s
4 - M203s	2 - M249s
LAWs	

* Dependent upon Dragon's effectiveness against building being attacked.

Table F-2. Alternative with an infantry rifle platoon.

F-5. CLEARING

Entry at the top and fighting downward is the preferred method of clearing a building (Figure F-2). Clearing a building is easier from an upper story since gravity and building construction become assets to the assault force when throwing hand grenades and moving from floor to floor. This method is only feasible, however, when access to an upper floor or rooftop can be gained from the windows or roofs of adjoining, secured buildings; or, when enemy air defense weapons can be suppressed and troops transported to the rooftops by helicopter. Helicopters should land only on those buildings that have special heliports on the roofs or parking garages. Soldiers can rappel onto the roof or dismount as the helicopter hovers a few feet above the roof. Troops then breach the roof or common walls with explosives and use ropes to enter the lower floors. Stairs are guarded by friendly security elements when not used.

Figure F-2. Helicopters used to clear buildings.

    a. Although the top-to-bottom method is preferred for clearing a building, assaulting the bottom floor and clearing upward is a common method in all areas except where buildings form continuous fronts. In this situation, the assault force attempts to close on the flank(s) or rear of the building. The assault team clears each room on the ground floor and then, moving up, begins a systematic clearance of the remaining floors.

    b. Preferably, entry is gained through walls breached by explosives or gunfire. Assault teams avoid windows and doors since they are usually covered by fire or are boobytrapped. If tanks are attached to the company, they can breach the wall by main gunfire for one entry point (Figure F-3).

Figure F-3. Main gun used to breach exterior.

c. Just before the rush of the assault force, suppressive fires on the objective should be increased by the support force and continued until masked by the advancing assault force. Once masked, fires are shifted to upper windows and continued until the assault force has entered the building. At that time, fires are shifted to adjacent buildings to prevent enemy withdrawal or reinforcement.

d. Assault parties quickly close on the building. Before entry through the breached wall, a hand grenade is cooked off (pin pulled, safety lever released, and held for two seconds before being thrown) and vigorously thrown inside. Immediately after the explosion, assault parties enter and spray the interior, using three-round bursts and concentrating on areas of the room that are possible enemy positions.

e. Once inside the building, the priority tasks are to cover the staircase leading to upper floors and the basement, and to seize rooms that overlook approaches to the building. These actions are required to isolate enemy forces within the building and to prevent reinforcement from the outside. The assault parties clear each ground floor room and then the basement.

(1) The assault team leader determines which room(s) to clear first.

(2) The support team provides suppressive fire while the assault team is systematically clearing the building. It also provides suppressive fire on adjacent buildings to prevent enemy reinforcements or withdrawal. The support team destroys any enemy trying to exit the building.

(3) After assault team 1 establishes a foothold in the building, a soldier from assault team 2 positions himself to provide security for the foothold. Assault team 1 proceeds to clear the first room.

(a) Soldier 1 throws a grenade into the room and yells, "Frag out," to alert friendly personnel that a grenade has been thrown toward the enemy.

> **WARNING**
>
> **If walls and floors are thin, fragments from hand grenades can injure soldiers outside the room.**

(b) After the grenade explodes, soldier 2 enters the room and positions himself to the left of the door up against the wall, spraying the room with automatic fire and scanning the room from left to right. (Soldiers 1 and 3 provide outside room security.) Soldier 2 will give a voice command of "All clear" before soldier 3 enters the room.

(c) Soldier 3 shouts, "Coming in," and enters the room. He positions himself to the right of the door up against the wall and scans the room from right to left. (Soldier 2 provides inside room security and soldier 1 provides outside room security.)

(d) Soldier 1 positions himself up against the hall wall so that he can provide security outside the room and can also observe into the room.

(e) Soldier 3 proceeds to clear the room while soldier 2 provides inside room security. Soldier 1 remains at his outside security position.

(f) After the room is cleared, the clearing team shouts, "Coming out," and proceeds to clear the next room(s). A soldier from the second assault team positions himself to cover the cleared room. The cleared rooms are marked IAW unit SOP.

(e) This procedure is continued until the entire floor is cleared.

f. If the assault force is preparing to clear a building from the top floor down, they should gain entrance through a common wall or the roof of an adjoining building. Accompanied by the company's attached engineer squad, the force uses a demolition charge to breach the wall and to gain entrance to the top floor. Access to lower floors and rooms may be gained by breaching holes in the floor and having the soldiers jump or slide down ropes to the lower floors. Stairs can be used if they are first cleared.

g. When using the top-to-bottom method of clearing, security requirements remain the same as for other methods (Figure F-4). After the floor is breached to gain access to a lower floor, a grenade is allowed to cook off and is dropped to the lower room. A soldier then sprays the lower room with gunfire using three-round bursts and drops through the mousehole.

Figure F-4. Upper floors secured.

h. Soldiers must avoid clearing rooms the same way each time by varying techniques so that the enemy cannot prepare for the assault (Figure F-5, page F-8). As rooms are cleared, doors should be left open and a predetermined mark (cloth, tape, spray paint) placed on the doorjamb or over the door.

i. If there is a basement, it should be cleared as soon as possible, preferably at the same time as the ground floor. The procedures for clearing a basement are the same as for any room or floor, but important differences do exist. Basements often contain entrances to tunnels such as sewers and communications cable tunnels. These should be cleared and secured to prevent the enemy from infiltrating back into cleared areas.

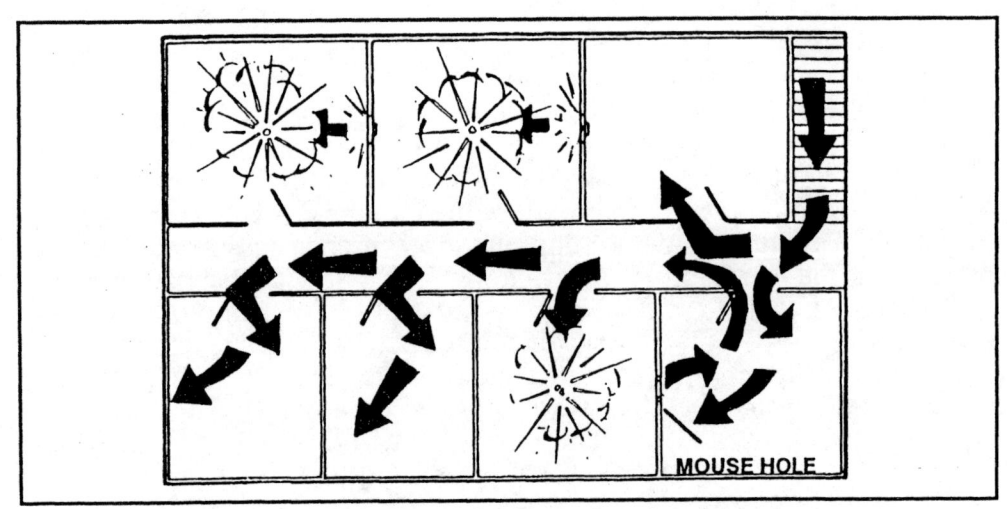

Figure F-5. Varying techniques for clearing rooms.

j. The most common types of buildings that must be cleared are brick buildings, brick houses, box-wall buildings, heavy-clad framed buildings, and light-clad framed buildings (Figure F-6). The best way to enter a brick building is to blow a breaching hole in the side with a tank firing HEAT ammunition. If tanks are not available, a door or window in the rear of the building usually provides better cover and concealment for entry than one in the front. If there is enough cover and concealment, the assault force should enter the rear of the building at an upper level, using a fire escape or grappling hook.

Figure F-6. Building being cleared.

(1) **Brick buildings.** To clear from building to building, the best method is to move from rooftop to rooftop since the roofs of brick buildings are usually easy to breach. The walls between buildings are at least three bricks thick (total of six bricks between buildings) and require large quantities of demolitions to breach. Walls are normally easier to breach on an upper floor than a lower floor, since the walls are thinner on upper floors. If rooftops are covered by fire and if there are not enough demolitions to breach walls between buildings, clearing from rear to rear of buildings is safer than clearing from front to front. The floor plans in brick buildings are different on ground floor levels than on upper levels (Figure F-7).

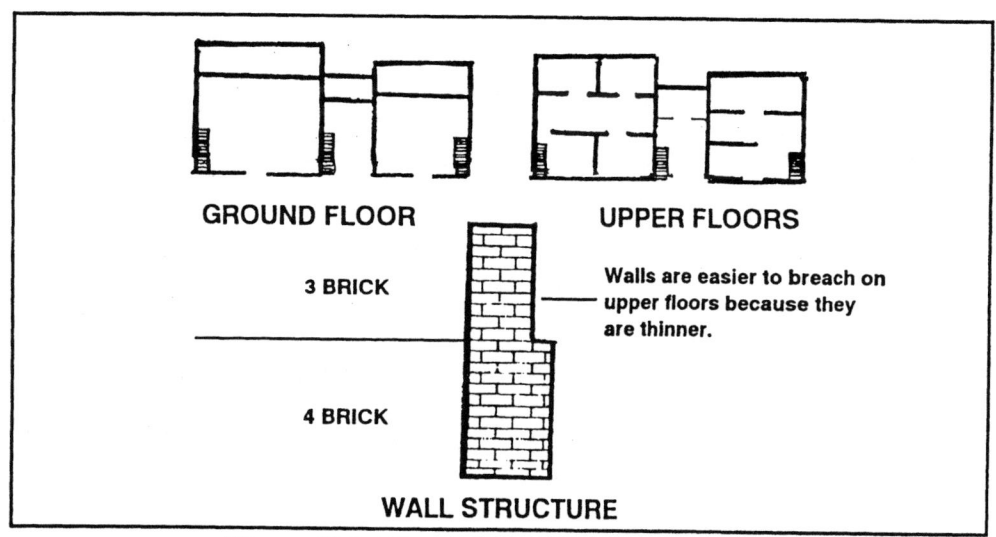

Figure F-7. Floor plans of brick buildings.

(2) **Brick houses.** Brick houses have similar floor plans on each floor (Figure F-8), therefore, ground floors are cleared the same way as upper floors.

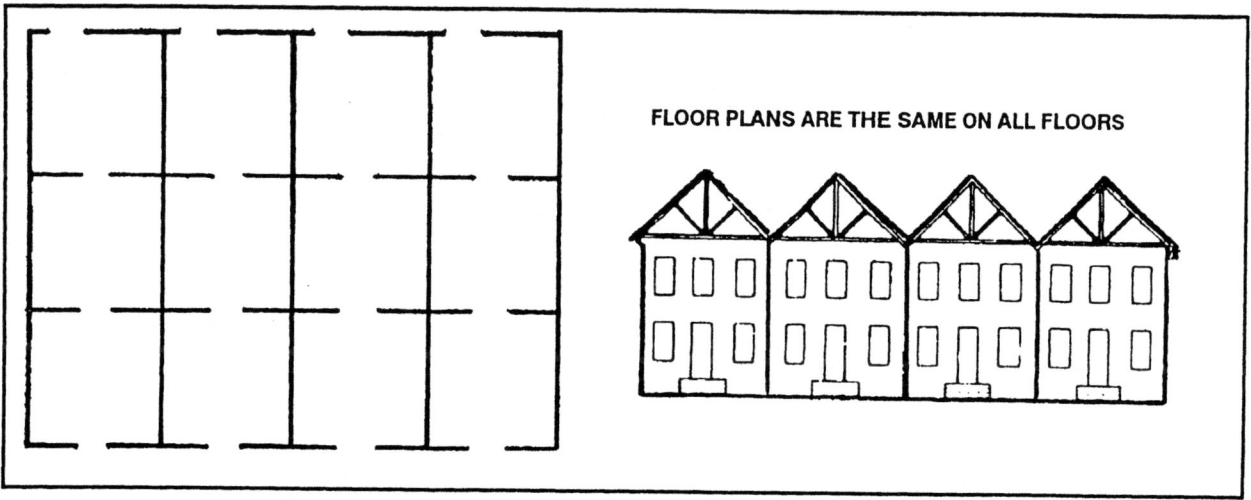

Figure F-8. Similar floor plans.

(3) *Box-wall buildings.* Box-wall buildings often have reinforced concrete walls (Figure F-9), which are difficult to breach due to the reinforcing bars. Therefore, the best way to enter is to blow down the door or to blow in one of the side windows. The floor plans of these buildings are predictable; clearing rooms is usually done from one main hallway. Interior walls are also constructed of reinforced concrete and are difficult to breach. The stairways at the ends of the building must be secured during clearing.

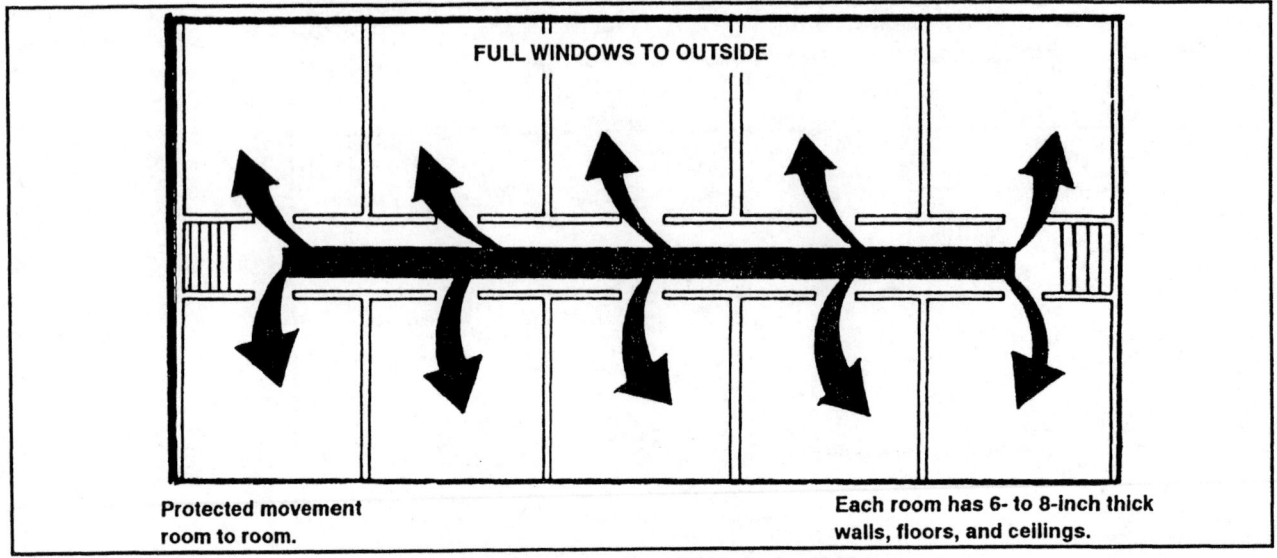

Figure F-9. Box-wall principle buildings.

(4) *Heavy-clad framed buildings.* Heavy-clad framed buildings are relatively easy to breach, because a tank can breach a hole in the cladding (Figure F-10). Their floor plans are oriented around a stairway or elevator, which must be secured during clearing. The interior walls of these buildings can be breached, although they may require use of demolitions.

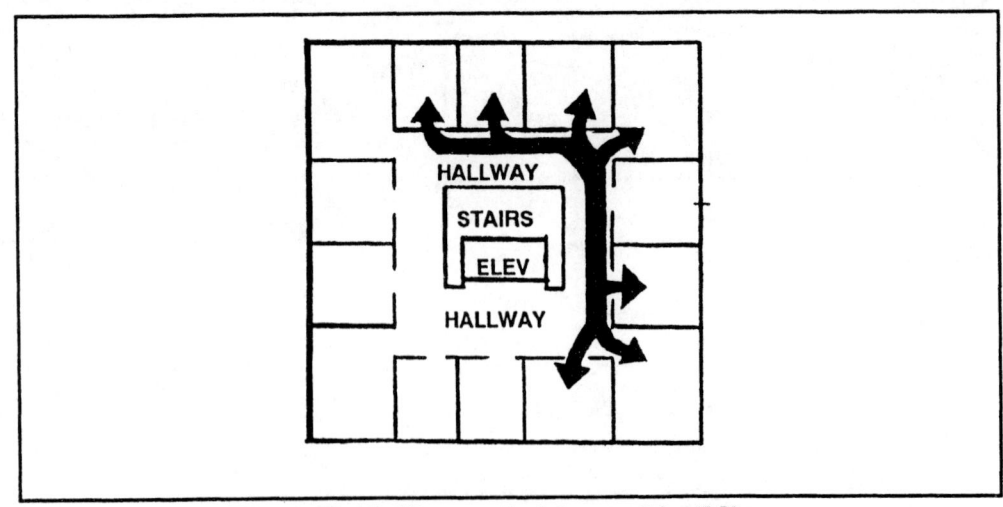

Figure F-10. Heavy-clad framed buildings.

(5) *Light-clad framed buildings.* On light-clad framed buildings (Figure F-11), the clearing tasks are usually the same: secure the central stairway and clear in a circular pattern. Walls are easier to breach since they are usually thin enough to be breached with an axe.

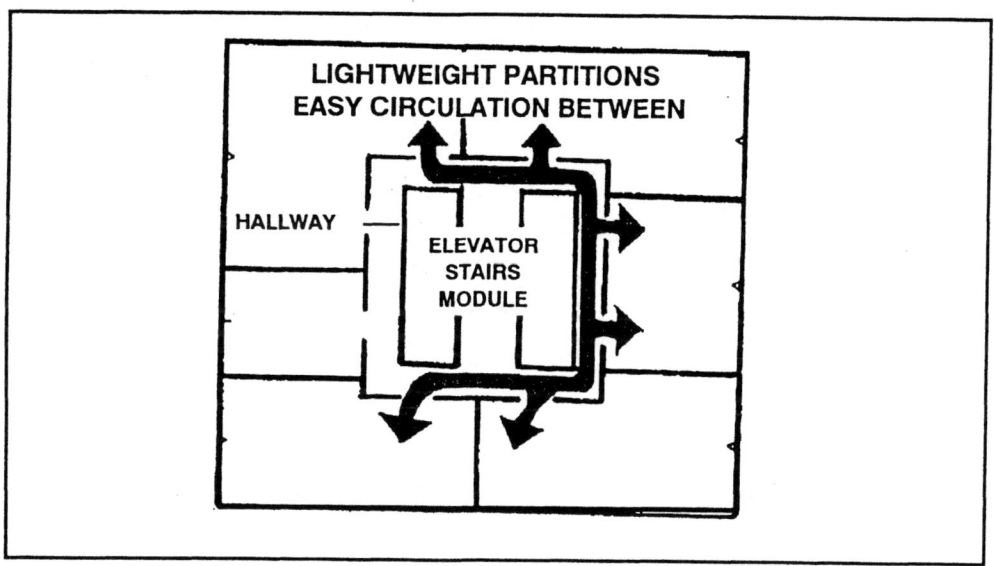

Figure F-11. Light-clad framed buildings.

F-6. REORGANIZATION

Reorganization of the assault force in a cleared building must be quick to repel enemy counterattacks and to prevent the enemy from infiltrating back into the cleared building. After securing a floor (bottom, middle, or top), selected members of the assault force are assigned to cover potential enemy counterattack routes to the building. Those sentinels alert the assault force and place a heavy volume of fire on enemy forces approaching the building. They guard—

- Enemy mouseholes between adjacent buildings.
- Covered routes to the building.
- Underground routes into the basement.
- Approaches over adjoining roofs.

As the remainder of the assault force completes search requirements, they are assigned defensive positions. After the building has been cleared, the following actions are taken:

- Resupplying and redistributing ammunition.
- Marking the building to indicate to friendly forces that the building has been cleared.
- Assuming an overwatch mission and supporting an assault on another building.
- Treating and evacuating wounded personnel.
- Developing a defensive position if the building is to be occupied for any period.

APPENDIX G

MILITARY OPERATIONS IN URBAN TERRAIN (MOUT) UNDER RESTRICTIVE CONDITIONS

Throughout the operational continuum, and especially during LIC operations, commanders can expect to encounter restrictions on their use of firepower, CS and CSS during MOUT. Basic doctrinal principles remain the same, but the tactics, techniques, and procedures may have to be modified to stay within established rules of engagement and to avoid unnecessary collateral damage.

G-1. PRECISION AND SURGICAL MOUT

Unlike MOUT under regular conditions, where the mission is to defeat the enemy while limiting noncombatant and collateral damage, precision and surgical MOUT require significant alterations in the METT-T and in political considerations. These alterations cause modifications to the way units fight.

    a. **Precision MOUT.** Under precision MOUT conditions, either the enemy is mixed with the noncombatants or political considerations require that the ROE be more restrictive than under regular MOUT conditions. Tightening the ROE requires strict accountability of individuals and unit actions. When preparing for precision MOUT operations, the commander must realize that not only is the ROE changing but the TTP also. These changes will require that the soldiers be given time to train for the specific operation. For example, when clearing a room, the established procedure of throwing a grenade into the room first must be modified to account for noncombatants interspersed with the enemy. Regular Army units are more likely to operate under precision MOUT than under surgical MOUT.

    b. **Surgical MOUT.** Operations conducted under surgical MOUT conditions include raids, strikes, or recovery operations in a MOUT environment and are usually conducted by joint special operating forces. Though regular units may not be involved in the actual operation, they may support the operation by isolating the area of operations.

G-2. RULES OF ENGAGEMENT

Rules of engagement for tactical forces come from the unified commander. They are based on NCA guidance, mission, threat, laws of war, and host nation or third-world country constraints on force deployment.

The political concerns used to develop ROE may conflict with the physical security needs of the force. Political needs should be weighed against the risks to the mission and force itself. They should be practical, realistic, and enforceable. Regardless of the situation, forces must operate in a highly constrained environment. This requires the patience, training, and dedication of the military force. An example of ROE used during Just Cause is shown in Figure G-1. It is not intended to be used as a sole source document for developing ROE.

ALL EMEMY MILITARY PERSONNEL AND VEHICLES TRANSPORTING THE ENEMY OR THEIR SUPPLIES MAY BE ENGAGED SUBJECT TO THE FOLLOWING RESTRICTIONS:

a. When possible, the enemy will be warned first and asked to surrender.

b. Armed force is the last resort.

c. Armed civilians will only be engaged in self defense.

d. Civilian aircraft will not be engaged without approval from above division level unless it is in self defense.

e. Avoid harming civilians unless necessary to save U.S. lives. If possible, try to arrange for the evacuation of civilians prior to any U.S. attack.

f. If civilians are in the area, do not use artillery, mortars, armed helicopters, AC-130s, tube- or rocket-launched weapons, or M551 main guns against known or suspected targets without the permission of a ground maneuver commander C or higher (for any of these weapons).

g. If civilians are in the area, all air attacks must also be controlled by a FAC or FO.

h. If civilians are in the area, close air support (CAS), white phosphorus, and incendiary weapons are prohibited without approval from above division level.

i. If civilians are in the area, infantry does not shoot except at known enemy locations.

j. If civilians are not in the area, you can shoot at suspected enemy locations.

k. Public works such as power stations, water treatment plants, dams, or other utilities may not be engaged without approval from above division level.

l. Hospitals, churches, shrines, schools, museums, and any other historical or cultural site will not be engaged except in self defense.

m. All indirect fire and air attacks must be observed.

n. Pilots must be briefed for each mission on the location of civilians and friendly forces.

o. No booby traps. No mines except as approved by division commander. No riot control agents without approval from above division level.

p. Avoid harming civilian property unless necessary to save U.S. lives.

q. Treat all civilians and their property with respect and dignity. Before using privately owned property, check to see if any publicly owned property can substitute. No requisitioning of civilian property without permission of a company level commander and without giving a receipt. If an ordering officer can contract for the property, then do not requisition it. No looting. Do not kick down doors unless necessary. Do not sleep in their houses. If you must sleep in privately owned buildings, have an ordering officer contract for it.

Figure G-1. Example of Just Cause ROE.

ALL EMEMY MILITARY PERSONNEL AND VEHICLES TRANSPORTING THE ENEMY OR THEIR SUPPLIES MAY BE ENGAGED SUBJECT TO THE FOLLOWING RESTRICTIONS:

r. Treat all prisoners humanely and with respect and dignity.

s. Annex R to the OPLAN provides more detail. Conflicts between this card and the OPLAN should be resolved in favor of the OPLAN.

DISTRIBUTION: 1 per every trooper deployed to include all ranks.

SUPPLEMENTAL RULES OF ENGAGEMENT FOR SELECTED RECURRING OPERATIONS

1. CONTROL OF CIVILIANS ENGAGED IN LOOTING.

    a. Senior person in charge may order warning shots.

    b. Use minimum force but not deadly force to detain looters.

    c. Defend Panamanian (and others) lives with minimum force including deadly force when necessary.

2. ROADBLOCKS, CHECKPOINTS AND SECURE DEFENSIVE POSITIONS:

    a. Mark all perimeter barriers, wires, and limits. Erect warning signs.

    b. Establish second positions to hastily block those fleeing.

    c. Senior person in charge may order warning shots to deter breach.

    d. Control exfiltrating civilians with minimum force necessary.

    e. Use force necessary to disarm exfiltrating military and paramilitary.

    f. Attack to disable, not destroy, all vehicles attempting to breach or flee.

    g. Vehicle that returns or initiates fire is presumed hostile. Fire to destroy hostile force.

    h. Vehicle that persists in breach attempt is presumed hostile. Fire to destroy hostile force.

    i. Vehicle that persists in flight after a blocking attempt IAW instruction 2b is presumed hostile. Fire to destroy hostile force.

Figure G-1. Example of Just Cause ROE (continued)..

> **3. CLEARING BUILDINGS NOT KNOWN TO CONTAIN HOSTILE FORCE:**
>
> a. Warn all occupants to exit.
>
> b. Senior person in charge may order warning shots to induce occupants to exit.
>
> c. Do not attack hospitals, churches, shrines, or schools, museums, and any historical or cultural sites except in self-defense.
>
> d. Respect and minimize damage to private property.
>
> e. Use minimum force necessary to control the situation and to ensure the area is free of hostile force.

Figure G-1. Example of Just Cause ROE (continued).

G-3. IMPACT OF CIVILIANS ON MOUT

The presence of large concentrations of civilians constrains the applications of combat power during tactical operations.

a. **Mobility.** Civilians attempting to escape over roads can block military movement. Commanders should plan routes to be used by civilians and should seek the assistance of the military and civil police in traffic control.

b. **Firepower.** The presence of civilians and the desire to limit collateral damage can restrict the use of fires and reduce the firepower available to a commander. Selected areas may be designated "no fire" areas to prevent civilian casualties and damage to urban structures. Other areas may be limited to small arms and grenades, with prohibitions on air strikes, artillery, mortars, and flame. Target acquisition and direction-of-fire missions will be complicated by the requirement for positive target identification. Detailed guidance on the use of firepower in the presence of civilians will be published by the division G3. When no guidance is available, the general rules of the law of land warfare apply.

G-4. CIVILIAN INFLUENCE ON ENEMY AND FRIENDLY OPERATIONS

Civilians in an urban environment, and the political setting, will influence both enemy and friendly operations.

a. **Enemy Operations.** These operations will cover the spectrum from terrorism to well-organized military operations. The enemy may be special purpose forces or insurgents that have the ability to operate freely throughout a city due to having the appearance of civilians. Conventional enemy forces may choose to occupy specific urban areas that civilians have not been able to evacuate. The swift occupation of a city may cause civilians to be trapped between opposing forces. This will enhance the enemy's ability to defend.

b. **Friendly Operations.** The most critical aspect of friendly operations will be the ROE. Examples of different ROEs that US forces used during urban battles are Aachen during WWII in 1944 and Panama City "Just Cause" in 1990. Aachen typified ROE that permitted the free use of any type munition to eliminate the enemy. Panama City, on the other hand, showed US forces operating under restrictive ROE.

(1) Offensive operations by friendly forces must be well planned. These plans will take into account the potential use of precision-guided munitions to achieve identified objectives while precluding unnecessary collateral damage. Precision operations will include sniper and countersniper operations by both special forces personnel and conventional forces.

(2) Tanks, CFVs, BFVs, and APCs can enhance a unit's ability to apply direct fire to specified portions of buildings occupied by the enemy. This highly accurate fire can suppress the enemy and create breach points for assault teams to enter. Direct fire can also induce the enemy to surrender. These vehicles can provide overwatching fire for infantry, but before moving into the city, infantry must clear ahead to provide protection from enemy antitank teams. These vehicles can also do the following:

(a) Isolate the objective building by occupying positions to prevent enemy withdrawal, reinforcements, or counterattack.

(b) Breach roadblocks, walls, or other obstacles en route to the objective.

(c) Establish a roadblock or barricade.

(d) Obscure the enemy's view using the BFV's and tank's smoke generators and smoke grenades.

(e) Evacuate casualties from the immediate battle area.

(f) Evacuate PWs to the unit collection point.

(g) Quickly resupply dismounted forces.

(3) When civilian personnel are present or are thought to be present in the objective area, the following room-clearing procedure is used:

(a) Rules of engagement must be identified and known by all personnel before entering or clearing a room or building. To preclude unnecessary collateral damage, ROE may dictate precision-guided munitions or weapons be used to eliminate the enemy.

(b) Nonstandard entrances to buildings or rooms should be used if available from a previous show of force or earlier conflicts. Close air support, tanks, CEVs, or direct fire artillery may facilitate these breach points.

(c) If the door must be used to enter the room, one member of the assault force tries to push open the door while the other two members provide security and prepare to enter the room.

(d) The team leader enters the room and positions himself to the left of the door up against the wall, scanning the room for weapons. He shouts, "Next man in, right" before the next man enters the room.

(e) The next soldier shouts, "Coming in, right" and enters the room. He positions himself to the right of the door up against the wall and scans the room from right to left. The team leader provides inside room security and the soldier left outside provides outside room security. If additional personnel are required, the team leader will shout, "Next man in," which will require the outside security man to enter the room.

(f) Psychological operations or civil affairs teams can help remove civilians before a battle starts. These teams can help plan a show of force. Assigned or attached heavy weapons can engage an unoccupied corner of a building. The noise and destruction can induce civilians to leave the building. Once the objective area has been isolated, PSYOP teams can also be used to induce enemy personnel to surrender.

G-5. FIRE SUPPORT
Fire support consists of field artillery, mortars, close air support, and naval gunfire. (See Chapter 6 and FM 7-20 for more details.)

   a. **Field Artillery.** Applying firepower must always reflect the principle of minimum-essential force. FA support normally provided to light infantry divisions consists of the towed 105-mm howitzer. However, FA units that augment division artillery can provide weapons of larger calibers. Their use in a MOUT environment must be carefully planned due to the great potential for killing noncombatants, unnecessary collateral damage, or the loss of a valuable piece of equipment. When in range of the objective area, FA units can be used to emplace FA-delivered FASCAM to enchance the security of the force. Copperhead rounds fired by 155-mm FA units can be terminally guided by the FOs. They can attack hardened point targets or enemy armored vehicles by using a man-portable laser target designator (Figure G-2). FA units can provide both indirect and direct fire in MOUT. Direct fire can accomplish the following:

   (1) Establish breach points.
   (2) Induce surrender of enemy personnel through a show of force.
   (3) Eliminate enemy defensive positions.
   (4) Create obstacles and rubble to restrict enemy freedom of movement.
   (5) Fire printed PSYOP product.

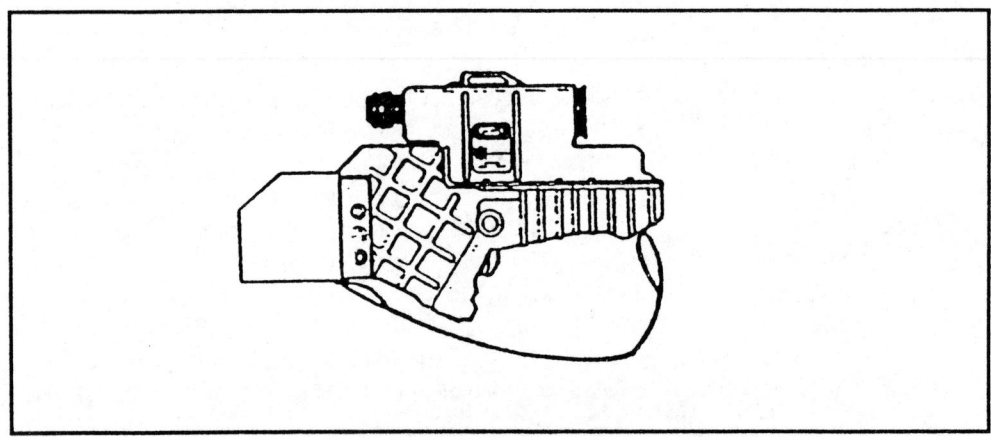

Figure G-2. AN/PAQ-1 laser target designator.

   b. **Mortars.** Mortars are the most accurate indirect fire weapon that can be used in an urban environment. These weapons have a high angle of fire, which enables placing fire accurately between or on buildings. Mortars do not have the destructive capability of FA due to their limited ability to penetrate most structures. Commanders should be aware that there will be no-fire areas, restrictive fire lines, and restrictions on the type of ammunition that can be fired. Illumination rounds will be greatly needed.

   c. **Close Air Support.** CAS assets should be used when other fire support means cannot fire on the target or the firepower of tactical fighter aircraft is necessary to obtain the desired results. If the ROE permits, tactical bombers capable of carrying more ordnance can be brought in to attack targets of tactical importance. CAS can also deliver laser-guided munitions into a target area. These munitions are capable of mass destruction and their employment could possibly be reduced or precluded by ROE.

G-6. AIR DEFENSE

Air defense combines all active and passive measures to counter hostile air operations. In a LIC, the hostile air threat may be none, minimal, or existing.

a. ADA weapons may remain in the rear staging area if the threat is none or minimal. Vulcans can be employed in a MOUT environment in the following missions:

(1) Establish breach points.
(2) Conduct show-of-force operations to induce surrender.
(3) Suppress or eliminate enemy positions.
(4) Isolate buildings.
(5) Establish base security.
(6) Establish convoy security.

b. Aerial fire support can be provided by either fixed-wing aircraft or helicopters.

(1) Fixed-wing aerial fire support may come from USAF, USN, or USMC units. The type of unit providing support, the aircraft, and the mix of ordnance carried all affect the fire support planning and coordination process. Some aircraft have a night and all-weather strike ability enabling them to support the force during any level of visibility. Operations during weather that limits visibility to less than 3 nautical miles are still somewhat restricted. The fire support coordinators must ensure that the correct aircraft are requested and employed effectively on the enemy. The tactical CP directs and adjusts aerial fires in the objective area.

(a) The unit can use ground laser target designators to pinpoint targets for air strikes, as well as electronic navigation aids to permit nonvisual air strikes (beacon bombing). The rifle company FIST or the tactical CP can control a laser-designated standoff air strike (Figure G-3).

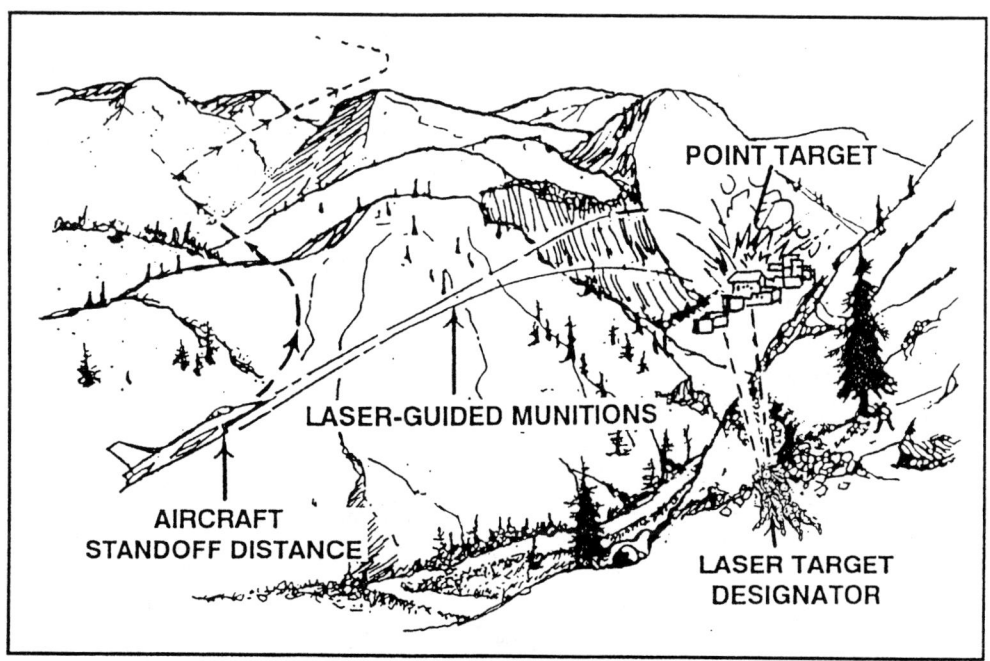

Figure G-3. Standoff air strike.

FM 90-10-1

(b) AC-130 aircraft (Figure G-4) provide an invaluable combination of firepower, night observation, and illumination, communications, and long loiter time. A well-planned and executed suppression of enemy air defense (SEAD) program, coupled with ECM directed against enemy ADA units, normally permits the use of AC-130 aircraft.

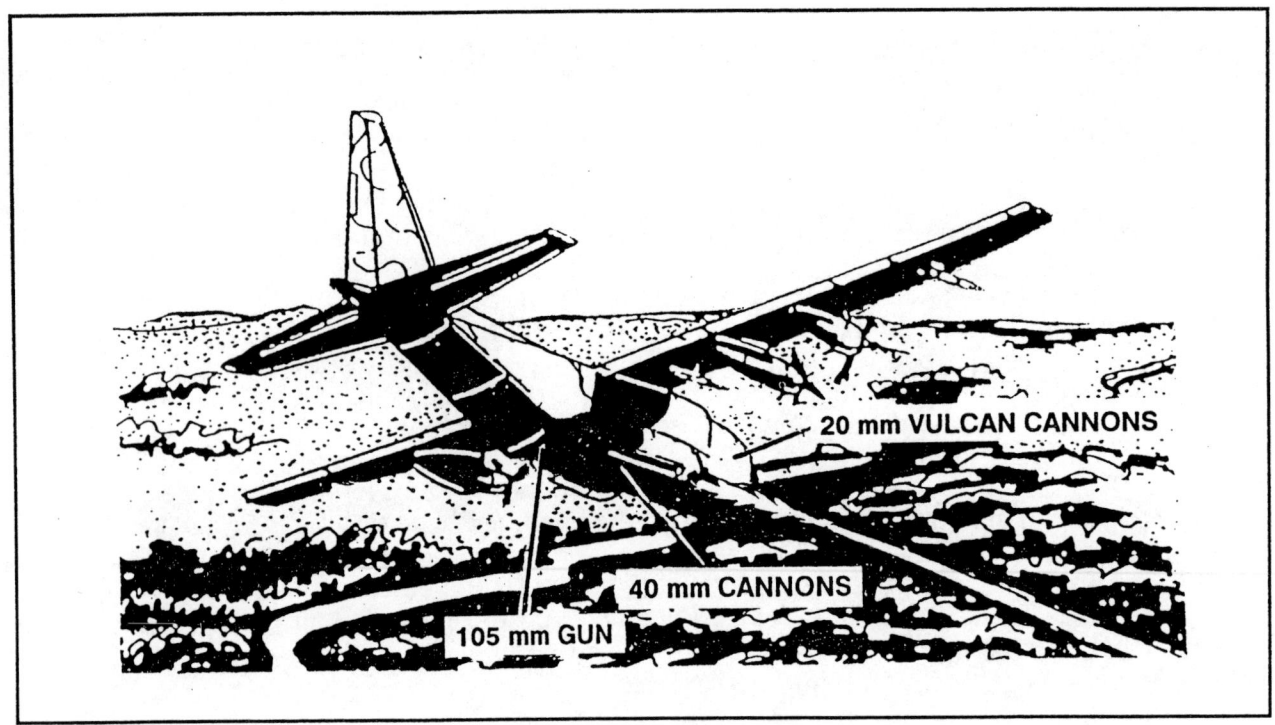

Figure G-4. AC-130 aircraft.

(2) If attack helicopters are used to support an operation, planned indirect fires are normally delivered along entry and exit corridors. Attack helicopters approach and depart the objective area using nap-of-the-earth flight profiles. Fires from armed helicopters are normally requested and controlled by the company FSO or one of his FOs, operating on a special ground-to-air net. The laser target designator may be used to precisely identify targets for the AH-64 Apache. Friendly unit locations may be marked by smoke, panels, lights, mirrors, or infrared sources.

c. Commanders must plan for the possibility that enemy may be supported by an outside air threat. Commanders must plan for such an attack by hostile or sympathetic forces.

G-7. COMMAND AND CONTROL

Leadership is a vital element of the command and control system, which includes communications. Performing assigned missions within the constraints of status of forces agreements and adhering to strict ROE require diplomatic leadership for success.

a. A brigade may consist of light infantry, mechanized infantry armor, air assault, or airborne units, or any combination of these. Commanders must task organize their available assets to exploit each unit's unique capabilities. All elements must have assigned missions and specified areas of

operation. To ensure success, the commander and his staff must concentrate on—
- Anticipating the enemy.
- Indirect approaches.
- Deception and effective OPSEC.
- Speed and violence.
- Flexibility and reliance on the initiative of junior leaders.
- Rapid decision making.
- Clearly designated main effort.

b. Communication, if effective in an urban environment, greatly enhances command and control. Communications sites are prime targets and personnel must employ all measures to protect and defend them.

(1) FM radio operations are reduced due to power output and the inability to maintain a line of sight for radio signals. However, communication with standard FM radios is difficult inside a large building during clearing operations.

(2) Retransmission stations can be used to help alleviate communication problems in any MOUT environment. Aerial platforms are the most effective means, if available.

(3) Though limited in number, AM radios within the infantry battalion or brigade are more desirable in a MOUT environment. AM assets should be used for those critical radio modes.

(4) Antennas must not be placed in obvious locations to avoid being targeted. Hanging antennas on the opposite side of the building from which the enemy is attacking will help to mask friendly communication from possible enemy interception. It will also make communication with adjacent units better.

(5) Wire is the most effective means of communications. Units should take the necessary wire to have all subordinate units hooked into the communication loop.

(6) Other means of communications include local telephone lines, messenger services, or visual signals.

G-8. INTELLIGENCE PREPARATION OF THE BATTLEFIELD

Low intensity conflict in an urban environment has an increased focus on urban terrorism, civil disturbance, and combat operations. No matter which situation applies, combat in urban terrain is expected to divide into many small-unit battles fought by battalions, companies, platoons, and squads, or into small assault groups in confined areas.

a. Battlefield area evaluation for urban LIC operations involves the analysis of the urban area and a definition of the actual area to be considered. Threat forces must be identified as conventional, an urban insurgency, a terrorist group, or a guerrilla war that has spread to the urban area.

b. Built-up areas are normally classified by size. Areas within cities and towns are classified by individual buildings and street layout patterns. These patterns have been categorized into five basic layouts, which are discussed in Chapter 2.

c. Key facilities and buildings are targets for enemy personnel. Their destruction can hinder the capabilities of a defending force.

NOTE: See FM 34-130 for a discussion of urban patterns, military aspects of urban terrain, cover and concealment, obstacles, and key terrain.

G-9. ENGINEERS
Engineer units provide needed mobility, countermobility, and survivability support to maneuver units. They can provide training, CS, and operational assistance to indigenous military and paramilitary forces, and can support military civic action programs that involve construction efforts.

Engineers have the necessary resources to help create breach points in buildings and to reduce obstacles. The CEV is capable of removing enemy positions and reducing obstacles with its demolition gun. Also, the armored combat earth mover and a small emplacement excavator can help emplace and reduce obstacles.

APPENDIX H

URBAN BUILDING ANALYSIS

As in other types of operations, success in urban combat depends largely on the ability to analyze the military aspects of soldiers' terrain. This appendix discusses in greater detail building analysis. Soldiers must be able to recognize certain terrain features when evaluating urban terrain. They must also be able to distinguish between mass-construction and framed buildings.

H-1. TYPES OF MASS-CONSTRUCTION BUILDINGS

Mass-construction buildings are those in which the outside walls support the weight of the building and its contents. Additional support, especially in wide buildings, comes from using load-bearing interior walls, strongpoints (called pilasters) on the exterior walls, cast-iron interior columns, and arches or braces over the windows and doors (Figure H-1). Modern types of mass-construction buildings are wall and slab structures such as many modern apartments and hotels, and tilt-up structures commonly used for industry or storage. Mass-construction buildings are built in many ways:

- The walls can be built in place using brick, block, or poured-in-place concrete.
- The walls can be prefabricated and "tilt-up" or reinforced-concrete panels.
- The walls can be prefabricated and assembled like boxes.

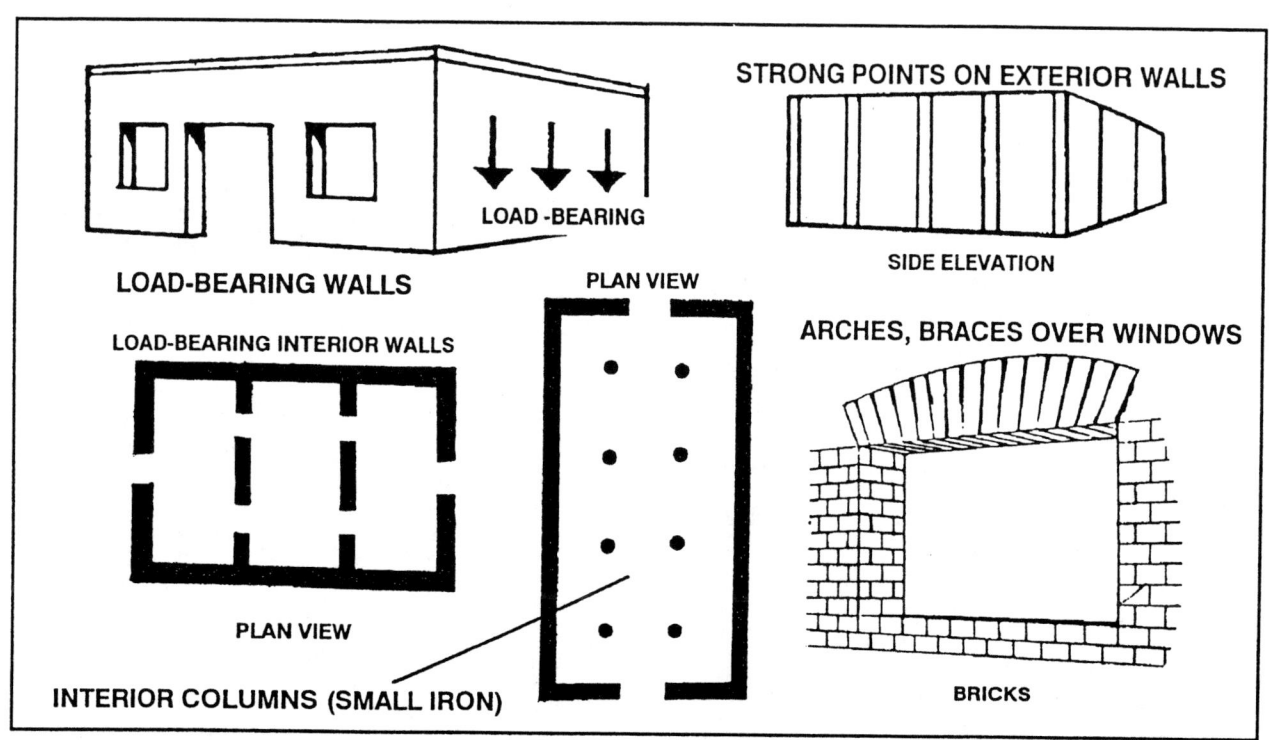

Figure H-1. Mass-construction buildings.

a. Brick buildings are the most common and most important of the mass-construction buildings. In Europe, brick buildings are commonly covered with a stucco veneer so that bricks do not show (Figure H-2). One of the most common uses of brick buildings is the small store. These buildings are found in all built-up areas but are most common in the core periphery (Figure H-3).

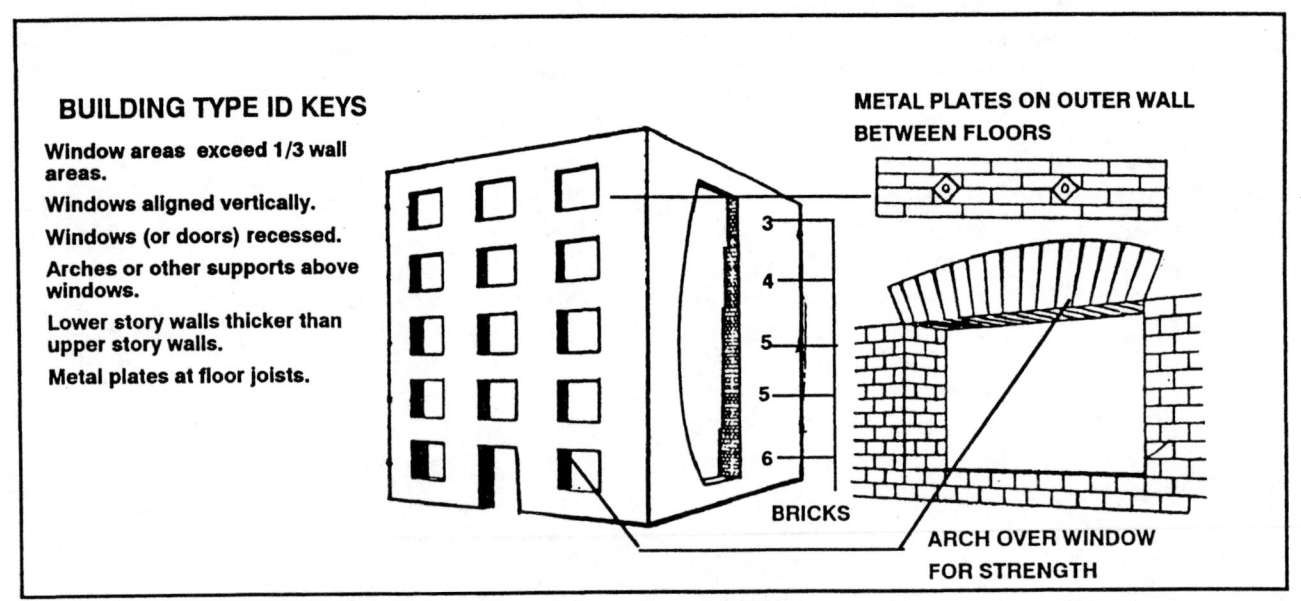

BUILDING TYPE ID KEYS
- Window areas exceed 1/3 wall areas.
- Windows aligned vertically.
- Windows (or doors) recessed.
- Arches or other supports above windows.
- Lower story walls thicker than upper story walls.
- Metal plates at floor joists.

Figure H-2. Brick buildings.

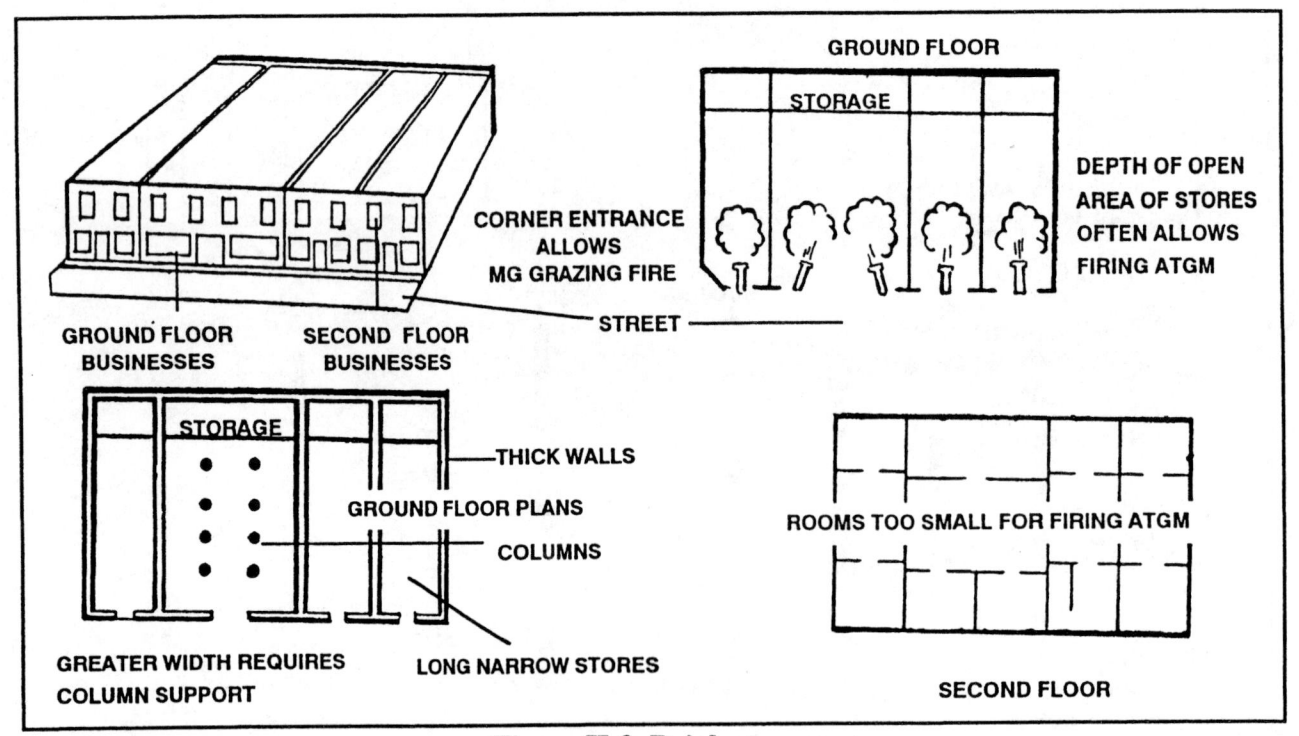

Figure H-3. Brick store.

b. Another common mass-construction building in industrial areas and along commercial ribbons is the warehouse. It is built of poured-in-place concrete reinforced with steel bars or of prefabricated walls that are "tilt-up." The walls of warehouses provide good cover, although the roof is vulnerable. The warehouses' large open bays permit firing of ATGMs and, because they are normally found in outlying areas, often afford adequate fields of fire for ATGMs. These buildings are built on slabs, which can normally support the weight of vehicles and can provide excellent cover and concealment for tanks (Figure H-4).

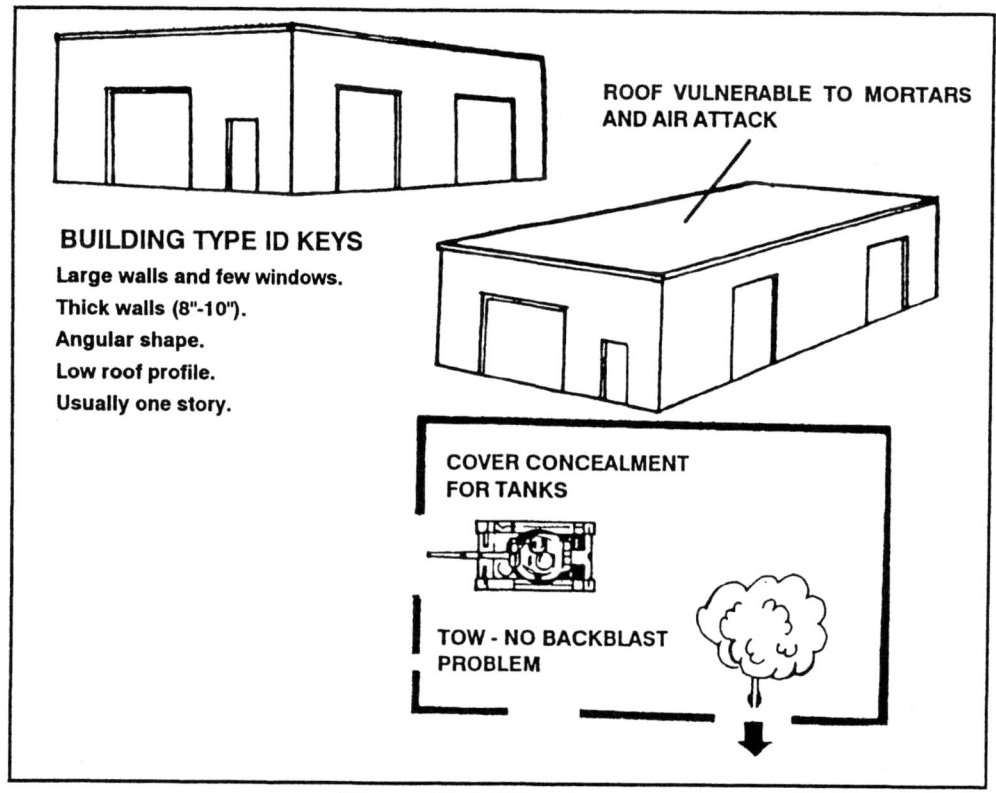

Figure H-4. Warehouse.

c. Another mass-construction building is the box-wall principle type. It is made from prefabricated concrete panels, which are made of 6- to 8-inch-thick reinforced concrete. The outside wall is often glass. The box-wall principle building provides good cover, except at the glass wall. The rooms are normally too small for ATGMs to be fired. A good circulation pattern exists from room to room and from floor to floor. These buildings are commonly used as hotels or apartments and are located in residential and outlying areas (Figure H-5, page H-4).

d. Public gathering places (churches, theaters) are mass-construction buildings with large, open interiors. The walls provide good cover, but the roof does not. The interior walls are not load-bearing and are normally easy to breach or remove. These buildings have adequate interior space for firing ATGMs. They are often located next to parks or other open areas and, therefore, have fields of fire long enough for ATGMs. Public gathering places are most common in core, core periphery, residential, and outlying high-rise areas (Figure H-6, page H-4).

FM 90-10-1

BUILDING TYPE ID KEYS

Uniform size cells (often fully vented).

Thick (6" - 8") floors, walls, ceilings (not always visible).

Windowless end walls.

FULL WINDOWS TO OUTSIDE

NO PROTECTION BUT USUALLY GOOD FIELDS OF FIRE

PROTECTED MOVEMENT ROOM TO ROOM

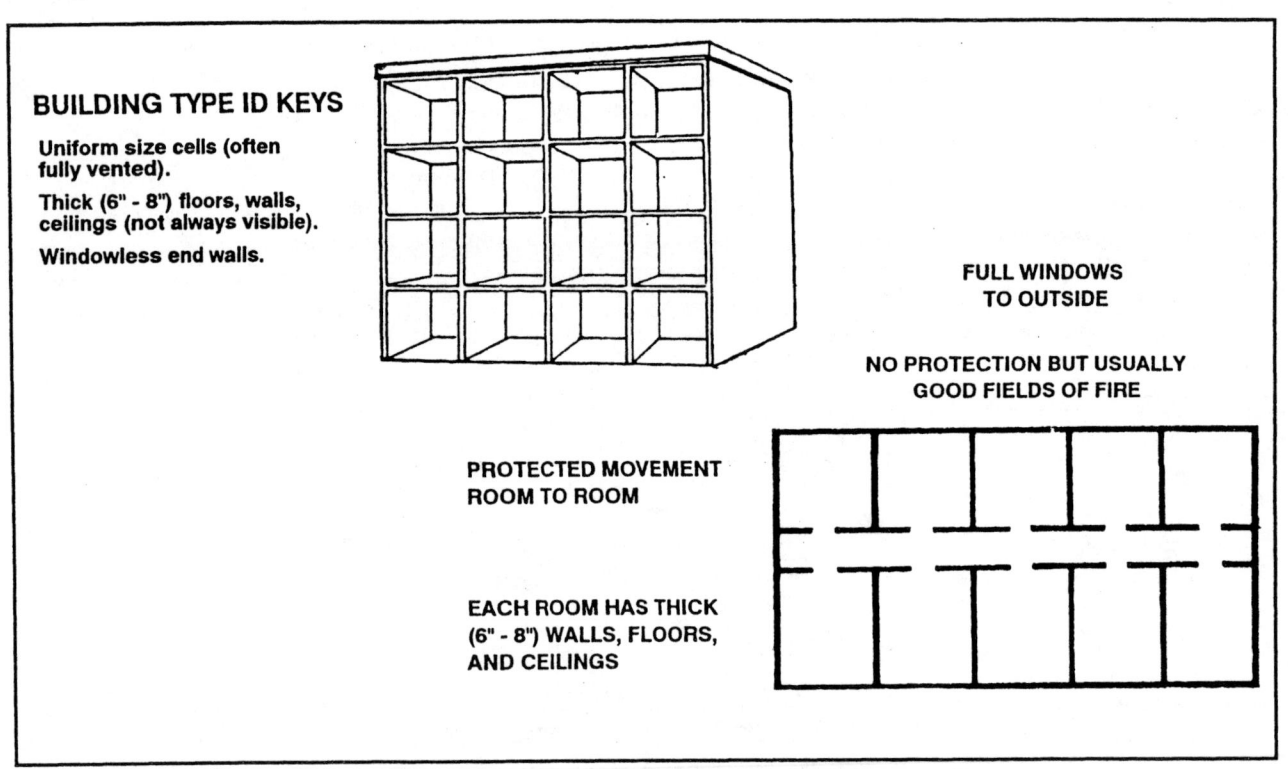

EACH ROOM HAS THICK (6" - 8") WALLS, FLOORS, AND CEILINGS

Figure H-5. Box-wall principle building.

STAGE, OFFICES

LARGE OPEN AREA

LOBBY, OFFICES

LARGE MAIN ENTRANCE

FEW, IF ANY, WINDOWS

THICK WALLS TO SUPPORT LONG ROOF SPANS

MASS-CONSTRUCTION

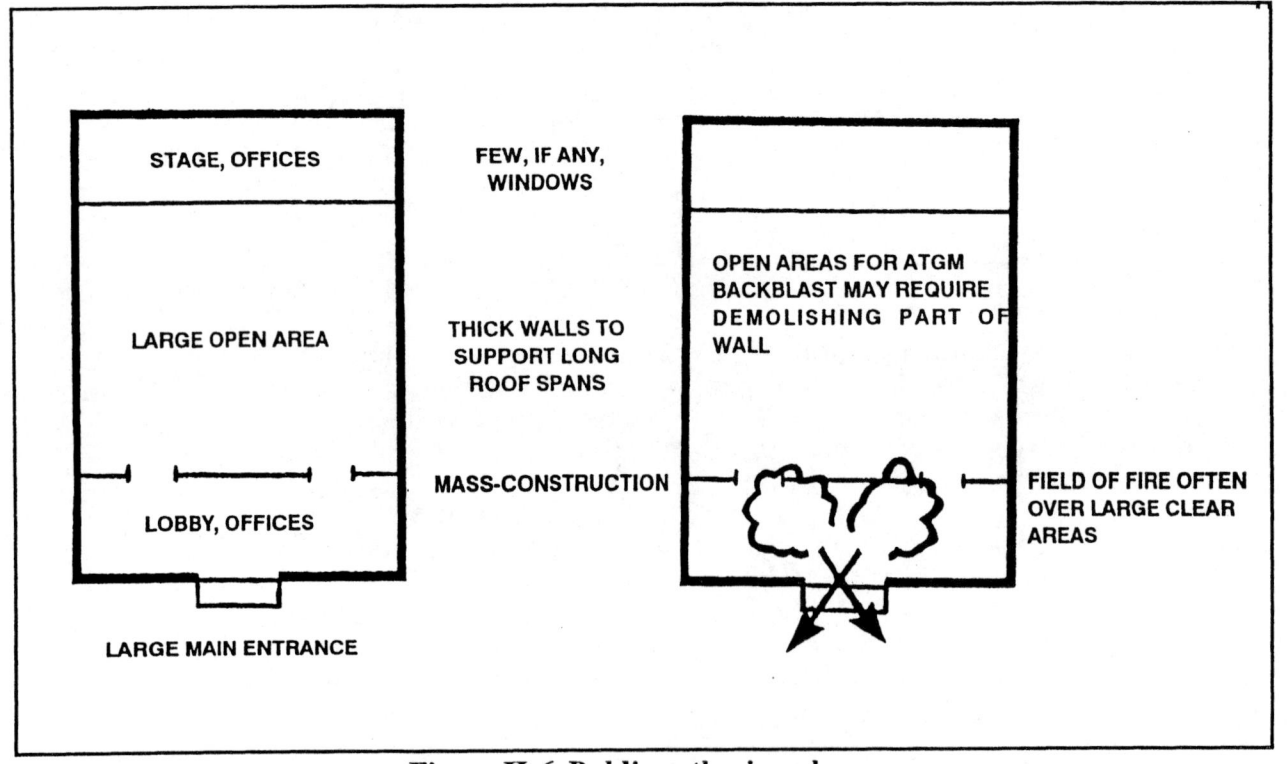

OPEN AREAS FOR ATGM BACKBLAST MAY REQUIRE DEMOLISHING PART OF WALL

FIELD OF FIRE OFTEN OVER LARGE CLEAR AREAS

Figure H-6. Public gathering places.

H-2. TYPES OF FRAMED BUILDINGS

Framed buildings are supported by a skeleton of columns and beams and are usually taller than frameless buildings (Figure H-7). The exterior walls are not load-bearing and are referred to as either heavy clad or light clad. Another type of framed building often found in cities is the garage, which has no cladding.

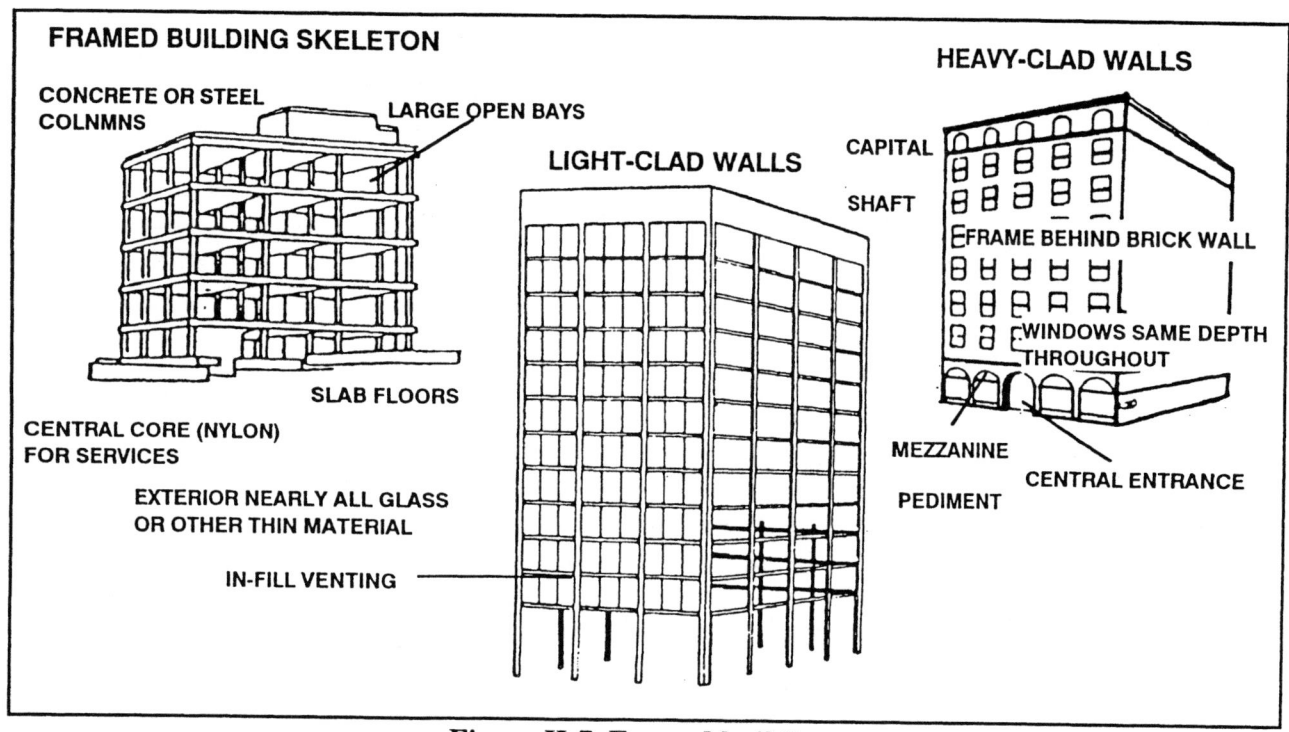

Figure H-7. Framed buildings.

a. Heavy-clad buildings were common when framed buildings were first introduced. Their walls are made of brick and block that are sometimes almost as thick as frameless brick walls, although not as protective. Heavy-clad framed buildings are found in core and core periphery areas. They can be recognized by a classic style or architecture in which each building is designed with three sections—the pediment, shaft, and capital. Unlike the brick building, the walls are the same thickness on all floors, and the windows are set at the same depth throughout. Often the frame members (the columns) can be seen, especially at the ground floor. The cladding, consisting of layers of terra cotta blocks, brick, and stone veneer, does not provide as good a cover as the walls of brick buildings. It protects against small-arms fire and light shrapnel but does not provide much cover against heavy weapons (Figure H-8, page H-6).

(1) The floor plans of these buildings depend upon their functions. Office buildings normally have small offices surrounding an interior hall. These offices have the same dimensions as the distance between columns (some large offices are as large as two times the distance between columns). These rooms are too small to permit firing of ATGMs but do provide some cover for snipers or machine gunners (Figures H-9 and H-10, page H-6).

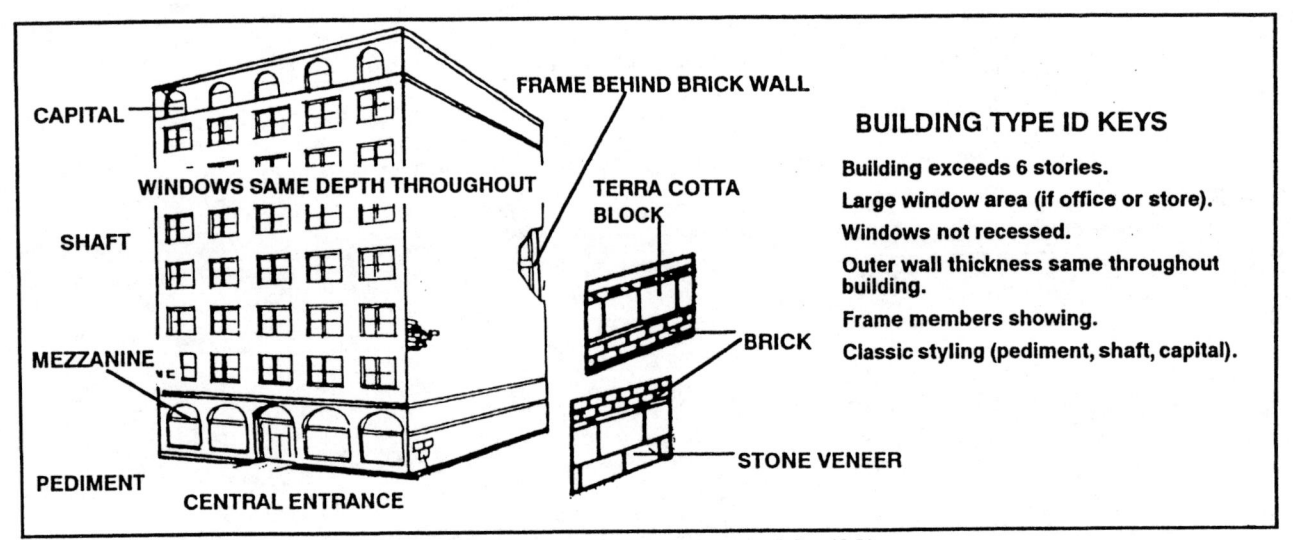

Figure H-8. Heavy-clad framed building.

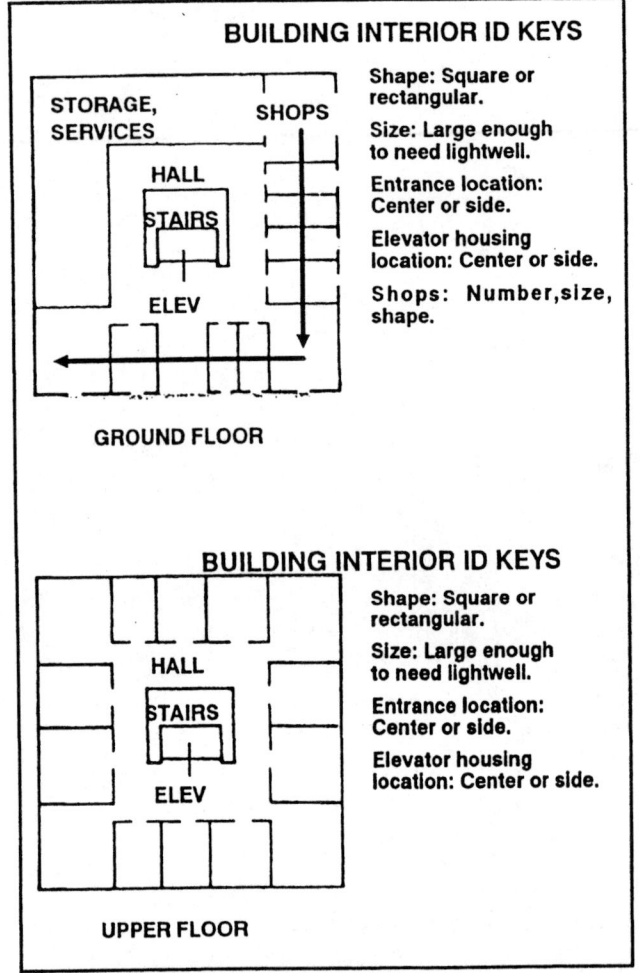

Figure H-9. Floor plan of heavy-clad framed office building.

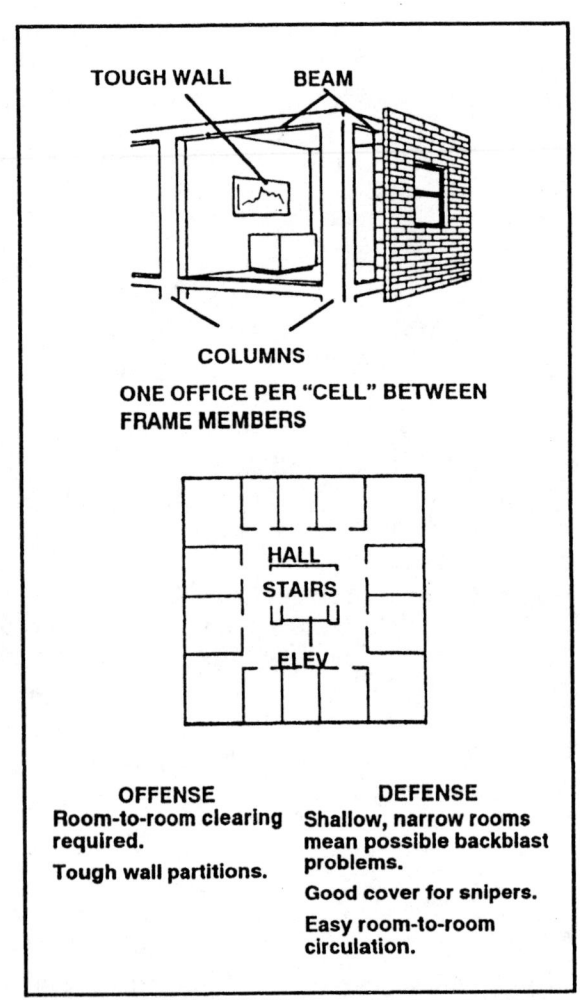

Figure H-10. Heavy-clad framed office.

(2) Department stores normally have large, open interiors (Figure H-11). Such areas permit firing ATGMs (if there are adequate fields of fire). Often a mezzanine level with a large backblast area permits firing down onto tanks. Steel fire doors often exist between sections of the store. The steel fire doors are activated by heat. Once closed, they are difficult to breach or force open, but they effectively divide the store into sections (Figure H-12).

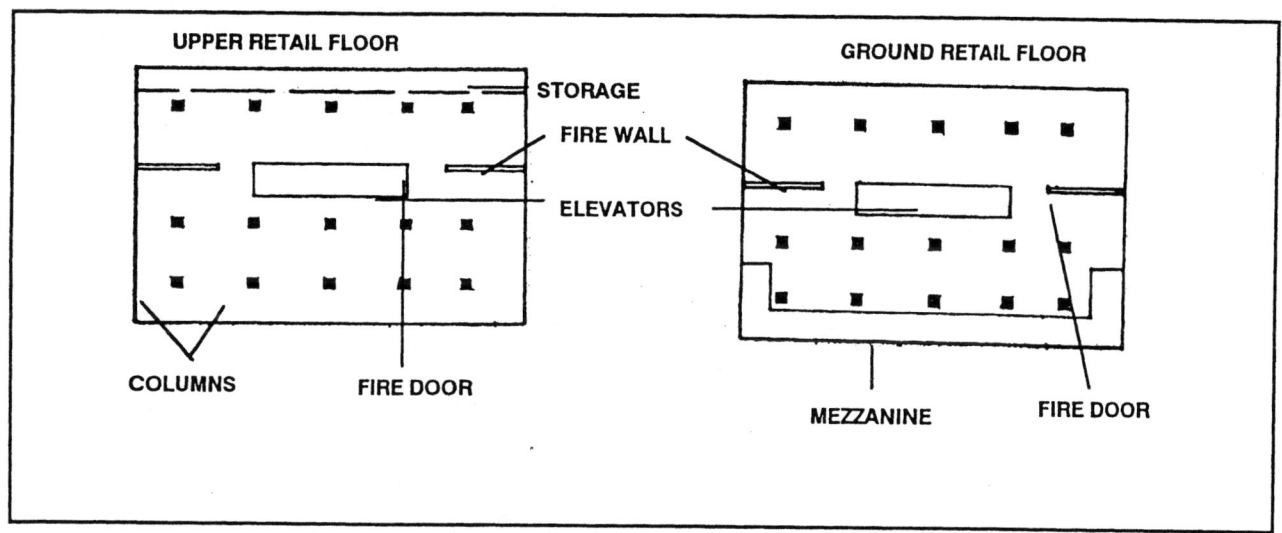

Figure H-11. Heavy-clad framed department store.

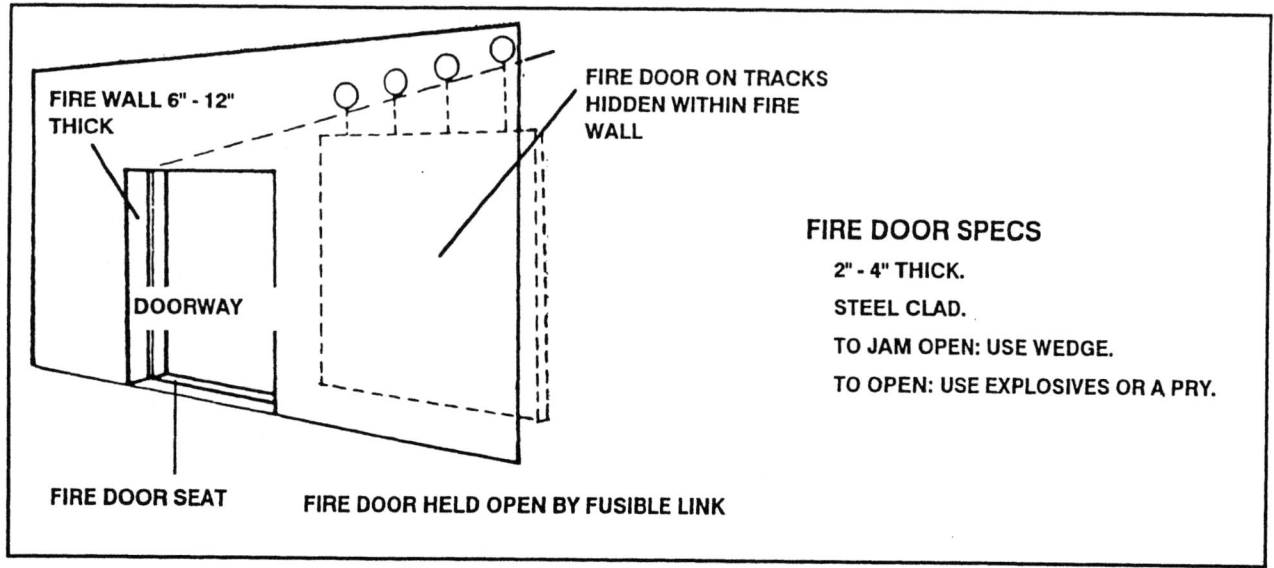

Figure H-12. Fire wall and fire door.

(3) Another type of heavy-clad framed building is used as a high-rise factory (Figure H-13, page H-8). Such buildings are normally easily recognized because the concrete beams and columns are visible from the outside. They are usually located in older industrial areas. The large windows and open interior favor the use of ATGMs. Because the floors are often made to support heavy machinery, this building provides good overhead cover.

FM 90-10-1

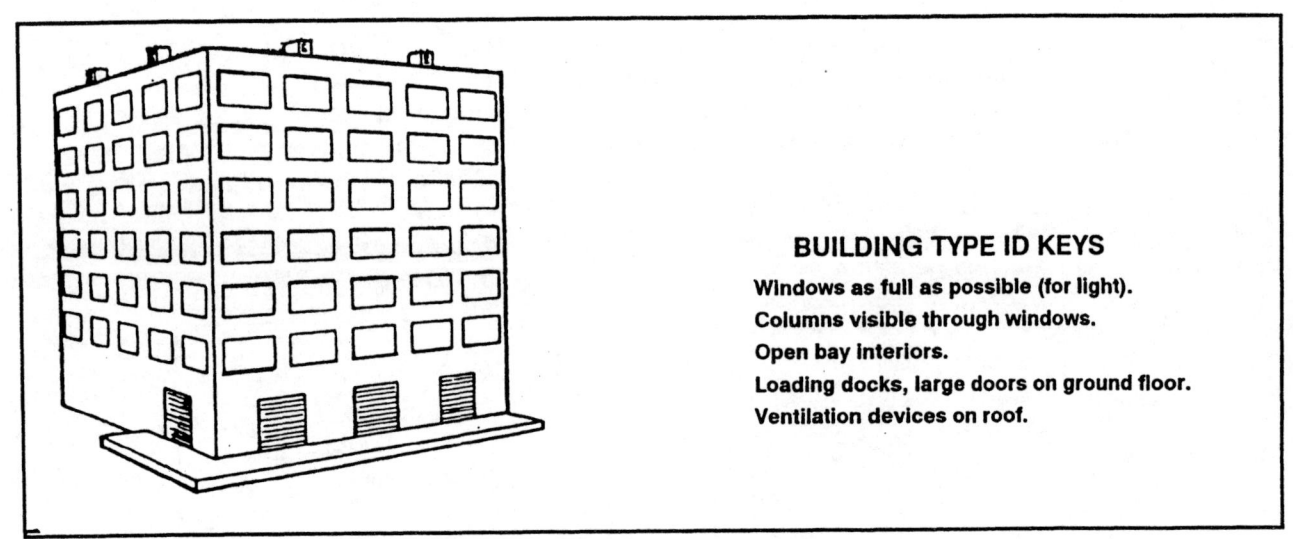

BUILDING TYPE ID KEYS

Windows as full as possible (for light).
Columns visible through windows.
Open bay interiors.
Loading docks, large doors on ground floor.
Ventilation devices on roof.

Figure H-13. High-rise factory.

b. Light-clad buildings are more modern and may be constructed mostly of glass (Figure H-7). Most framed buildings built since World War II are light-clad buildings. They are found in both core and outlying high-rise regions. Their walls consist of a thin layer of brick, lightweight concrete, or glass. Such materials provide minimal protection against any weapon. However, the floors of the buildings are much heavier and provide moderate overhead cover (Figure H-14). The rooms in light-clad framed buildings are much bigger than those in heavy-clad. This feature, along with the fact that the buildings usually stand detached from other buildings, favors the employment of ATGMs. The interior partitions are thin, light, and easy to breach (Figure H-15).

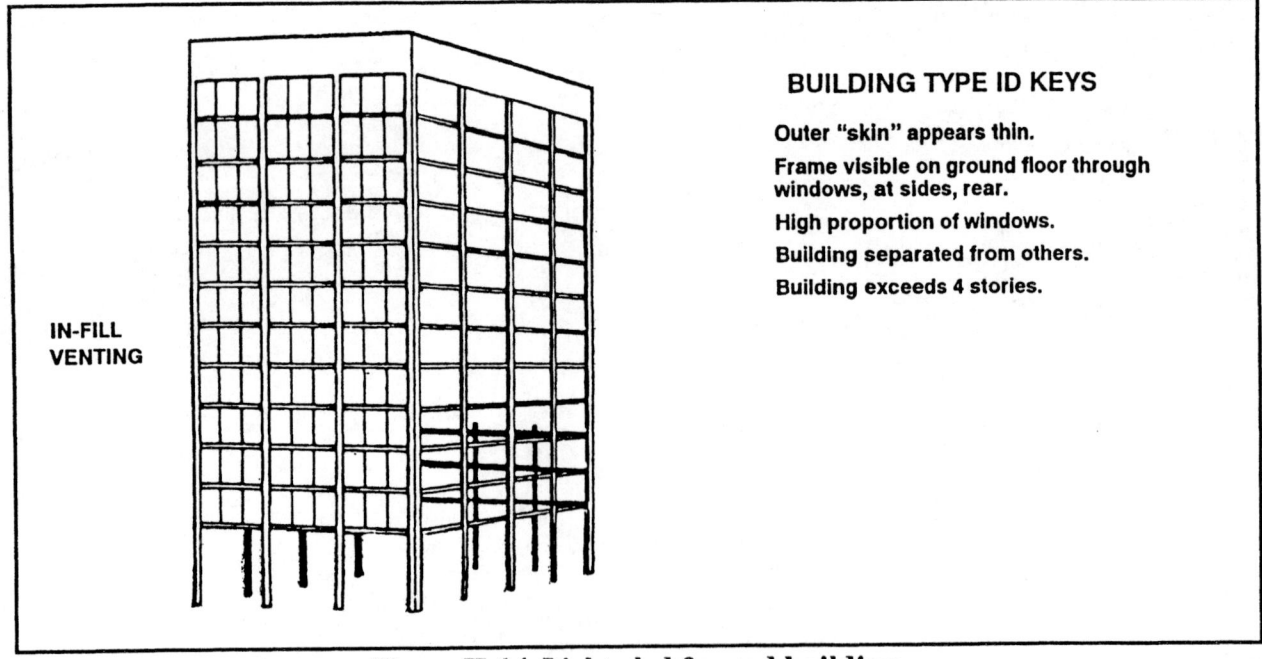

BUILDING TYPE ID KEYS

Outer "skin" appears thin.
Frame visible on ground floor through windows, at sides, rear.
High proportion of windows.
Building separated from others.
Building exceeds 4 stories.

IN-FILL VENTING

Figure H-14. Light-clad framed building.

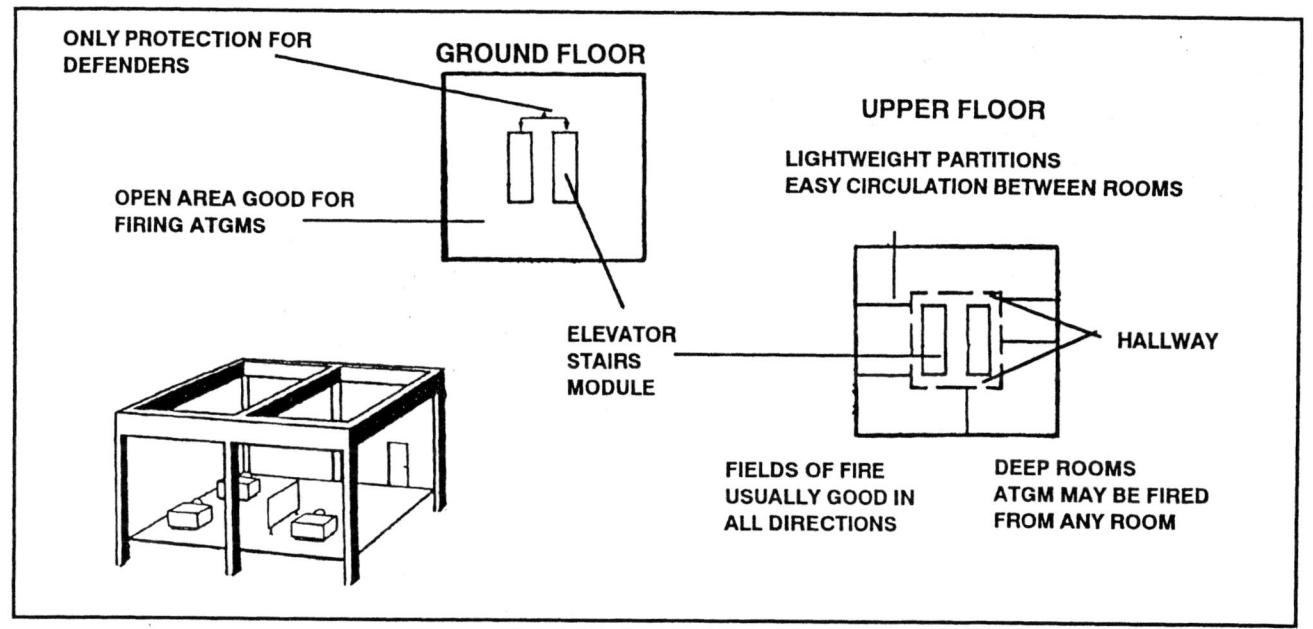

Figure H-15. Light-clad framed room.

c. The garage is one of the few buildings in an urban area in which all floors support vehicles. It provides a means to elevate vehicle-mounted TOWs, and the open interiors permit firing of ATGMs. Garages are normally high enough to provide a 360-degree field of fire for antiaircraft weapons. For example, a Stinger could hide under the top floor of the garage, come out to engage an aircraft, and then take cover again (Figure H-16).

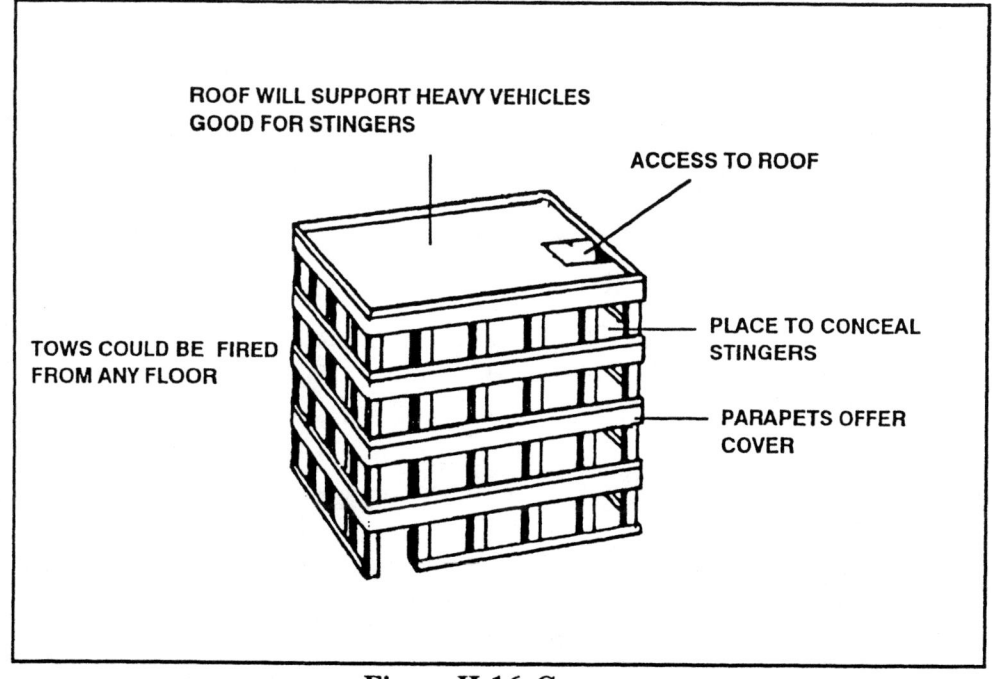

Figure H-16. Garage.

H-3. FLOOR PLANS

Floor plans in buildings follow predictable patterns. One of the factors that determines floor plans is building shape (Figure H-17). The basic principle governing building shape is that rooms normally have access to outside light. This principle helps to analyze and determine the floor plans of large buildings.

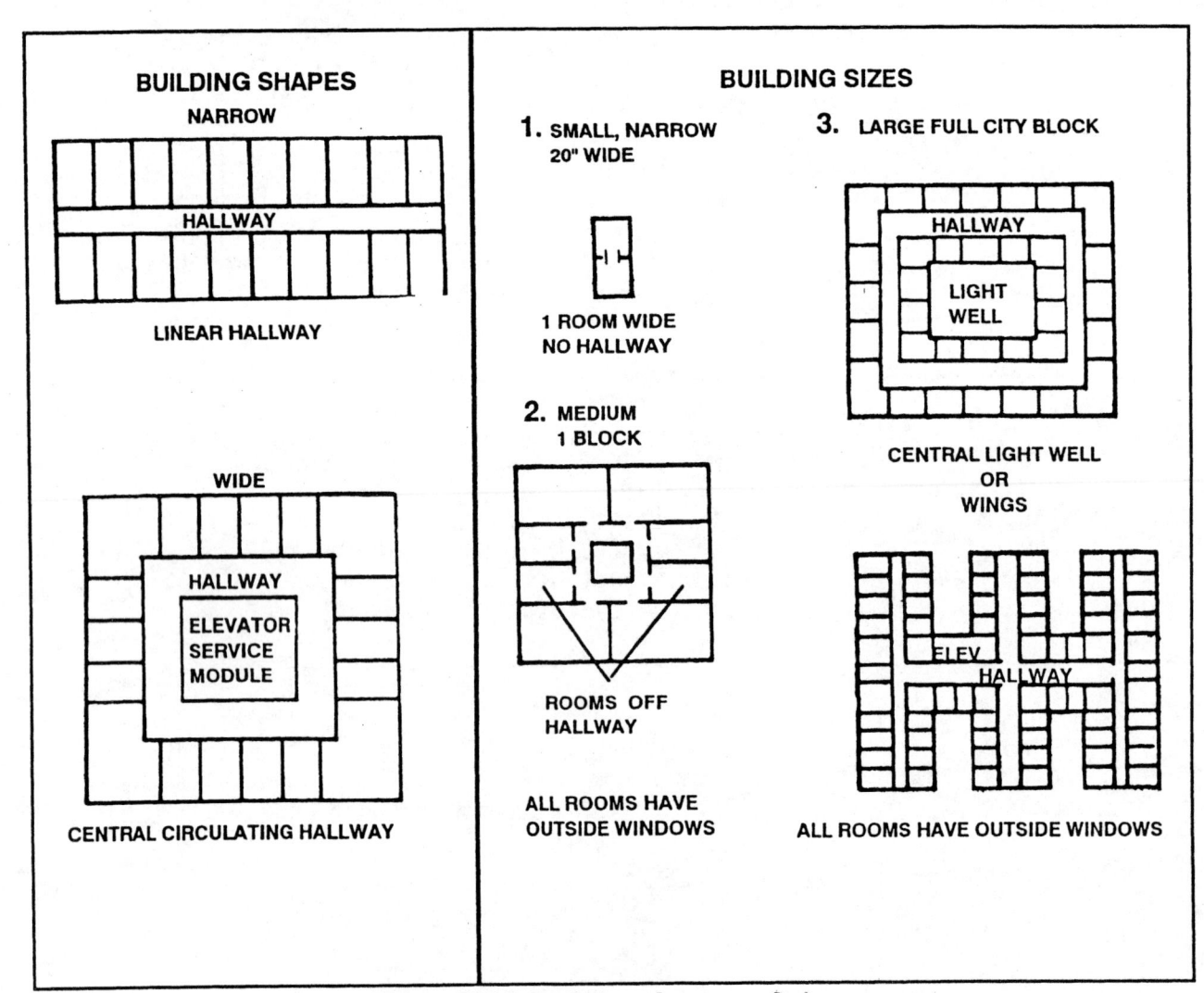

Figure H-17. Building shapes and sizes.

H-4. RESIDENTIAL AREAS

The two basic types of houses in the western world are located in and around cities and in rural areas. City houses are normally mass-construction brick buildings. Rural buildings in the continental US, South America, and Southeast Asia are commonly made of wood. In continental Europe, Southwest Asia, and sub-Saharan Africa, where wood is extremely scarce, rural buildings are normally constructed of concrete blocks (Figure H-18).

    a. Another common type of building structure in cities with European influences is called the Hof-style apartment building (Figure H-19).

URBAN

RURAL

CHARACTERISTICS

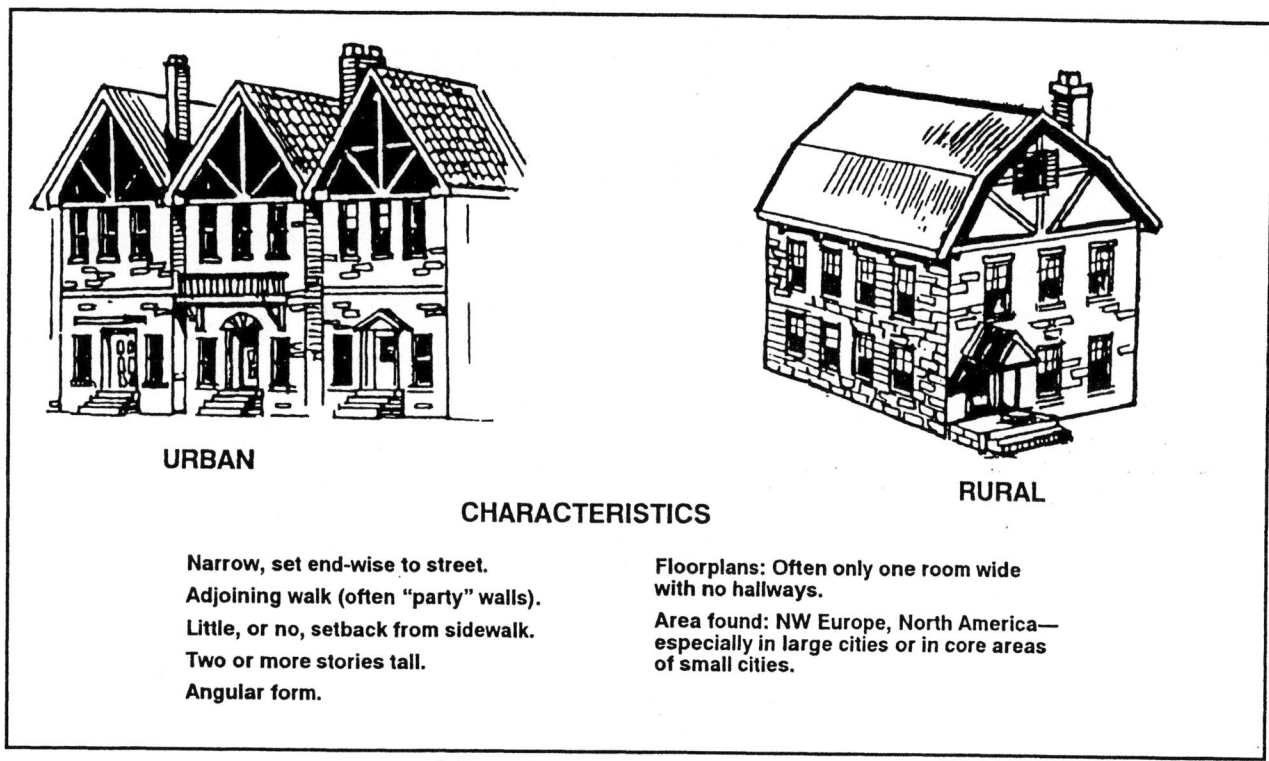

Narrow, set end-wise to street.
Adjoining walk (often "party" walls).
Little, or no, setback from sidewalk.
Two or more stories tall.
Angular form.

Floorplans: Often only one room wide with no hallways.
Area found: NW Europe, North America— especially in large cities or in core areas of small cities.

Figure H-18. Types of housing.

PERSPECTIVE VIEW

FLOOR PLAN

COURTYARD TUNNEL

CHARACTERISTICS

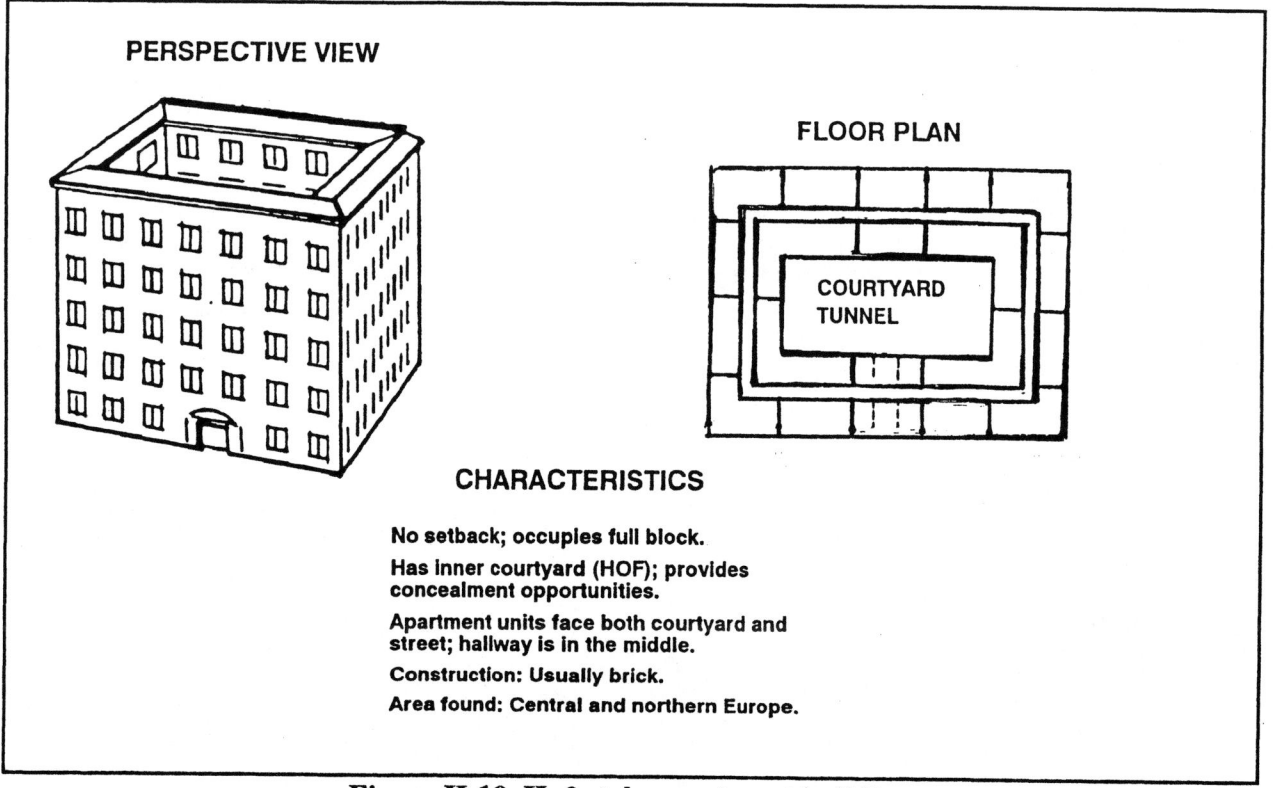

No setback; occupies full block.
Has inner courtyard (HOF); provides concealment opportunities.
Apartment units face both courtyard and street; hallway is in the middle.
Construction: Usually brick.
Area found: Central and northern Europe.

Figure H-19. Hof-style apartment building.

b. In the Mideast and tropical regions, the most common housing is the enclosed courtyard. Houses are added one to another with little regard to the street pattern. The result is a crooked, narrow maze, which is harder to move through or fire in than dense European areas (Figure H-20).

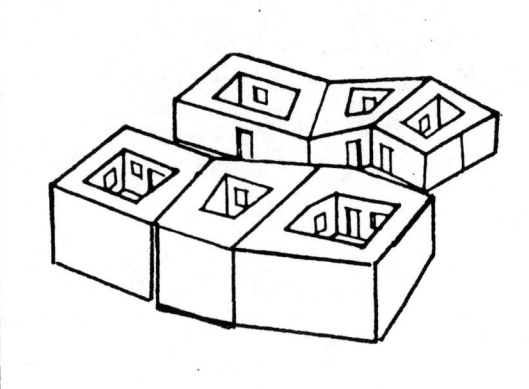

CHARACTERISTICS

Windowless outer walls, inner courtyards.
Varying size, dimensions.
No setbacks.
One to two stories tall.
Flat roofs.
Floor plan: All rooms open onto courtyard.
Location: On narrow, curving streets with short horizontal lines of sight.
Area found: Middle East, north Africa, and Mediterranean.

Figure H-20. Enclosed courtyard.

H-5. CHARACTERISTICS OF BUILDINGS

Certain characteristics of both mass-construction and framed buildings can be helpful in analyzing a built-up area. Leaders can use Table H-1 to determine how to defend or attack a certain building given the unit's available weapon systems.

TYPE OF CONSTRUCTION	BUILDING MATERIAL	HEIGHT (STORIES)	AVERAGE WALL THICKNESS (CM)
Mass	Stone	1 to 10	75
Mass	Brick	1 to 3	22
Mass	Brick	3 to 6	38
Mass	Concrete block	1 to 5	20
Mass	Concrete wall and slab	1 to 10	22 to 38
Mass	Concrete "tilt-ups"	1 to 3	18
Framed	Wood	1 to 5	3
Framed	Steel (heavy cladding)	3 to 50	30
Framed	Concrete/steel (light cladding)	3 to 100	2 to 8

Table H-1. Characteristics of buildings.

H-6. DISTRIBUTION OF BUILDING TYPES

Certain types of buildings dominate certain parts of a city, which establishes patterns within a city. Analysis of the distribution and nature of these patterns has a direct bearing on military planning and weapon selection (Figure H-21).

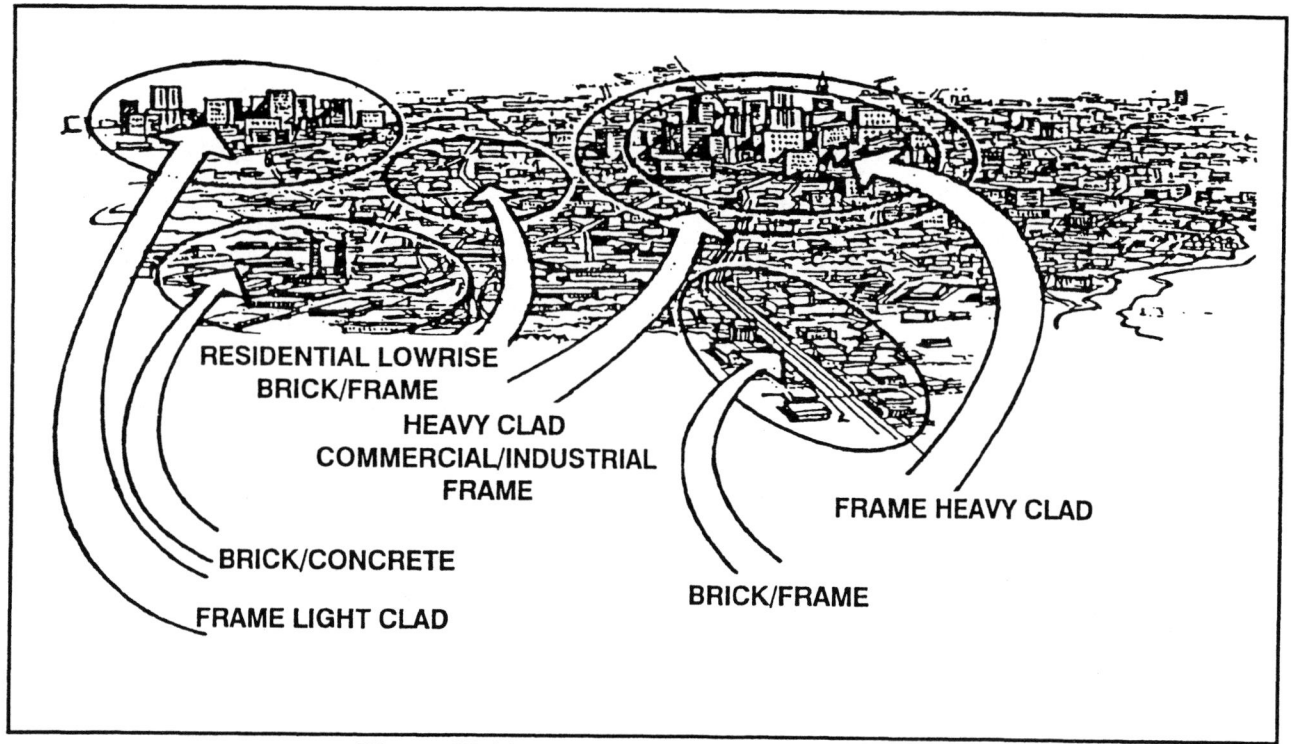

Figure H-21. Distribution of building types.

a. Mass-construction buildings are the most common structures in built-up areas, forming about two-thirds of all building types. Brick structures account for nearly 60 percent of all buildings, especially in Europe.

b. Steel and concrete framed multistory buildings have an importance far beyond their one-third contribution to total ground floor area. They occupy core areas—a city's most valuable land—where, as centers of economic and political power, they have a high potential military significance.

c. Open space accounts for about 15 percent of an average city's area. Many open spaces are grass-covered and are used for parks, athletic fields, and golf courses; some are broad, paved areas. The largest open spaces are associated with suburban housing developments where large tracts of land are recreation areas.

d. Streets serving areas consisting of mostly one type of building normally have a common pattern. In downtown areas, for example, high land values result in narrow streets. Street widths are grouped into three major classes: 7 to 15 meters, located in medieval sections of European cities; 15 to 25 meters, located in newer planned sections of most cities; and 25 to 50 meters, located along broad boulevards or set far apart on large parcels of land. When a street is narrow, observing or firing into windows of a building across the street can be difficult because an observer is forced to look along the building rather than into windows. When the street is wider, the observer has a better chance to look and fire into the window openings (Figure H-22).

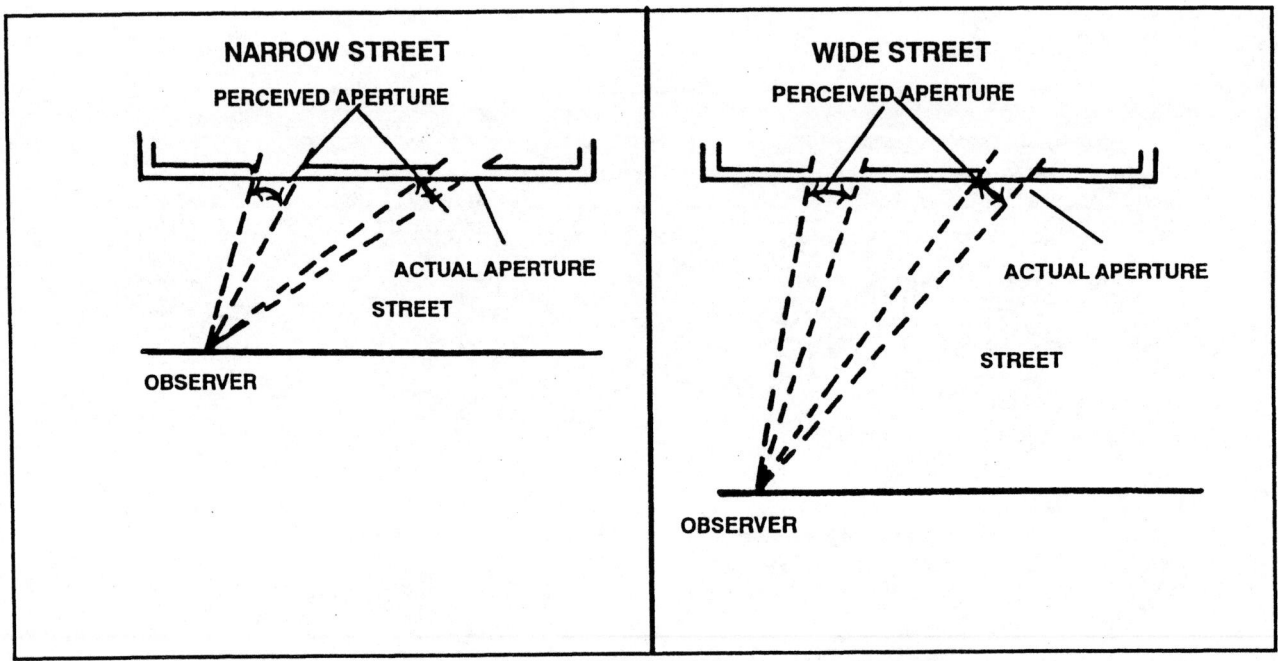

Figure H-22. Line-of-sight distances and angles of obliquity.

APPENDIX I

LIMITED VISIBILITY OPERATIONS UNDER MOUT CONDITIONS

With the rapid development of night vision devices throughout the world and AirLand operations doctrine that mandates continuous operations, US forces will continue to fight in built-up areas regardless of the weather or visibility conditions. To be successful, leaders must anticipate the effects of limited visibility conditions on operations and soldiers.

I-1. ADVANTAGES

When fighting in built-up areas during periods of limited visibility, attacking or defending forces have several advantages.

  a. In most cases, US forces have a technological advantage in thermal imagery and light intensification over their opponents. This enables US forces to identify, engage, and destroy enemy targets before detection by the enemy.

  b. AirLand operations stress continuous operations, day and night. This allows the attacking forces to conclude the battle decisively in a shorter period of time. It also allows the attacker to retain the initiative.

  c. Direct-fire target ranges in the MOUT environment are greatly reduced. During periods of limited visibility, effective target acquisition ranges are even further reduced. This enables attacking forces to close to shorter ranges, thus increasing the lethality and accuracy of weapons. Attacking forces can also take advantage of the enemy's reduced visibility and can engage before being detected with thermal imagery or light intensification devices.

  d. Air assault operations are best conducted during periods of limited visibility, since the enemy's air defenses are degraded.

  e. Attacking during periods of limited visibility gives the attacker a greater chance of surprise.

I-2. DISADVANTAGES

When fighting in built-up areas during limited visibility, attacking and defending forces also face some disadvantages.

  a. Command and control is difficult in any operation in a built-up area, and periods of limited visibility increase this difficulty.

  b. Soldiers have an instinctive tendency to form groups during limited visibility. Constant attention must be given to prevent the soldiers from "bunching up."

  c. Due to the low visibility and the characteristics of built-up areas, soldiers become disoriented easily.

  d. Target identification becomes difficult in limited visibility conditions. Depending on the individual, the soldier may fire at anything he sees, or he may hesitate too long before firing. This is one of the leading causes of fratricide, so leaders must pay close attention to soldiers' target engagement.

I-3. FRATRICIDE AVOIDANCE

The risk of fratricide is much greater during periods of limited visibility. The key to avoiding fratricide is situational awareness by leaders and individuals coupled with training. Other considerations include:

a. Graphic control measures should be clearly defined and obvious. Examples include distinct buildings, large boulevards, rivers, and so forth.

b. Leaders must exercise firm control when engaging targets. Movements should also be tightly controlled.

c. Cleared rooms and buildings should be distinctly marked to identify cleared areas and friendly troops to any base of fire supporting the maneuver.

d. Visible markers (for example, glint tape or thermal strips) should be attached to individual soldiers.

e. Far and near recognition symbols should be used properly.

f. Units using close air support must exercise firm control and direct their firing. Failure to do so may lead to the pilot becoming disoriented and engaging friend and foe alike.

I-4. URBAN ENVIRONMENTAL EFFECTS ON NIGHT VISION DEVICES

The characteristics of built-up areas affect standard US NVDs and sights differently than do open areas. This may cause some confusion for soldiers operating during limited visibility, since the images they receive through their NVDs are unusual.

a. Since most built-up areas have electric power, street lights and or building lights "white out" any light intensification devices unless the power is disrupted.

b. The chance that fires will be burning in the area of operations is high. This causes problems not only for light intensification devices, but possibly for thermal devices as well.

c. Subterranean areas and the interiors of buildings will not have ambient light if the power is off. Passive NVDs must have an artificial light source, such as infrared, to provide enough ambient light for the devices to work.

d. The many reflective surfaces found in built-up areas may cause false images, especially for laser range finders and laser target designators.

e. Large amounts of dust particles suspended in the air prevent thermal imaging devices from seeing through the dust cloud.

f. Smoke also affects NVDs similar to the way dust does.

g. Fog degrades long-range target acquisition from thermal sights, which may cause problems for any overwatching or supporting elements.

h. Weapons flashes within enclosed areas appear to be much brighter. This causes soldiers to lose their night vision and washes out light intensification devices.

I-5. CONSIDERATIONS

The environment of built-up areas presents special challenges and considerations during periods of limited visibility.

a. The use of glint tape, thermal tape, budd lites, or chemlites is an important consideration. These can be used to mark the forward line of troops, casualties, cleared buildings and rooms, weapons positions, and individual soldiers. Their use must be clearly addressed in the unit's TAC SOP. When markers are used for extended periods, their meanings should change since the enemy may be able to capture or manufacture and use these marking devices.

b. The use of tracer and incendiary ammunition may be restricted to prevent fires. Also, the light of the fires "whites out" light intensification night vision devices and may interfere with thermal devices.

c. The control of power stations may be essential to operations during limited visibility. This enables friendly forces to control, to a degree, background illumination. Shutting off the power to the street lights is preferable to shooting the lights out. Commanders must balance the trade-off between force protection and maintaining law and order after the battle is over. During cold weather, the control of power stations may be critical for the welfare of the civilian population.

d. The identification between friendly soldiers, noncombatant civilians, and enemy troops becomes more difficult during limited visibility operations.

e. The location of the source of sounds becomes more difficult due to the natural echoing in built-up areas and the tendency of sounds to carry farther at night.

f. The location of booby traps and obstacles also becomes more difficult at night. Movement rates are slower than during normal visibility.

I-6. SPECIAL EQUIPMENT

Fighting during periods of limited visibility requires some specialized equipment to maximize maneuver and target engagement.

a. As a rule, thermal imaging devices, such as the AN/PAS-7 IR viewer (LIN Y03104) and the Dragon IR sight AN/TAS-5 (LIN N23721), are better for limited visibility operations than light intensification devices such as the AN/PVS-7 (LIN N05482). Light intensification devices are easily washed out from background light, weapons flashes in enclosed areas, and fires. Thermal devices, while also affected by fires, are not as easily washed out.

b. The AN/PAQ-4 infrared aiming device (LIN A34938) is similar to its civilian laser aiming sight counterparts except it is not visible to the naked eye. Pen lights can also be attached to weapons to provide a quick sight picture, illuminate rooms and hallways, identify obstacles and booby traps, and identify friendly forces.

c. Other night sights for weapons include the AN/TVS-5 (LIN N04596) crew-served weapon night vision sight, the AN/PVS-4 (LIN N04734) individual weapons night vision sight, and the AN/UAS-11 (LIN N05050) night vision set. The AN/UAS-11, while not easily man-portable, has the advantage of an integral laser range finder. It is also a thermal sight similar to the TOW 2 AN/TAS-4 night sight.

d. Trip flares, flares, illumination from mortars and artillery, and spotlights (visible light or infrared) can be used to blind enemy NVDs or to artificially illuminate the battlefield (Figure I-1). (See FM 7-90 for more information on illumination from mortars and artillery.)

e. Spare batteries for the NVDs should be carried to keep the devices operational. Soft, clean rags should be used to clean the lenses.

Figure I-1. Use of indirect fire illumination during MOUT.

I-7. COMBAT SUPPORT

Loss of synchronization is one of the major concerns to commanders and leaders during limited visibility operations under MOUT conditions. The concentration of forces and fires at the point of decision is faciliated by US forces' technological edge and by clear orders.

a. Any degradation of artillery fire will be due to the limited target acquisition assets. While the field artillery FOs and combat observation and lasing teams (COLTs) have thermal sights and laser range finders, most soldiers on the battlefield do not have devices that will enable them to accurately call for fire. The following are some devices and techniques to acquire targets for indirect fires.

(1) The AN/UAS-11 determines accurate coordinates with its thermal imaging sight coupled with a laser range finder. For the AN/UAS-11 to obtain accurate coordinates, the crew must first have an accurate location. The same technique can be used by any attached armor unit. The BFV can be used as well if it is equipped with a laser range finder.

(2) Preregistered TRPs are effective only if the TRPs can be observed and the observer has clear communications to the firing unit.

(3) Reflective surfaces found in built-up areas may affect laser designators.

(4) Counterfire radar should be employed to cover likely areas of enemy mortar, cannon, and rocket use. Because of the masking effect of built-up areas, counterfire radars are not normally emplaced within the built-up area.

b. Fixed-wing aviation assets face a lower ADA threat during periods of limited visibility. However, the need for command and control is greater to prevent fratricide. The best fixed-wing aircraft available for fire support is the AC-130 because of its target acquisition capabilities, deadly and accurate fire, and long loiter time.

c. Army aviation operates on similar limitations and considerations as fixed-wing aircraft. Most US Army attack helicopters have a forward-looking infrared (FLIR) night sight. Coupled with the slower speed and hover capability of the helicopter, Army aviation assets can deliver highly accurate and responsive fire on enemy targets. However, helicopters are more susceptible to enemy air defense artillery assets and, therefore, should only be employed where the enemy air defense threat is light. Commanders must identify clear land marks for the pilots to navigate to and from the objective.

d. ADA is significantly degraded during periods of limited visibility. Visual detection, identification, and range estimation are all difficult, if not impossible. Radar guidance systems have difficulty determining the target from ground clutter.

e. The lack of thermal imaging devices may hamper engineer units. Locating and clearing mines and booby traps also become more dangerous and difficult. The method of marking cleared lanes should be determined and coordinated in advance to avoid confusion with other limited visibility markers (glint tape, infrared strobe lights [budd lites], chemlites, and so forth).

f. Military intelligence relies primarily on human intelligence assets to gain information about the enemy in urban environments.

(1) GSR and REMs have limited use in the center of built-up areas. They are best employed on the outskirts to monitor traffic into and out of the built-up area. If necessary, GSR can be used to cover large open areas such as parks and public squares. REMs can be used in subterranean areas such as sewers and utility tunnels.

(2) Military intelligence units equipped with the AN/UAS-11 can use it in a variety of target acquisition and surveillance roles.

(3) Based on the time available before the operation or the urgency of need, satellite photographs of the built-up area may be available.

(4) Military intelligence officers at brigade level and below should obtain city maps of the area of operations. The normal 1:50,000 map scale is virtually useless to soldiers fighting in a built-up area. The Defense Mapping Agency maintains various city maps with either a 1:10,000 or 1:12,500 scale. These maps are created based on the need for contingency operations and

noncombatant evacuation operations (NEO). If no maps are available for the area, the S2 at battalion level can request that the division topographic section produce some products based on Terrabase and satellite imagery. If the division topographic section cannot produce the map, the request should be forwarded through channels to corps.

I-8. COMBAT SERVICE SUPPORT
Maneuver unit commanders and their soldiers are not the only individuals that must adjust to combat under limited visibility conditions in MOUT operations. Logisticians at every level must anticipate requirements for this unique environment.

   a. Units conducting resupply operations during periods of limited visibility should remember the following:

   (1) Drivers and vehicle commanders should be issued night vision devices so the vehicles going to and from logistic release points do not need any illumination. This also prevents the enemy from acquiring resupply locations by following vehicles with blackout lights on.

   (2) Strict noise and light discipline should be maintained.

   (3) Vehicles should follow a clearly marked route to avoid any obstacles and prevent the resupply vehicle(s) from becoming disoriented.

   (4) Radios should be redistributed to resupply vehicles whenever possible, in case of further instructions.

   (5) Each vehicle should have a map of the area of operations (preferably a city map with the street names).

   b. Combat units operating for extended periods during limited visibility should have enough batteries to keep the NVDs functioning at optimum power and sensitivity.

   c. A large operational readiness float of night vision devices, especially thermal sights such as the AN/TAS-4 or AN/UAS-11, should be maintained.

   d. Casualty collection during periods of limited visibility is much more difficult. Clear methods for marking any casualties must be established before the operation begins.

   e. CSS operations in existing structures at night must not be visible from a long distance. This includes limiting vehicle traffic to an absolute minimum, sealing doors and windows to prevent light leakage, and dispersing assets as much as possible.

I-9. OFFENSIVE CONSIDERATIONS
US forces conduct attacks during periods of limited visibility to gain or sustain the momentum of the attack. Before conducting a limited visibility attack, the commander must balance the risks and ensure that every soldier understands the intent and control measures. Rehearsals and strict command and control reduce casualties and greatly enhance the chances for mission accomplishment.

   a. Soldiers should clear buildings and rooms using the same techniques they use during periods of unlimited visibility to reduce confusion. The soldiers are well rehearsed in these techniques and, therefore, more confident. The only major difference is in equipment used. (See the paragraph on special equipment in this appendix.)

   b. Movement rates are slower. Each soldier must be alert for mines, booby traps, and enemy positions. Although thermal imaging devices can detect the difference in the temperature of the soil, light intensification devices are usually better for detecting recently disturbed dirt. Thermal imaging devices are better for identifying personnel; however, light intensifiers can identify friendly soldiers, noncombatants, and enemy troops better than the thermals.

    c. Squads and fire teams should be equipped with a mixture of both thermal imaging and light intensifying devices whenever possible. This enables the squads and fire teams to obtain a better picture of the night environment and enables the soldiers to balance the strengths and weaknesses of each type of night vision device.

    d. When moving through buildings, the assault teams must mark cleared rooms and buildings, and communicate with the support team(s). This communication is critical if more than one assault team is in the same building.

    e. Soldiers should maximize the use of ambient light whenever possible for two reasons: to conserve the batteries of the night vision devices, and to make detection of attacking US forces harder for the enemy.

    f. If flashlights or chemlites are used, they should be held away from the head or chest area. This will make it harder for enemy soldiers firing at the light to kill the soldier holding the flashlight or chemlite.

    g. The assault team must have clear communications with all supporting elements, whether they are organic, in DS, under OPCON, or attached. Supporting units should not fire unless they have good communications with the assaulting elements and are sure the targets they are engaging or suppressing are the enemy.

    h. Units must know where everyone is during offensive operations. Not only does this reduce the risk of fratricide, but it also increases the time of identifying, locating, and treating casualties. Also, it greatly reduces the chance of soldiers becoming disoriented and separated from the unit.

    i. Assault teams should be aware of adjacent fires that diminish the effectiveness of night vision devices. Weapons flashes within small rooms cause soldiers to lose their night vision and wash out light intensification devices. Also, enemy soldiers may use flares inside and outside of buildings to render night vision devices ineffective.

    j. Leaders must ensure that all soldiers follow the rules of engagement and the laws of land warfare. This is critical if the enemy is intermixed with the local civilian population. Also, soldiers and leaders must follow all control measures, especially graphic control measures.

I-10. DEFENSIVE CONSIDERATIONS

Enemy forces can be expected to use periods of limited visibility for the same reasons US forces do. (See the paragraph on advantages in this appendix.) Enemy forces may have access to sophisticated night vision devices manufactured in Europe, the United States, Japan, Korea, and the former Soviet Union. (See Chapter 4 for more information on defensive techniques.)

GLOSSARY

ADA	air defense artillery
ANGLICO	air naval gunfire liaison company
AP	armor-piercing, antipersonnel
APC	armored personnel carrier
APDS-T	armor-piercing discarding sabot—tracer
APFSDS	armor-piercing, fin-stabilized, discarding sabot
AT	antitank
ATGMs	antitank guided missiles
ATTN	attention
AXP	ambulance exchange point
BDAR	battle damage assessment and repair
BDU	battery display unit; battle dress uniform
BFV	Bradley fighting vehicle
BMNT	beginning morning nautical twilight
BMPs	Threat fighting vehicles
BTRs	Threat fighting vehicles
CA	civil affairs
CAS	close air support
CEV	combat engineer vehicle
CFV	combat fighting vehicle
CI	configuration item; command information; counterintelligence
CN	chloroacetothenone
COLT	combat observation and lasing team
CP	command post
CS	combat support, chemical smoke, O-chlorobenzylidene malononitrile
CSS	combat service support
DA	Department of the Army
DPRE	displaced persons, refugee, and evacuee
DS	direct support

ECM	electronic countermeasures
EOD	explosive ordnance disposal
EPW	enemy prisoner of war
FA	field artillery
FAC	forward air controller
FAE	fuel air explosives
FASCAM	family of scatterable mines
FAST	Freight Automated System for Traffic Management, forward area support team
FCL	final coordination line
FEBA	forward edge of battle area
FIST	fire support team
FLIR	forward-looking infrared
FM	field manual; frequency modulation
FO	forward observer
FPF	final protective fire
FPL	final protective line
FRAGO	fragmentary order
FSB	final staging base
FSE	fire support element
FSO	fire support officer
G3	Assistant Chief of Staff, G3 (Operations and Plans)
G5	Assistant Chief of Staff, G5 (Civil Affairs)
GS	general support
GSR	ground surveillance radar
GRREG	graves registration
HC	hydrogen chloride
HE	high explosive
HEAT	high-explosive antitank
HEAT-MP	high-explosive antitank, multipurpose
HEI-T	high-explosive incendiary—tracer
HMMWV	high-mobility, multipurpose, wheeled vehicle

IAW	in accordance with
ICM	improved capabilities missile
ID	identification
IPB	intelligence preparation of the battlefield
ITOW	improved TOW
ITV	improved TOW vehicle
J5	Plans and Policy Directorate
LAW	light antitank weapon
LD	line of departure
LIC	low-intensity conflict
LOGPAC	logistics package
LOS	line of sight
LTC	lieutenant colonel
MBA	main battle area
MCOO	modified, combined obstacle overlay
MDP	meteorological datum plane, main defensive position
MEDEVAC	medical evacuation
METT-T	mission, enemy, terrain, troops, and time available
MOPP	mission-oriented protective posture
MOUT	military operation on urbanized terrain
MP	military police
MRB	motorized rifle battalion
MRR	motorized rifle regiment
MSR	main supply route
NATO	North Atlantic Treaty Organization
NBC	nuclear, biological, chemical
NCA	national command authority
NCO	noncommissioned officer
NEO	noncombatant evacuation operations
NOE	nap-of-the-earth
NVG	night vision goggles
OP	observation post

OPCON	operational control
OPLAN	operation plan
OPORD	operation order
OPSEC	operational security
PA	public affairs
PAC	Personnel and Administration Center; plastic ammunition container
PDDA	power-driven decontaminating apparatus
PE	probable error
PEWS	platoon early warning system
POL	petroleum, oils, and lubricants
PSYOP	psychological operations
PW	prisoner of war
RAAWS	Ranger antiarmor weapon system
RCLR	recoilless rifle
REMs	remote sensors
ROE	rules of engagement
RP	reference point, red phosphorus
RPG	Threat antiarmor weapon
S1	Adjutant (US Army)
S2	intelligence officer
S3	Operations and Training Officer (US Army)
S4	Supply Officer (US Army)
S5	Civil Affairs Officer (US Army)
SALT	supporting arms liaison team
SEAD	suppression of enemy air defenses
SIDPERS	Standard Installation/Division Personnel System
SMAW	shoulder-launched, multipurpose, assault weapon
SOP	standing operating procedure
STB	supertropical bleach
TAACOM	Theater Army Area Command
TF	task force

TI	technical inspection; technical intelligence
TM	technical manual, team (graphics only)
TNT	trinitrotoluene
TOC	tactical operations center
TOW	tube-launched, optically tracked, wire-guided missile
TP-T	target practice-tracer
TRADOC	Training and Doctrine Command
TTP	tactics, technics, and procedures
UCMJ	Uniform Code of Military Justice
US	United States
USAF	United States Air Force
USMC	United States Marine Corps
USN	United States Navy
VT	variable time
WP	white phosphorous

REFERENCES

Documents Needed

These documents must be available to the intended users of this publication.

FIELD MANUALS (FMS)

7-7	The Mechanized Infantry Platoon and Squad (APC). 15 March 1985.
7-7J	The Mechanized Infantry Platoon and Squad (Bradley). 18 February 1986.
7-8	The Infantry Platoon and Squad (Infantry, Airborne, Air Assault, Ranger). 22 April 1992.
7-10	The Infantry Rifle Company. 14 December 1990.
7-20	The Infantry Battalion (Infantry, Airborne and Air Assault). 06 April 1992.
7-30	Infantry, Airborne and Air Assault Brigade Operations. 24 April 1981.
34-130	Intelligence Preparation of the Battlefield. 23 May 1989.
71-1	Tank and Mechanized Infantry Company Team. 22 November 1988.
71-2	The Tank and Mechanized Infantry Battalion Task Force. 27 September 1988.
71-3	Armored and Mechanized Infantry Brigade. 11 May 1988.
100-5	Operations. 05 May 1986.
101-5	Staff Organization and Operations. 25 May 1984.
101-5-1	Operational Terms and Symbols. 21 October 1985.

Readings Recommended

These readings contain relevant supplemental information.

ARMY REGULATIONS

385-62	Regulations for Firing Guided Missiles and Heavy Rockets for Training, Target Practice and Combat. 05 January 1977.
385-63	Policies and Procedures for Firing Ammunition for Training, Target Practice and Combat. 15 October 1983.
600-8-1	Army Casualty and Memorial Affairs and Line of Duty Investigations. 18 September 1986.

FIELD MANUALS (FMS)

1-100	Doctrinal Principles for Army Aviation in Combat Operations. 28 February 1989.
1-112	Tactics, Techniques, and Procedures for the Attack Helicopter Battalion. 21 February 1991.

3-3	NBC Contamination Avoidance. 30 September 1986.
3-4	NBC Protection. 29 May 1992.
3-5	NBC Decontamination. 23 July 1992.
3-6	Field Behavior of NBC Agents (Including Smoke and Incendiaries). 3 November 1986.
3-9	Potential Military Chemical/Biological Agents and Compounds. 12 December 1990.
3-11	Flame Field Expedients. 19 September 1990.
3-100	NBC Defense, Chemical Warfare, Smoke, and Flame Operations. 23 May 1991.
5-33	Terrain Analysis. 11 July 1990.
5-101	Mobility. 23 January 1985.
5-102	Countermobility. 14 March 1985.
5-103	Survivability. 10 June 1985.
5-105	Topographic Operations. 09 September 1987.
5-250	Explosives and Demolitions. 15 June 1992.
6-20	Fire Support in the AirLand Battle. 17 May 1988.
6-20-40	Tactics, Techniques, and Procedures for Fire Support for Brigade Operations (Heavy). 05 January 1990.
6-20-50	Tactics, Techniques, and Procedures for Fire Support for Brigade Operations (Light). 05 January 1990.
7-90	Tactical Employment of Mortars. 09 October 1992.
7-91	Tactical Employment of Antiarmor Platoons, Companies, and Battalions. 30 September 1987.
10-63-1	Graves Registration Handbook. 17 July 1986.
11-50	Combat Communications Within the Division (Heavy & Light). 04 April 1991.
17-95	Cavalry Operations. 19 September 1991.
17-98	Scout Platoon. 07 October 1987.
19-1	Military Police Support for the AirLand Battle. 23 May 1988.
19-15	Civil Disturbances. 25 November 1985.
19-30	Physical Security. 01 March 1979.
19-40	Enemy Prisoners of War, Civilian Internees, and Detained Persons. 27 February 1976.
20-32	Mine/Countermine Operations. 09 December 1985.
21-10	Field Hygiene and Sanitation. 22 November 1988.

21-11	First Aid for Soldiers. 27 October 1988.
21-60	Visual Signals. 30 September 1987.
21-75	Combat Skills of the Soldier. 03 August 1984.
23-1	Bradley Fighting Vehicle Gunnery. 01 March 1991.
23-9	M16A1 Rifle and M16A2 Rifle Marksmanship. 03 July 1989.
23-30	Grenades and Pyrotechnic Signals. 27 December 1988.
24-1	Signal Support in the AirLand Battle. 15 October 1990.
27-10	The Law of Land Warfare. 18 July 1956.
32-1	Signal Intelligence (SIGINT)(U)
33-1	Psychological Operations. 31 July 1987.
34-1	Intelligence and Electronic Warfare Operations. 02 July 1987.
34-3	Intelligence Analysis. 15 March 1990.
41-10	Civil Affairs Operations. 17 December 1985.
44-3	Air Defense Artillery Employment: Chaparral/Vulcan/Stinger. 15 June 1984.
44-8	Small Unit Self-Defense Against Air Attack. 30 December 1981.
44-16	Chaparral/Vulcan/Stinger Platoon Combat Operations. 20 May 1987.
63-1	Combat Service Support Operations, Separate Brigade. 30 September 1983.
63-2	Division Support Command, Armored, Infantry, and Mechanized Infantry Divisions. 20 May 1991.
71-100	Division Operations. 16 June 1990.
71-101	Infantry, Airborne, and Air Assault Division Operations. 26 March 1980.
90-2	Battlefield Deception. 03 October 1988.
100-2-1	Soviet Army Operations and Tactics. 16 July 1984.
100-2-2	Soviet Army Specialized Warfare and Rear Area Support. 16 July 1984.
100-2-3	The Soviet Army Troops Organization and Equipment. 06 June 1991.
100-10	Combat Service Support. 18 February 1988.
100-20	Military Operations in Low Intensity Conflict. 05 December 1990.
100-26	The Air-Ground Operations System. 30 March 1973.
101-10-1/1	Staff Officers Field Manual - Organizational, Technical, and Logistical Data (Volume 1). 07 October 1987.

101-10-1/2	Staff Officers Field Manual - Organizational, Technical, and Logistical Data Planning Factors (Volume 2). 07 October 1987.
700-80	Logistics. 15 August 1985.

TRAINING CIRCULARS (TCS)

24-20	Tactical Wire and Cable Techniques. 03 October 1988.
34-40-6	Signal Intelligence (SIGINT). 11 April 1989.
90-6-1	Military Mountaineering. 26 April 1989.

INDEX

25-mm automatic gun, 8-31
   obliquity, 8-31
   penetration, 8-32
   target types, 8-31

aerial weapons, 8-40
   fixed-wing aircraft, 8-41
     AC-130, 8-41
   rotary-wing aircraft, 8-40
     Hellfire, 8-40

air defense, 6-5, G-7
   Stingers, 6-5
   Vulcan, 6-5

ambush, 4-32

antitank guided missiles, 8-19
   employment, 8-19
     backblast, 8-21
     dead space, 8-19
     obstacles, 8-19
     penetration, 8-23
     wall breaching, 8-23

Army aviation, 6-6
   defensive missions, 6-6
   offensive missions, 6-6

BFV, B-1
   defense, B-8
   Echo Company, B-10
   employment, B-1
   offense, B-2

building analysis, H-1

built-up areas, 1-1
   categories, 1-5
   characteristics, 1-4, 2-1, 2-2
     regional, 2-1
     specific, 2-1
     urban, 2-2
   navigation, 5-39
   special considerations, 1-5
   threat, 1-3

camouflage, 5-40

casualties, 7-2, 7-3

city core, 2-1 (illus), 2-2, 2-3 (illus)

clearing, 5-19, F1

combat service support, 7-1, I-6
   guidelines, 7-1

combat support, 6-1, I-4

command and control, 3-17, 4-16, G-8

commercial ribbon, 2-2, 2-4 (illus)

communications, 1-6, 4-15
   restrictions, 4-4
   support, 6-9

core periphery, 2-2, 2-4 (illus)

counterguerilla, 2-9

counterinsurgency, 2-9

counterterrorist, 2-9

cover and concealment, 4-4

defensive operations, 4-1
   characteristics, 4-2
   considerations, 4-1
   obstacles, 4-2

deliberate attack, 3-3, 3-17
   foothold, 3-4, 3-5 (illus), 3-18

demolitions
   bulk, 8-41
   cratering charges, 8-42
   defensive use, C-13
   offensive use, C-12
   safety, C-18
   satchel charges, 8-42
   shaped charges, 8-42

direct fire, 3-13, 3-14 (illus), 3-16

engineers
   employment of, 4-12
   support, 6-8, G-10

field artillery, 6-3

fighting positions, 4-14, 4-15, E-1
    preparation, E-2

fire support
    employment of, 4-14, 4-15, E-1

firing positions
    hasty, 5-23
    prepared, 5-26

flame operations, 5-37

flame weapons, 8-23
    effects, 8-24
    employment, 8-24

fratricide, 3-2
    avoidance, 1-7, I-2

grenade launchers, 8-7
    employment, 8-7
    penetration, 8-8

hand grenades, 5-19
    effects, 8-27
    employment, 8-27
    fragmentation, 8-27
    riot control, 8-27
    smoke, 8-27

hasty attack, 3-3

helicopters, 6-6
    assaults, 6-6, 6-7 (illus)

high-rise areas, 2-2, 2-3 (illus)

loopholes, 5-26, 5-27 (illus), 5-32 (illus), 5-33, E-5

M16 rifle, 8-2
    employment, 8-2
    penetration, 8-3, 8-4
    protection, 8-3

M249, 8-2
    employment, 8-2
    penetration, 8-3, 8-4
    protection, 8-3

machine guns, 8-4
    employment, 8-5
    penetration, 8-5, 8-6

medical, 7-5
    evacuation, 7-6
    supplies, 7-5

METT-T factors
    defensive considerations, 4-5 to 4-15
    offensive considerations, 3-7 to 3-17

military maps, 5-39

military police, 6-8, 6-9

mines, C-6
    types, C-7

mortars, 6-1
    employment, 8-28
    effects, 8-30

MOUT, 1-1
    precision, 1-2, G-1
    restrictive conditions, G-1
    surgical, 1-1, G-1

NBC, A-1
    decontamination, A-2
    detection, A-2

naval gunfire, 6-4

OPCON, 3-22

observation post, 5-35, 5-36 (illus)

obstacles, 4-2, 4-15, 7-8
    field-expedient, 4-3 (illus)
    types, C-1

offensive operations, 3-1
    characteristics, 3-2
    considerations, 3-1, 3-2
    equipment, 3-2, 3-3
    limited visibility, I-6
    maneuver, 3-2
    troop requirements, 3-2
    types, 3-3

open city, 3-1

outlying industrial area, 2-2, 2-5 (illus)

recoilless weapons, 8-9
    backblast, 8-12
    employment, 8-9
    penetration, 8-16

reconnaissance, 3-17, 3-27, 3-28 (illus)
    subterranean, D-3

residential sprawl, 2-2, 2-5 (illus)

roit control agents, A-3

rules of engagement, 1-1, G-1

smoke operations, A-2

snipers, 3-16, 5-38

supply, 7-4

tactical air, 6-4

tactics, techniques, and procedures, 1-7

tank cannon, 8-34
    ammunition, 8-34
    characteristics, 8-35
    effects, 8-36
    employment, 8-36
    obliquity, 8-34

task force, 3-17 to 3-21

techniques, 5-1
    entry, 5-11
    movement, 5-1 to 5-11

terrain, 2-6, 3-8
    analysis, 2-6
    special considerations, 2-6

threat, 1-3, 2-8
    evaluation, 2-8, 2-9
    integration, 2-8
    projected capabilities, 2-10, 2-11

weather, 2-6
    analysis, 2-6
    special considerations, 2-7